INDICATORS FOR
WATERBORNE PATHOGENS

Committee on Indicators for Waterborne Pathogens

Board on Life Sciences

Water Science and Technology Board

Division on Earth and Life Studies

NATIONAL RESEARCH COUNCIL
OF THE NATIONAL ACADEMIES

THE NATIONAL ACADEMIES PRESS
Washington, D.C.
www.nap.edu

THE NATIONAL ACADEMIES PRESS   500 Fifth Street, N.W.   Washington, DC 20001

This study was supported by Grant X-82928901 between the National Academies and the United States Environmental Protection Agency. Any opinions, findings, conclusions, or recommendations expressed in this publication are those of the author(s) and do not necessarily reflect the views of the organizations or agencies that provided support for the project.

International Standard Book Number 0-309-09122-5 (Book)
International Standard Book Number 0-309-52941-7 (PDF)
Library of Congress Control Number 2004108875

Cover:
The Committee on Indicators for Waterborne Pathogens thanks the following organizations and individuals for use of their images on the cover of this report:

- City of Wilsonville, Oregon
- Isle Royale National Park (Michigan), National Park Service, U.S. Department of the Interior
- Randal Kath, The State University of West Georgia, Carrollton
- National Center for Infectious Diseases, Centers for Disease Control and Prevention, U.S. Department of Health and Human Services
- National Institute of Allergy and Infectious Diseases, National Institutes of Health, U.S. Department of Health and Human Services
- Southern California Coastal Water Research Project Authority, Westminster
- James Terry, Jr., Concord, Massachusetts

Additional copies of this report are available from the National Academies Press, 500 Fifth Street, N.W., Lockbox 285, Washington, DC 20055; (800) 624-6242 or (202) 334-3313 (in the Washington metropolitan area); Internet, http://www.nap.edu.

THE NATIONAL ACADEMIES
Advisers to the Nation on Science, Engineering, and Medicine

The **National Academy of Sciences** is a private, nonprofit, self-perpetuating society of distinguished scholars engaged in scientific and engineering research, dedicated to the furtherance of science and technology and to their use for the general welfare. Upon the authority of the charter granted to it by the Congress in 1863, the Academy has a mandate that requires it to advise the federal government on scientific and technical matters. Dr. Bruce M. Alberts is president of the National Academy of Sciences.

The **National Academy of Engineering** was established in 1964, under the charter of the National Academy of Sciences, as a parallel organization of outstanding engineers. It is autonomous in its administration and in the selection of its members, sharing with the National Academy of Sciences the responsibility for advising the federal government. The National Academy of Engineering also sponsors engineering programs aimed at meeting national needs, encourages education and research, and recognizes the superior achievements of engineers. Dr. Wm. A. Wulf is president of the National Academy of Engineering.

The **Institute of Medicine** was established in 1970 by the National Academy of Sciences to secure the services of eminent members of appropriate professions in the examination of policy matters pertaining to the health of the public. The Institute acts under the responsibility given to the National Academy of Sciences by its congressional charter to be an adviser to the federal government and, upon its own initiative, to identify issues of medical care, research, and education. Dr. Harvey V. Fineberg is president of the Institute of Medicine.

The **National Research Council** was organized by the National Academy of Sciences in 1916 to associate the broad community of science and technology with the Academy's purposes of furthering knowledge and advising the federal government. Functioning in accordance with general policies determined by the Academy, the Council has become the principal operating agency of both the National Academy of Sciences and the National Academy of Engineering in providing services to the government, the public, and the scientific and engineering communities. The Council is administered jointly by both Academies and the Institute of Medicine. Dr. Bruce M. Alberts and Dr. Wm. A. Wulf are chair and vice chair, respectively, of the National Research Council.

www.national-academies.org

COMMITTEE ON INDICATORS FOR WATERBORNE PATHOGENS[1]

MARY JANE OSBORN, *Chair*, University of Connecticut Health Center, Farmington

R. RHODES TRUSSELL, *Vice Chair*, Trussell Technologies, Inc., Pasadena, California

RICARDO DELEON, Metropolitan Water District of Southern California, La Verne

DANIEL Y.C. FUNG, Kansas State University, Manhattan

CHARLES N. HAAS, Drexel University, Philadelphia, Pennsylvania

DEBORAH A. LEVY, Centers for Disease Control and Prevention, Atlanta, Georgia

J. VAUN MCARTHUR, University of Georgia, Savannah River Ecology Laboratory, Aiken, South Carolina

JOAN B. ROSE, Michigan State University, East Lansing

MARK D. SOBSEY, University of North Carolina, Chapel Hill

DAVID R. WALT, Tufts University, Boston, Massachusetts

STEPHEN B. WEISBERG, Southern California Coastal Water Research Project Authority, Westminster

MARYLYNN V. YATES, University of California, Riverside

Staff

MARK C. GIBSON, Study Director

JENNIFER KUZMA, Co-Study Director *(until January 2003)*

SETH H. STRONGIN, Project Assistant

---

[1]The activities of the Committee on Indicators for Waterborne Pathogens were organized and supported by the National Research Council's Board on Life Sciences (lead) and Water Science and Technology Board. Biographical sketches of committee members are contained in Appendix E and current rosters of the two parent boards are contained in Appendix D.

WORKSHOP ON INDICATORS FOR WATERBORNE PATHOGENS PARTICIPANTS

September 4, 2002

JOHN M. COLFORD, JR., University of California, Berkeley
CHRISTOPHER S. CROCKETT, Philadelphia Water Department
RAYMOND P. MARIELLA, Lawrence Livermore National Laboratory
PETER MARSDEN, Drinking Water Inspectorate of the United Kingdom
J. MICHAEL RAMSEY, Oak Ridge National Laboratory
GEOFFREY I. SCOTT, National Oceanic and Atmospheric Administration
MARK D. SOBSEY, University of North Carolina, Chapel Hill
TIMOTHY J. WADE, University of California, Berkeley

Preface

The use of bacterial indicator organisms to signal the possible presence of human pathogens in drinking water began more than a century ago in the United States, at a time when contamination of drinking and source waters by enteric bacterial pathogens, such as the typhoid bacillus, was a major public health threat. In subsequent decades, the use of bacterial indicators, predominantly coliforms, has been expanded to U.S. ambient, recreational, and shellfish waters and continues to focus on identification of fecal contamination, principally of human origin. Although these approaches have been extremely effective in reducing outbreaks of waterborne human disease, significant numbers of such outbreaks are still reported annually, many of unknown etiologic origin, and it is generally agreed that a substantial number of occurrences of waterborne human disease go unrecognized or unreported.

Recent advances in understanding the diversity and ecology of waterborne human pathogens as well as the ongoing rapid development of new techniques for detection and identification of waterborne microbes make it timely to reevaluate the standard indicators and indicator approaches employed to determine the microbiological quality of waters used for recreation or as sources of drinking water. Accordingly, the National Research Council (NRC) formed the Committee on Indicators for Waterborne Pathogens in 2002 at the request of the U.S. Environmental Protection Agency (EPA) to report on candidate indicators and/or indicator approaches (including technologies for detection) for assessing contamination of U.S. recreational waters and source water (including groundwater) for drinking water. The original charge to the committee excluded coastal marine and marine-estuarine waters, but these were added after subsequent discussion with EPA, and it was agreed that the study would then give less emphasis to some other aspects of the charge as described in Chapter 1. For example, the committee did not explicitly address indicators of water treatment performance. Furthermore, the report does not specifically address the threat of bioterrorism or the protection of vulnerable subpopulations such as infants and immunocompromised persons regarding microbial water quality.

To address its charge, the Committee on Indicators for Waterborne Pathogens met four times, starting in April 2002. The committee quickly concluded that it is not possible to identify a single, unique indicator or even a small set of indicators that is capable of identifying all classes of waterborne pathogens of public health concern for all applications and water media. Rather, priority should be given to the development of a phased monitoring approach for assessing microbial water quality that relies on a flexible "tool box" containing a spectrum of indicators and indicator approaches (to include direct monitoring of pathogens) that can be matched according to specific circumstances and needs. Thus, the committee did not conduct a comprehensive evaluation of candidate indicators or specific pathogens per se.

The committee would like to thank the many experts who contributed to this report by participating and/or speaking at committee meetings, including Rita Schoeny, Betsy Southerland, Ephraim King, Alfred Dufour, and Rebecca Calderon, EPA; and Roger Fujioka, University of Hawaii.

The committee also sponsored a one-day public workshop on candidate indicators and indicator approaches for waterborne pathogens on September 4, 2002, in Washington, D.C. This workshop provided insight on a wide variety of subjects related to the committee's charge, ranging from epidemiology to emerging detection technologies. The names and affiliations of the workshop presenters are listed in the front of this report.

This report has been reviewed in draft form by individuals chosen for their diverse perspectives and technical expertise, in accordance with procedures approved by the NRC's Report Review Committee. The purpose of this independent review is to provide candid and critical comments that will assist the NRC in making its published report as sound as possible and to ensure that the report meets institutional standards for objectivity, evidence, and responsiveness to the study charge. The review comments and draft manuscript remain confidential to protect the integrity of the deliberative process. We thank the following individuals for their review of this report: Jennifer Clancy, Clancy Environmental Consultants, Inc.; James Crook, Consultant; Mark Gold, Heal the Bay; Robert Haselkorn, University of Chicago; Mark LeChevallier, American Water; Laura Leff, Kent State University; Daniel Lim, University of South Florida; Christine Moe, Emory University; Erik Olson, Natural Resources Defense Council; David Relman, Stanford University; and Gary Toranzos, University of Puerto Rico.

Although the reviewers above have provided many constructive comments and suggestions, they were not asked to endorse the conclusions or recommendations nor did they see the final draft of the report before its release. The review of this report was overseen by Edward Bouwer, Johns Hopkins University. Appointed by the NRC, he was responsible for making certain that an independent examination of this report was carried out in accordance with institutional procedures and that all review comments were carefully considered. Responsibility for the final content of this report rests entirely with the authoring committee and the NRC.

This report would not have been possible without the hard work and dedication of Mark Gibson, study director and staff officer for the NRC's Water Science and Technology Board. The committee would like to thank project assistant Seth Strongin from the Board on Life Sciences (BLS) for logistical support throughout the study. We would also like to thank former co-study director Jennifer Kuzma and research associate Laura Holliday of the BLS for their early contributions to this report.

Finally, I would like to thank the 12 members of this committee for bringing this report together. Their diverse backgrounds and perspectives provided for lively and insightful discussions throughout the course of the entire study.

Mary Jane Osborn, *Chair*

Contents

Executive Summary

BACKGROUND

The establishment of sanitary practices in the late nineteenth century for the disposal of sewage and the increasing use of filtration and chlorination of drinking water throughout the twentieth century resulted in a dramatic decrease in waterborne diseases such as cholera and typhoid fever in the United States. Despite these historical efforts, ongoing advances in water and wastewater treatment, and several layers of federal, state, and local government laws, regulations, and guidance designed to protect public water supplies from contamination, waterborne disease outbreaks still occur every year in the United States. Furthermore, epidemiologists generally agree that these documented outbreaks represent only a fraction of the total that actually occur because many go undetected or unreported.

In order to protect public health, and as mandated in the Clean Water Act and the Safe Drinking Water Act (SDWA), it is important to have accurate, reliable, and scientifically defensible methods for determining whether source waters for drinking water and recreational waters are contaminated by pathogens and to what extent. For more than 100 years, U.S. public health personnel have relied extensively on an indicator organism approach to assess the microbiological quality of drinking water. More specifically, these enteric bacterial indicator microorganisms (predominantly "coliforms"[1]) are typically used to detect the possible

---

[1]Coliforms include several genera of bacteria belonging to the family Enterobacteriaceae, of which *Escherichia coli* is the most important member. The historical definition of this group is based on the method (lactose fermentation) used for its detection (see Chapter 1 for further information).

1

presence of microbial contamination of drinking water from human waste. The use of coliforms was later expanded and adopted for ambient, recreational, and shellfish waters and continues to focus on identification of fecal contamination.

Over the long history of their development and use, the current bacterial indicator approaches have become standardized, are relatively easy and inexpensive to use, and constitute a cornerstone of local, state, and federal monitoring and regulatory programs. An increased understanding of the diversity of waterborne pathogens, their sources, physiology, and ecology, however, has resulted in a growing understanding that the use of bacterial indicators may not be as universally protective as was once thought. For example, the superior environmental survival of pathogenic viruses and protozoa raised serious questions about the suitability of relying on relatively short-lived coliforms as an indicator of the microbiological quality of water. That is, while the presence of coliforms could still be taken as a sign of fecal contamination, the absence of coliforms could no longer be taken as assurance that the water was uncontaminated. Thus, existing bacterial indicators and indicator approaches do not in all circumstances identify all potential waterborne pathogens. Furthermore, recent and forecasted advances in microbiology, molecular biology, and analytical chemistry make it timely to reassess the current paradigm of relying predominantly or exclusively on traditional bacterial indicators for waterborne pathogens. Nonetheless, indicator approaches will still be required for the foreseeable future because it is not practical or feasible to monitor for the complete spectrum of microorganisms that may occur in source waters for drinking water and recreational waters, and many known pathogens are difficult to detect directly and reliably in water samples.

This report was written by the National Research Council (NRC) Committee on Indicators for Waterborne Pathogens—jointly overseen by the NRC's Board on Life Sciences and Water Science and Technology Board—and comprised of 12 volunteer experts in microbiology, waterborne pathogens (bacteriology, virology, parasitology), aquatic microbial ecology, microbial risk assessment, water quality standards and regulations, environmental engineering, biochemistry and molecular biology, detection methods, and epidemiology and public health. This report's contents, conclusions, and recommendations are based on a review of relevant technical literature, information gathered at four committee meetings, a public workshop on indicators for waterborne pathogens (held on September 4, 2002), and the collective expertise of committee members. Furthermore, because of space limitations, this Executive Summary includes only the major conclusions and related recommendations of the committee in the general order of their appearance in the report. More detailed conclusions and recommendations can be found within individual chapters and are summarized at the end of each chapter.

The committee was formed in early 2002 at the request of the U.S. Environmental Protection Agency (EPA) Office of Water and originally charged to report on candidate indicators and/or indicator approaches (including detection technologies) for microbial pathogen contamination in U.S. recreational waters (*excluding* coastal marine water and marine-estuarine water) and source water (including groundwater) for drinking water. It is important to note that the committee's original charge was slightly but substantively altered after its first meeting and subsequent discussions with EPA, most notably to *include* coastal and marine-estuarine recreational waters that were originally excluded (see Box ES-1). The consequences of this important change regarding the emphasis and content of this report are summarized in Chapter 1. For example, it was agreed that the report would give less space and emphasis to defining currently known waterborne pathogen classes and anticipating those emerging waterborne patho-

---

BOX ES-1
Statement of Task

The NRC will convene a committee to report on candidate indicators and/or indicator approaches (including detection technologies) for microbial pathogen contamination in U.S. recreational waters (including coastal marine water and marine/estuarine water) and source water (including groundwater). Specifically, the committee will:

1. Review and provide perspective on the importance and public health impacts of waterborne pathogens as discussed in previous National Academies' reports and other seminal reports.
2. Develop candidate lists or sets of appropriate and scientifically defensible indicators and/or indicator approaches. In doing so, the committee will:

- define currently known waterborne pathogen classes and anticipate those emerging waterborne pathogens that are likely to be of public health concern;
- evaluate the strengths and weaknesses of the candidates for reflecting the presence, quantity, and viability of these important pathogens;
- explore whether a selected subset of indicators, a unique indicator, and/or specific indicator approaches can help to identify the source or sources of water contamination (including discharges from municipal publicly owned treatment works);
- assess the practicality of using these candidates at local, state, and federal levels given current technology, personnel, and water quality monitoring programs;
- comment on data, research, and information needs for short- and long-term validation of candidates; and
- consider how the list of recommended candidates might change with future technological developments.

gens that are likely to be of public health concern (although Appendix A provides a brief summary discussion and table of new and [re]emerging waterborne pathogens).

While this is the first NRC study to focus specifically on indicators for waterborne pathogens, issues surrounding their use have been discussed in several recent and historical NRC reports, as summarized in chronological order in Appendix B. In addition, many federal, state, and local government and nongovernmental organizations, including the water industry and academia, have addressed the issue of the microbiological quality of drinking water and recreational water and its association with various adverse human health effects. Thus, Appendix B also includes summaries of some major reports that have been published addressing these concerns.

HEALTH EFFECTS ASSESSMENT: EPIDEMIOLOGY OF WATERBORNE HUMAN DISEASE

The ultimate objective for determining the microbiological quality of water is to identify and then minimize the public health risk from consuming water intended for drinking and from exposure to recreational water. Health effects assessments for waterborne pathogens can be based on a number of approaches. Each approach has strengths and weaknesses (see Chapter 2 for further information), and all have been or are being used to document and quantify the health risks from microbes in water. Many of these approaches involve the use of epidemiology, which is a well-established, essential tool for determining the linkage between the presence of identified waterborne pathogens and their indicators and human disease. However, the significant cost and methodological difficulty of designing, conducting, and interpreting epidemiologic studies have limited their use.

One health effects assessment approach that is being used increasingly for waterborne pathogens is quantitative microbiological risk assessment (QMRA). Developed in the 1980s following the traditional framework for chemical risk assessment, QMRA has several substantive differences, and most applications to date have focused on its use to predict primary infections or illnesses resulting from exposure to a contaminated medium such as water. QMRA is a useful tool for identifying the potentially most influential parameters in the waterborne disease transmission process for which there are data gaps, especially models that include infectious disease parameters such as immunity. However, some of the key needs for QMRA include dose-response and exposure information (e.g., intensity and duration of contagion), which are often lacking. In some cases, impacts from such population level phenomena may dramatically alter projected estimates of human risk.

The comprehensiveness of investigations of waterborne disease outbreaks in the United States varies by the type of outbreak and by state, and results are compiled in the Centers for Disease Control and Prevention's (CDC's) surveillance system. However, this system has low sensitivity and does not consistently provide information that links indicator and pathogen data with adverse health outcomes. This gap occurs because most outbreak investigations include primarily the epidemiologic component which concentrates on linking illness to water (e.g., through determination of the agent in clinical specimens), and tend to neglect the environmental component (e.g., determination of water quality through measurement of indicator and pathogen occurrence in the water). This gap occurs more frequently with outbreaks associated with drinking water than with those associated with recreational water. In addition, 40-50 percent of the identified waterborne disease outbreaks remain of unknown etiology.

A substantial effort to determine the potential health risks associated with consumption of drinking water has been going on since the late 1990s. As required by the SDWA Amendments of 1996, epidemiologic studies of drinking water and endemic disease have focused on establishing associations between water consumption and gastrointestinal illness. Thus far, these studies have not established a good correlation between indicators of waterborne pathogens, the pathogens themselves, and adverse human health effects, although some earlier studies have shown an association between tap water and endemic gastrointestinal illness. To have adequate statistical power to address the epidemiologic association of health outcomes with specific indicators and specific waterborne pathogens, the study sample has to be large, which leads to significant costs. In addition, methodologic complexities as well as difficulty in interpretation of results have limited the use of some of the studies.

In contrast, epidemiologic studies involving recreational bathing waters have shown predictive associations between several swimming-associated health effects and various microbial indicators or pathogens. A recent systematic review and meta-analysis of recreational waterborne studies (both freshwater and marine) undertaken at the specific request of the committee confirmed that indicators can provide reliable estimates of water quality that are predictive of human health risks under some but not all water quality conditions (see Chapter 2 for further information).

Building on these conclusions, the committee provides several recommendations regarding future directions for epidemiologic and microbiological research as related to health effect assessment for waterborne pathogens and their indicators. In this regard, the committee first recommends that EPA and CDC take a greater leadership role in such efforts, and fund and work with stakeholders and academic researchers in the following areas:

• CDC should actively work with state and local health departments to encourage testing for pathogens (especially viruses and parasites) in clinical specimens during waterborne outbreak investigations.
• CDC and EPA should actively work with state and local health depart-

ments to encourage collection and testing of environmental data (i.e., water quality data for source, finished, and distribution system waters that include indicators and pathogens) during waterborne outbreak investigations.

• Standardized protocols and definitions are needed for outbreak investigations and epidemiologic studies, especially to help ensure a comprehensive investigation or study that includes the collection of clinical, laboratory, and environmental data (including co-occurrence of pathogens and indicators).

• Fewer but more comprehensive epidemiologic studies should be conducted rather than multiple small-scale studies that do not adequately address multiple risk factors and health outcomes when working within a fixed or constrained budget. More specifically, the link between pathogens and their potential indicators, and among pathogens, indicators, and adverse health outcomes, would be strengthened by including in comprehensive and adequately funded studies, epidemiologic measurements of health outcomes, measurements of pathogens in clinical specimens, as well as measurements of pathogens and their potential indicators in relevant water samples.

• Additional epidemiologic studies are needed to look at the association between water consumption and gastrointestinal illness in groundwater systems, and to correlate water quality data (pathogens and indicators) with health outcomes. Furthermore, these studies should include the collection of epidemiologic, clinical, laboratory, and environmental data whenever feasible.

• Health outcomes studied in association with drinking water exposure should not be limited to gastrointestinal illness (e.g., should consider including respiratory and dermatological illnesses).

• Additional epidemiologic studies should be conducted to determine the occurrence of chronic/recurrent disease attributable to waterborne pathogens in habitual users of recreational waters (e.g., surfers) from point and nonpoint sources of contamination.

• Studies on recreational waters should be carried out on a broader range of geographical and ecological sites, including tropical and sub-tropical waters, and ocean beaches.

ECOLOGY AND EVOLUTION OF WATERBORNE PATHOGENS AND INDICATOR ORGANISMS

The nature and abundance of waterborne microbial assemblages and communities varies widely from location to location and over time, according to various environmental and anthropogenic factors. In general, most waterborne pathogens of public health concern discussed in this report (see also Appendix A) are not native to the types of waterbodies addressed here, but are introduced from either point sources (e.g., sewage discharge) or nonpoint sources (e.g., agriculture, rainfall).

Past efforts to develop and implement indicators of microbial contamination have often given little or no consideration to the role of evolution in the ecology and natural history of such waterborne pathogens. Their ecology and evolution, however, have important implications for their emergence and reemergence in the aquatic environment and for subsequent public health concerns. Furthermore, the concept of using indicators for waterborne pathogens implies that certain characteristics of microorganisms such as genes and gene products remain constant under varying environmental conditions. However, this assumption does not always hold because the effectiveness of indicator technologies that are based on the detection of some aspect of the biology or chemistry of a living organism (whether a pathogen or an indicator microorganism) may decrease over time due to evolutionary changes in the target organism. Therefore, it is important to understand the effects of the environment on these targets and organisms. For these reasons, existing and candidate indicator organisms should have ecologies and responses to environmental variations that are similar to those of the pathogenic organisms whose presence they are supposed to be indicating.

The committee provides the following recommendations to improve understanding of the ecology and evolution of waterborne pathogens, as related to the development of new and effective indicators of microbial contamination:

- Bacteria, viruses, and protozoa have evolved mechanisms that facilitate their rapid response to environmental changes and may influence their infectivity and pathogenicity. Therefore, additional research is needed on microbial evolutionary ecology to address long-term public health issues.
- Genetic and phenotypic characterizations of pathogenic viral, bacterial, and protozoan parasites are needed to elucidate zoonotic relationships with their animal hosts and the factors influencing waterborne transmission to humans.
- The ecology of waterborne pathogens should be assessed in relation to modern agricultural practices and other anthropogenic activities, such as urbanization. Animal wastes from agriculture and urban sewage, runoff and storm water are major contributors of both human pathogenic and non-pathogenic strains of microbes and the wide use of antibiotics in animal agriculture and in human and veterinary therapy leads to selection for antibiotic-resistant phenotypes.
- Advanced analytical methods should be used to help discriminate between introduced pathogenic and naturally occurring non-pathogenic strains of waterborne microorganisms and to characterize the emergence of new strains of pathogens as a result of genetic change.
- Natural background density of waterborne pathogens should be established to differentiate between native opportunistic pathogens and introduced pathogens. Efforts should be made to differentiate between indicators and pathogens that are native to the environment and those that are introduced from external sources, such as human and animal wastes.
- Research is need to develop a better understanding of the ecology and natural history of both the environmental and infectious stages of pathogens and

the parallel stages of indicator organisms to grasp how the organisms are distributed in nature; how they persist and accumulate in water, other environmental media, and animal reservoirs; and how dissemination of the environmental form occurs, especially human exposure.

ATTRIBUTES AND APPLICATION OF INDICATORS

Microbial water quality indicators are used in a variety of ways within public health risk assessment frameworks, including assessment of potential hazard, exposure assessment, contaminant source identification, and evaluating effectiveness of risk reduction actions. As noted previously, however, no single indicator or analytical method (or even a small set of indicators or analytical methods) is appropriate to all applications. A suite of indicators and indicator approaches is required for different applications and different geographies.

For almost 40 years, Bonde's attributes of an ideal indicator have served as an effective model of how a fecal contamination index of public health risk and treatment efficiency should function (see Box 4-1). However, Bonde's attributes must be refined to continue their relevance to public health protection because the development and increasing availability of new measurement methods necessitates the separation of criteria for evaluating indicators and detection methods. Historic definitions of microbial indicators, such as coliforms, have been tied to the methods used to measure them. Newly available methods (particularly molecular methods) allow more specificity in the taxonomic grouping of microorganisms that are measured. More importantly, a variety of new methods are becoming increasingly available, providing several options for measuring each indicator group. Thus, separate criteria allow one to choose the indicator with the most desirable biological attribute for a given application and then match this with a measurement method that best meets the need of the application. Box ES-2 provides a summary listing of desirable biological attributes of indicators and desirable attributes of indicator methods (see Boxes 4-2 and 4-3 for further information).

The most important biological attribute is a strong quantitative relationship between indicator concentration and the degree of public health risk. One of the most important method attributes is its specificity, or ability to measure the target indicator organism in an unbiased manner. The speed of the method (processing time and rapidity of results) is also an important characteristic in many applications.

Several factors limit the effectiveness of current recreational water warning systems, the most prominent of which is the delay in warnings caused by long laboratory sample processing time. One approach that is increasingly being used to address this problem is predictive models intended to prevent exposure. Another shortcoming of present warning systems is the poorly established relationship between presently used indicators and health risk. Present studies do not

---

BOX ES-2
Summary of Desirable Attributes of Indicators and
Indicator Methods

Biological Attributes of Indicators
- Correlated to health risk
- Similar (or greater) survival to pathogens
- Similar (or greater) transport to pathogens
- Present in greater numbers than pathogens
- Specific to a fecal source or identifiable as to source of origin

Attributes of Methods
- Specificity to desired target organism
- Broad applicability
- Precision
- Adequate sensitivity
- Rapidity of results
- Quantifiable
- Measures viability or infectivity
- Logistical feasibility

---

address all sources of contamination, have not been conducted in enough geographic locations, and do not address chronic exposure. Many reported failures of beach water quality standards are associated with nonpoint source contamination, but the epidemiologic studies used to establish recreational bathing water standards have been based primarily on exposure to point source contamination dominated by human fecal material. A final problem with present water contact warning systems is that bacterial indicator concentrations are spatiotemporally variable and most sampling is too infrequent to transcend this granularity. There are many promising source identification techniques that can help in deciding whether a health warning should be issued or in identifying the best approach for fixing the problem. However, these techniques have not yet been standardized or fully tested.

Groundwater quality monitoring is rare, despite data that show the majority of waterborne outbreaks of disease in the United States result from groundwater systems. Viral contamination of groundwater is a particular concern because the small size and considerable environmental persistence of viruses makes it more likely they will reach and contaminate groundwater. The known risks from viruses in fecally contaminated groundwater, and evidence that human enteric viruses are detectable in fecally contaminated groundwater, suggest that coliphage or direct virus monitoring would enhance the assessment of groundwater microbiological quality and would make a better indicator of human health risk.

The committee makes the following recommendations related to the consideration and use of indicator attributes for different applications:

• The link between potential indicators and pathogens, and among indicators, pathogens, and adverse health outcomes, would be strengthened by including measurements of both indicators and pathogens in comprehensive epidemiologic studies. In particular, studies should be conducted to better assess the role of nonpoint sources in occurrence of human pathogens and indicator organisms, disease outbreaks, and endemic health risks in recreational waters. Use of alternative indicators need to be included in these studies.

• Improved indicators for viruses in groundwater sources of drinking water need to be developed.

• New paradigms for reporting water contact health risk, such as "letter grades" for public beaches, need to be developed. The present all-or-none closure decisions can misinform the public because of large spatiotemporal heterogeneity in indicator concentrations. Letter grades are one option that would effectively address the granularity issue by integrating data over a longer time period and are readily understandable.

• Investment should be made in developing rapid analytical methods because the most commonly used warning systems involve laboratory methods that are too time consuming to achieve the best possible public health protection.

• There are several promising source identification (i.e., microbial source tracking) techniques on the horizon that should be incorporated into monitoring systems when they have been adequately validated. Public health risk from exposure to fecally contaminated water is likely to vary depending on whether high indicator concentrations resulted from animal or human sources, and microbial source tracking tools will allow public health managers to incorporate that distinction into their decision making.

• Models that predict future water quality conditions, based on factors such as rainfall, are potentially valuable tools for warning the public before exposure occurs, but the scientific foundation for these models has to be enhanced before they can be widely used.

NEW BIOLOGICAL MEASUREMENT OPPORTUNITIES

Although classic microbiological culture methods for detection of pathogens and indicator microorganisms have proved effective over many decades, the advent of increasingly sophisticated and powerful molecular biology techniques provides new opportunities to improve upon present indicators and pathogens by both culture and non-culture methods. Regardless of the indicator or detection method or approach used, it is essential that collection, sample processing or preprocessing, measurement, and data processing all be considered for accurate analysis of microbial water quality (i.e., not merely the measurement itself).

The collection of representative samples requires careful consideration of the objectives or purpose of sampling in the context of the need to obtain a reliable estimate of microbial exposure in a timely fashion. At present, most water quality measurement methods are single-parameter based. Ongoing research in the micro/nano-technology field, combined with efforts in array sensing and intelligent processing, should provide the tools for creating inexpensive, ubiquitous universal sensing and detection systems now and over the next several decades. This development is essential because the committee recognizes the lack of technical, infrastructure, and financial resources required to implement advanced water quality monitoring methods in many parts of the United States. The microbiological community needs to develop and implement multiparameter approaches in which many technologies and methods are integrated to provide the best possible information regarding the microbial quality of water.

The funding of methods development in the United States has been relatively poor to date for many pathogens, for new and emerging methods, and for new and innovative indicators. Development of new and improved methods has been funded substantially for only a few pathogens, specifically those targeted for regulation in drinking water. Greater and more consistent efforts should be made to support methods development for new and emerging microbial detection technologies, for many more pathogens, and for new and improved candidate indicators of waterborne pathogens.

Newer methods involving immunofluorescence techniques and nucleic acid analysis are proving their value, and novel microtechnologies are evolving rapidly, spurred partly by recent concerns about bioterrorism. However, problems associated with sample concentration, purification, and efficient (quantitative) recovery remain and will require significant effort to be resolved. One technology area that will enable significant reductions in sample preparation and separation time is the field of microfluidics and microelectromechanical systems. Thus, the introduction of molecular techniques for nucleic acid analysis is viewed by the committee as a growth opportunity for waterborne pathogen detection.

With the prospect for such an enormous amount of data to be collected from the many sensors disposed on arrays, the potentially large numbers of sensor arrays deployed for water monitoring, and the continuous data streams coming from these sensor networks, greater attention must be paid to the fields of data analysis, intelligent decision making, and archiving. There is a need for a database that compiles and serves as a clearinghouse for all microbiological methods that have been utilized and published for studying water quality. Research methods, in particular those that have great potential for becoming accepted as conventional methods, will have to be documented.

Recent developments in molecular and microbiology methods and their application to public health-related water microbiology have necessitated a new approach for the rapid assessment, standardization, and validation of such methods. It is clear that a major effort is needed for accessible methods to examine microbial water quality for health decisions. To move new methods into the main-stream, a process is required that not only allows for standardization and validation, but also facilitates widespread acceptance and implementation. In this regard, the committee concludes that the Association of Analytical Communities (AOAC International) Peer-Verified approach or its equivalent may be the best way forward.

Based on these conclusions, the committee provides the following recommendations regarding the development of new biological measurement opportunities in the field of microbial water quality assessment:

- A specific program on promising research methodologies for waterborne microorganisms of public health concern should be supported by EPA and other organizations concerned with microbial water quality. Such methodologies need not be microorganism specific, but should be application specific, focusing on the desirable attributes of the method.

- Ongoing research should be supported and expanded to develop and validate rapid, ultimately inexpensive, sensitive, and robust methods for detection and measurement of all classes of waterborne pathogens and their indicators. Such expanded research should go beyond pathogenic bacteria and indicators to include improved methods for the detection of pathogenic viruses and protozoa.

- Additional research is needed to develop improved methods for rapid sample concentration and effective, reproducible microbial recovery.

- The adoption of new molecular techniques should be accelerated for waterborne pathogen detection. New methods undergoing validation should be tested using whole microorganisms, rather than just extracted DNA or RNA targets.

- Research should be funded to develop approaches to the detection of infectious or viable microbes by nucleic acid detection methods.

- Focused efforts should be made to support the development of inexpensive and rapid fieldable methods for testing microbial water quality. This will require the concurrent development of reagents, methods, and the attendant portable instruments that can survive repeated transport and use in the field.

- Issues of sensitivity, reproducibility, and representativeness of miniaturized detection methods should be addressed as one of the most important technological challenges to analysis of waterborne pathogens and indicators.

- EPA should reinvigorate its role with standard-setting organizations, including the American Society for Testing and Materials International (ASTM), AOAC, and the International Organization for Standardization (ISO), that focus on new and innovative methods. Regular input by professional organizations such as the American Society for Microbiology should also be encouraged.

- EPA should support the design, development, and maintenance of a nationwide database that compiles and serves as a clearinghouse for all microbiological methods used and published for studying water quality. Guidance on appropriate data needed for methods studies and a process for on-line iterative development of consensus methods should be included in this database.

EVALUATION OF RISK:
A PHASED APPROACH TO MONITORING
MICROBIAL WATER QUALITY

Many factors, including differences in the purpose for which indicator data will be used, environmental conditions, and the availability of technology to detect the microorganism(s), profoundly affect the choice of appropriate indicators and approaches to assessing microbial water quality. The selection of a suitable indicator or indicator strategy may also differ significantly depending on the water type or waterbody under consideration and the intended use of the information. Often, indicators are used to provide an early warning of potential microbial contamination, an application for which a rapid, simple, broadly applicable technique is appropriate. Indicators are used for confirmation of health risk, where resulting actions can be costly and time consuming. They are also used to identify and ameliorate the source of a microbial contamination problem. In both of the latter applications, the time frame and investment in indicators, indicator approaches, and methods must be greater than those typically used in routine monitoring.

Because a single, unique indicator or even a small set of microbial water quality indicators cannot meet this diversity of needs and applications, what is required is development and use of a "tool box" in which the indicator(s) and method(s) are matched to the requirements of a particular microbial water quality application. In this regard, the committee recommends the use of a phased, three-level monitoring framework, as illustrated in Figure ES-1, for selecting indicators and indicator approaches for waterborne pathogens. Chapter 6 describes the potential application of this framework to three typical monitoring situations in the near-term future (including the proposed Ground Water Rule and Interim Enhanced Surface Water Rule provisions) and in the long-term future at each level of investigation.

The first phase of this framework is screening or routine monitoring (Level A). The objective of this phase is early warning of a health risk or of a change from background condition that could lead to a health risk. This is the most frequent type of monitoring and is conducted routinely throughout the country. In general, the most important indicator attributes at this level are speed, low cost (logistical feasibility), broad applicability, and sensitivity.

Once screening has identified a potential problem, the second phase involves more detailed studies to confirm a health risk (Level B). The aim of such investigations is to assess the need for further management actions and/or expanded specific data gathering efforts. The confirmation phase often involves measurement of new indicators, including direct measurement of pathogens. Such studies

Phases of investigation

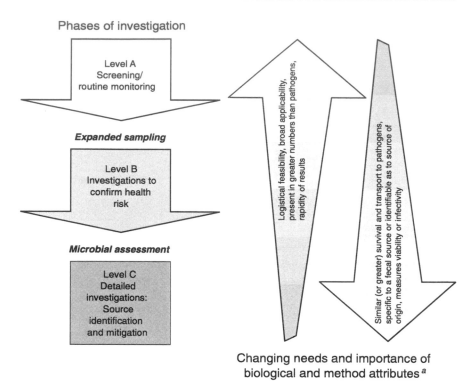

Changing needs and importance of
biological and method attributes [a]

FIGURE ES-1 Recommended three-level phased monitoring framework for selection and
use of indicators and indicator approaches for waterborne pathogens.
[a]Not all biological and method attributes summarized in Box ES-2 are included in this
figure, nor are they listed in order of importance.

are not initiated on a routine basis, but would typically be undertaken when screening indicators persist at high levels without a clearly identifiable contamination
source. Many of the new and emerging biological measurement methods and
detection technologies described in Chapter 5 and Appendix C will be useful at
this stage of investigation. Since confirmation studies focus on assessing health
risk, the most important indicator biological attributes during this phase are correlation with contamination sources and transport or survival behavior similar to
pathogens.

The third phase (Level C) involves studies to determine the sources of microbial contamination so that the health risk can be abated through a variety of engineering and policy solutions. However, this report focuses on the identification of
sources of microbial contamination rather than their mitigation. In some cases,
source identification is accomplished through expanded spatial sampling to look
for gradients in indicator organisms or pathogens. Where recreational waters are

concerned, new indicator strategies based on molecular signatures are increasingly used in place of screening indicators. For these detailed investigations, essential indicator attributes include specificity to a fecal source and quantifiability. Level C studies often overlap in goal with Level B health risk confirmation studies, since identifying the source helps to identify health risk. Depending on the overlap, the ability to measure infectiousness may also be important.

The preceding discussion is focused on microbial water quality monitoring to support risk management, not the risk management actions themselves which are largely beyond the scope of this report. Appropriate risk management decisions depend not only on the results of monitoring but on the application the monitoring is designed to address. For example, when Level A monitoring identifies a potential risk in a drinking water supply that already receives complete treatment before use, management actions might be limited until Level B and/or C investigations enable the solidification of that risk assessment, at which time a significant upgrade in treatment might be in order. On the other hand, if the Level A monitoring outcomes are on water to which the public is directly exposed such as at recreational beaches, significant and immediate management action may be called for in parallel with Level B and C monitoring activities, depending on the risks implied by the Level A results. Under any circumstances, time is of the essence in all three levels of microbial water quality monitoring whenever there exists a possibility that the public may be at risk.

In general, the standardization and regulation of monitoring methodology decrease as indicator use moves through the three phases. Methods used in the screening phase are typically "standard" and can be accomplished by almost all county health department laboratories. This also holds true for many Level B studies. However, the techniques used during the latter phases may be more specialized and require the expertise of a research laboratory—this is especially true for Level C studies. The responsibility for study costs may also shift through the three phases, with parties potentially responsible for contamination sources typically more involved with the latter two levels of studies.

Microbial measurement technology is evolving rapidly and there is an opportunity to leverage these advances toward water quality needs. If sufficient investment is made in the coming decade, indicator systems will undergo a comprehensive evolution, and the correct and rapid identification of waters that are contaminated with pathogenic microorganisms will be substantially enhanced. Historically, EPA has focused much of its investment on indicators and indicator systems that are used at the screening level (A), but there is an increasing need for national leadership and guidance for the subsequent phases of microbial investigation that follow screening. Within the limited context of screening, EPA's guidance has been of mixed value, and in this regard, the committee concludes that (1) the selection of enterococci for screening at marine recreational beaches is appropriate because enterococci have shown to have the best relationship to health risk; (2) existing and proposed monitoring requirements for surface water sources of drinking water are irregular and are not supported by adequate research; and (3) proposed monitoring requirements for groundwater are not adequately protective for viral pathogens.

Based on these conclusions, the committee provides the following recommendations regarding the development of a phased approach to monitoring microbial water quality:

- EPA should invest in a long-term research and development program to build a flexible tool box of indicators and methods that will serve as a resource for all three phases of investigation identified in this report.
- This tool box should include the following: (1) the development of new indicators, particularly direct measures of pathogens that will enhance health risk confirmation and source identification; (2) the use of coliphages, as suggested by EPA's Science Advisory Board, in conjunction with bacterial indicators as indicators of groundwater vulnerability to fecal contamination; and (3) the use of routine microbiological monitoring of surface water supplies of drinking water before as well as after treatment.
- A significant portion of that investment should be directed toward concentration methods because existing technology is inadequate to measure pathogens of concern at low concentrations.
- Consistent with previous related recommendations, EPA should invest in comprehensive epidemiologic studies to (1) assess the effectiveness and validity of newly developed indicators or indicator approaches for determining poor microbial water quality and (2) assess the effectiveness of the indicators or indicator approaches at preventing and reducing human disease.
- EPA should develop a more proactive and systematic process for addressing microorganisms on the Drinking Water Contaminant Candidate List (CCL). The EPA should (1) prepare a review of published methods for each CCL microorganism and groups of related microorganisms, (2) publish those reviews on the Internet so researchers and practitioners can use them and comment on how to improve them, and (3) promote their use in special studies and monitoring efforts.

These conclusions and recommendations should not be taken as an excuse to either cling to or abandon current indicator systems until research develops new approaches. On the contrary, the committee recommends a phased approach to monitoring, as both a means to make existing indicator systems more effective, and to encourage the successive adoption of new, more promising indicator systems as they become available.

1

Introduction and Historical Background

INTRODUCTION

The first outbreak of a waterborne disease to be scientifically documented in modern Western society occurred in London, England, in 1854. This early epidemiology study by John Snow, a prominent local physician, determined that the consumption of water from a sewage-contaminated public well led to cholera (Snow, 1854a,b). This connection, decades before the germ theory of disease would be hypothesized and proven, was the first step to understanding that water contaminated with human sewage could harbor microorganisms that threaten public health. Since then, epidemiology has been the major scientific discipline used to study the transmission of infectious diseases through water (NRC, 1999a).

In the late nineteenth century and throughout the twentieth century, sanitary practices were established in the UnhÚed States regarding the handling and disposal of sewage, while filtration and chlorination systems were increasingly used to disinfect drinking water. Through these historical efforts and owing to ongoing advances in water and wastewater treatment and source water protection, the United States has secured and maintains one of the cleanest and safest supplies of drinking water in the world. Starting in 1920, national statistics on waterborne disease outbreaks caused by microorganisms, chemicals, or of unknown etiology have been collected by a variety of researchers and federal agencies (Lee et al., 2002). These data demonstrate that several outbreaks still occur every year in this country. Moreover, epidemiologists generally agree that these reported outbreaks represent only a fraction of the total that actually occur because many go undetected or unreported (NRC, 1999a). Thus, continued vigilance to protect the public from waterborne disease remains a necessity.

For more than 100 years, U.S. public health personnel have relied extensively on an indicator organism approach to assess the microbiological quality of drinking water. These bacterial indicator microorganisms (particularly "coliforms," described later) are typically used to detect the possible presence of microbial contamination of drinking water by human waste. More specifically, fecal indicator bacteria provide an estimation of the amount of feces, and indirectly, the presence and quantity of fecal pathogens in the water. Over the long history of their development and use, coliform test methods have been standardized, they are relatively easy and inexpensive to use, and enumeration of coliforms has proven to be a useful method for assessing sewage contamination of drinking water. In conjunction with chlorination to reduce coliform levels, this practice has led to a dramatic decrease in waterborne diseases such as cholera and typhoid fever. Furthermore, the use of bacterial indicators has been extended to U.S. "ambient" waters in recent decades—especially freshwater and marine-estuarine waters used for recreation. However, an increased understanding of the diversity of waterborne pathogens, their sources, physiology, and ecology has resulted in a growing understanding that the current indicator approach may not be as universally protective as was once thought. In this regard, several limitations of bacterial indicators for waterborne pathogens have been reported and are discussed throughout this report.

To protect public health, it is important to have accurate, reliable, and scientifically defensible methods for determining when water is contaminated by pathogens and to what extent. Furthermore, recent and forecasted advances in microbiology, biology, and analytical chemistry make it timely to assess the current paradigm of relying predominantly or exclusively on traditional bacterial indicators for waterborne pathogens in order to make judgments concerning the microbiological quality of water to be used for recreation or as a source for drinking water supply.

Committee and Report

This report was prepared by the National Research Council (NRC) Committee on Indicators for Waterborne Pathogens—jointly overseen by the NRC's Board on Life Sciences and Water Science and Technology Board. The committee consists of 12 volunteer experts in microbiology, waterborne pathogens (bacteriology, virology, parasitology), aquatic microbial ecology, microbial risk assessment, water quality standards and regulations, environmental engineering, biochemistry and molecular biology, detection methods, and epidemiology and public health. The report's conclusions and recommendations are based on a review of relevant technical literature, information gathered at four committee meetings, a public workshop on indicators for waterborne pathogens (held on September 4, 2002), and the collective expertise of committee members.

The committee was formed in early 2002 at the request of the U.S. Environ-

mental Protection Agency (EPA) Office of Water to report on candidate indicators and/or indicator approaches (including detection technologies) for microbial
pathogen contamination in U.S. recreational waters (*excluding* coastal marine
water and marine-estuarine water) and source water (including groundwater) for
drinking water.[1] It is important to note that the committee's charge, as outlined in
its statement of task (see Box ES-1), was slightly but substantively altered after
its first meeting and subsequent discussions with EPA, most notably to *include*
coastal and marine-estuarine recreational waters that were originally excluded.
As a result, it was agreed that the committee's report would give less space and
emphasis to the importance and public health impacts of waterborne pathogens;
place less emphasis on defining currently known waterborne pathogen classes
and anticipating those emerging waterborne pathogens that are likely to be of
public health concern (although Appendix A provides a brief summary discussion and table of new and [re]emerging waterborne pathogens); exclude consideration of blue-green algae and their toxins; and not specifically consider how the
use of candidate indicators might allow for determination of an appropriate level
of water treatment needed to protect public health. It is also important to state that
although an assessment of suitable indicators for shellfish waters is beyond the
scope of this report, some discussion of shellfish experience is included because
of the (especially historical) interrelatedness of the various microbial indicator
standards and their development. Lastly, this report does not address public swimming and wading pools that are regulated by state and local health departments
whose disinfection practices vary widely from place to place.

This chapter provides an introduction to the public health importance of
waterborne pathogens; a brief summary of key federal laws, regulations, and programs concerning microbial water quality monitoring and especially the use of
indicator organisms; the historical development and current use of microbial indicators for waterborne pathogens; and the current status of waterborne disease
outbreaks and endemic disease. The chapter ends with a summary of its contents
and conclusions. Chapter 2 provides an overview of health effects assessment as
related to the current and future use of indicators of waterborne pathogens to help
protect public heath. Chapter 3 focuses on the ecology and evolution of waterborne pathogens and indicator organisms by major classes (i.e., viruses, bacteria,
protozoa). Chapter 4 assesses the development and uses of indicators and indicator approaches according to their applications and attributes, while Chapter 5
reviews some emerging and innovative approaches for measuring indicator organisms and waterborne pathogens. Lastly, Chapter 6 provides a recommended

---

[1]For the purposes of this report, surface water sources for drinking water and recreational waters
can be considered a subset of U.S. "ambient waters" and "waters of the United States" (see footnote
2). As such and per the statement of task, unless noted otherwise all discussion of "water" in this
report refers to source water for drinking water (including groundwater) and freshwater, coastal, and
marine-estuarine recreational waters.

phased monitoring framework for selection and use of indicators, along with examples of how to use such a framework.

Relevant Laws and Regulations

It is beyond the scope of this report to systematically review and discuss all
federal, state, or local laws, regulations, and programs that concern the microbiological quality of source water for drinking water and ambient recreational waters. Regarding the latter, state and local governments have primary authority for
maintaining the quality and safety of recreational waters (both freshwater and
marine). However, given their nationwide application, importance to this report,
and direct relevance to the committee's charge, a brief discussion of the Safe
Drinking Water Act (SDWA), Clean Water Act (CWA), Beaches Environmental
Assessment and Coastal Health (BEACH) Act of 2000, and several related regulations and programs follows (see also Tables 1-1 and 1-2).

Safe Drinking Water Act

The SDWA, enacted in 1974 and administered by EPA, is the most important and comprehensive law designed to protect the public from man-made or
naturally occurring contaminants in drinking water. It has been amended regularly, including significant changes in 1986 and 1996. Prior to passage of the

TABLE 1-1 Microbiological and Other Indicators Used Under EPA's Drinking
Water Regulations[a,b]

Rule or Program	Indicator	Use	URLs and Notes
Total Coliform Rule	Total coliforms (TC)	Determine treatment efficiency and distribution system integrity	http://www.epa/gov/ safewater/tcr/tcr.html
	Fecal coliforms (FC) *Escherichia coli*	Determine or verify presence of fecal contamination if PWSs obtain sample(s) positive for TC	
Surface Water Treatment Rule (SWTR), as amended by the following rules	Turbidity	Measure of filter efficiency and source water quality	http://www.epa.gov/ safewater/mdbp/ ieswtrfr.pdf
	Disinfectant residual	Nondetection of a disinfectant residual indicates a distribution system problem	

TABLE 1-1 Continued

Rule or Program	Indicator	Use	URLs and Notes
Interim Enhanced SWTR	Heterotrophic plate count	Measure of drinking water quality in distribution system	http://www.epa.gov/ OGWDW/mdbp/ ieswtr.html
Long-Term 1 ESWTR	TC FC	Measure of source water quality (for unfiltered PWSs)	http://www.epa.gov/ safewater/mdbp/ lt1eswtr.html
Information Collection Rule (1996-1998)	*Cryptosporidium* *Giardia* Total culturable viruses TC FC	Results provided information to facilitate development of the Long-Term 2 ESWTR	http://www.epa.gov/ safewater/icr.html Applicable for PWSs serving ≥ 100,000 persons to provide treatment data and monitor disinfection by-products and source water quality parameters
Long-Term 2 ESWTR (proposed rule)	*Cryptosporidium* *E. coli*	Determine minimum treatment level needed by surface water system	www.epa.gov/ safewater/lt2/ st2eswtr.html
Groundwater Rule (final rule expected in late 2004)	*E. coli* Enterococci Coliphage	Determine presence of fecal contamination in source groundwater	http://www.epa.gov/ safewater/gwr/ gwrprop.pdf GWR does not apply to privately owned wells that serve <25 person (e.g., household wells)
Drinking Water Contaminant Candidate List (CCL)	Virulence-factor activity relationships (under consideration)	Assess potential pathogenicity (virulence) of waterborne pathogens as recommended in *Classifying Drinking Water Contaminants for Regulatory Consideration* (NRC, 2001)	www.epa.gov/safewater/ ccl/ccl_ fr.html www.epa.gov/safewater/ ndwac/ mem_ccl_cp.html First CCL published in 1998 as required by the SDWA Amendments of 1996, includes 10 pathogens and groups of related pathogens (EPA,1998a)

[a]As of August 11, 2003.
[b]Refer to actual rules (URLs) for a description of the monitoring requirements.
SOURCES: EPA, 2002d; Lee et al., 2002.

TABLE 1-2 Microbiological and Other Indicators Used Under Select CWA
Regulations and Related Programs

Activity or Program	Indicator	Use	URLs and Notes
Ambient Water Quality Criteria for Bacteria	Freshwater: *Escherichia coli* Enterococci Marine water: Enterococci	Determine presence of fecal contamination in ambient and recreational waters	http://www.epa.gov/ost/ pc/ambientwqc/ bacteria1986.pdf
Beaches Environmental Assessment and Coastal Health Act of 2000	*E. coli* Enterococci Proposed rapid methods such as Bioluminometer Fiber optics System flow cytometry	Rapidly determine presence of fecal contamination in freshwater and marine recreational waters	http://www.epa.gov/ ORD/WebPubs/ beaches/
Shellfish Program	Total coliforms (TC) Fecal coliforms (FC) *E. coli*	Help ensure shellfish waters are adequately protected from microbial contamination	http://www.epa.gov/ waterscience/shellfish/
Biosolids (Treated Sewage Sludge) Program	FC Salmonella Enteric viruses Viable helminth ova	Adequacy of sludge treatment practices to protect human and environmental health	http://www.epa.gov/ owmitnet/mtb/ biosolids/ See also (NRC, 2002) http://cfpub.epa.gov/ npdes/
National Pollutant Discharge Elimination System (NPDES Permitting Program)	TC FC Fecal streptococci	Ensure ambient water quality standards are maintained despite pollutant discharges	
305(b) Water Quality Assessment Report Program	Varies by state	Determine if waters meet state-determined ambient water quality standards	http://www.epa.gov/ owow/monitoring/ guidelines.html http://www.epa.gov/ owow/tmdl/ 2002wqma.html See also Table 1-3
303(d) Impaired Waters List and Total Maximum Daily Load (TMDL) Program	Varies by state	Determine if waters meet state-determined ambient water quality standards	http://www.epa.gov/ owow/tmdl/ http://www.epa.gov/ owow/tmdl/ pathogen_all.pdf http://www.epa.gov/ owow/tmdl/examples/ pathogens.html http:///www.epa.gov/ owow/tmdl/ 2002wqma.html

SOURCE: EPA, 2002d.

SDWA, the only enforceable federal drinking water standards were for water-borne pathogens in water supplies used by interstate carriers such as buses and trains. Interested readers should refer to *Safe Water from Every Tap: Improving Water Service to Small Communities* (NRC, 1997) for an overview of the development of drinking water supply regulations in the United States to include the SDWA, or to Pontius and Clark (1999) for a more thorough discussion of the SDWA and its subsequent amendments.

Under the SDWA, microbial contamination is regulated primarily under the Total Coliform Rule (TCR) and the Surface Water Treatment Rule (SWTR), both originally promulgated in 1989 (EPA, 1989a,b, 1990). Under the TCR, all public water systems (PWSs) are required to routinely collect total coliform samples at sites that are considered representative of water throughout the distribution system. The SWTR covers all drinking water systems using surface water or groundwater systems that rely on surface water, requiring them to disinfect their water, while most must also filter (unless they meet EPA-stipulated filter avoidance criteria). The SWTR is intended to protect the public from exposure to the intestinal protozoan parasite *Giardia lamblia* and viruses through a combination of removal (filtration) and inactivation (disinfection) (EPA, 1989a).

In 1998, EPA promulgated the Interim Enhanced Surface Water Treatment Rule (IESWTR; EPA, 1998c), which builds on the SWTR and includes more stringent requirements related to the performance of filters used in drinking water treatment to protect against the protozoan parasite *Cryptosporidium* and other pathogens for systems that serve more than 10,000 persons. Similarly, EPA promulgated and finalized the Long-Term 1 Enhanced Surface Water Treatment Rule (LT1ESWTR) requiring PWSs that serve less than 10,000 persons (EPA, 2002a) to meet more stringent filtration requirements. In addition, EPA recently proposed a Long-Term 2 Enhanced Surface Water Treatment Rule (LT2ESWTR) that will provide additional protection against *Cryptosporidium* and will apply to all systems using surface water or groundwater under the influence of surface water (EPA, 2003b). All PWSs will be assigned to a water treatment category ("bin") based on *Cryptosporidium* concentrations in their source water; the category determines how much additional treatment is required. In 2000, EPA proposed the Ground Water Rule (GWR) in response to the SDWA Amendments of 1996 that mandate the development of regulations for the disinfection of groundwater systems as necessary to protect public health (EPA, 2000b). The GWR had not yet been finalized as this report neared publication in early 2004. Table 1-1 summarizes these and other existing and proposed rules and programs concerning the use of pathogens under the auspices of the SDWA and EPA.

Clean Water Act

Growing public awareness of and concern for controlling water pollution nationwide led to enactment of the Federal Water Pollution Control Act (FWPCA;

originally enacted in 1948) Amendments of 1972. Together with the Clean Water Act of 1977 and the Water Quality Act of 1987—both of which amended and reauthorized the FWPCA—it provides the foundation for protecting the nation's surface waters. Collectively, they are referred to as the Clean Water Act, and that usage is maintained throughout this report. The CWA is of central importance to this report in that it is a comprehensive statute intended to restore and maintain the chemical, physical, and biological integrity of the waters of the United States.[2] To accomplish this, the CWA sought to attain a level of water quality that "provides for the protection and propagation of fish, shellfish, and wildlife, and provides for recreation in and on the water" by 1983 and to eliminate the discharge of pollutants into navigable waters by 1985. Primary authority for implementation and enforcement of the CWA rests with the EPA. In addition to measures authorized before 1972, the CWA authorizes water quality programs; requires federal effluent limitations for wastewater discharges to surface waters and publicly owned treatment works (i.e., municipal sewage treatment plants) and ambient water quality standards;[3] requires permits for discharge of pollutants[4] into waters of the United States; provides enforcement mechanisms; and authorizes funding for wastewater treatment works construction grants and state revolving loan programs, as well as funding to states and tribes for their water quality programs. Provisions have also been added to address water quality problems in specific regions and specific waterways, and the CWA has been amended almost yearly since its inception. Due consideration must be given to the improvements necessary to conserve these waters for the protection and propagation of fish and

---

[2]As defined in the CWA, "waters of the United States" applies only to surface waters, rivers, lakes, estuaries, coastal waters, and wetlands. However, not all surface waters are legally waters of the United States, and the exact division between waters of the United States and other waters can be difficult to determine. In addition, it is important to note that the CWA does not deal directly with groundwater or water quantity issues; see http://www.epa.gov/r5water/cwa.htm or http://www.epa.gov/watertrain/cwa/ for further information about the CWA.

[3]Ambient water quality standards (AWQSs) are determined by each state (collectively includes territories, American Indian tribes, the District of Columbia, and interstate commissions of the United States) and consist of (1) designated beneficial uses (e.g., aquatic life support, drinking water supply, primary contact recreation); (2) narrative and numeric criteria (ambient water quality criteria, or AWQC; discussed later) for biological, chemical, and physical parameters to meet designated use(s); (3) antidegradation policies to protect existing uses; and (4) general policies addressing implementation issues (e.g., low flows, variances). State water quality standards have become the centerpiece around which most surface water quality programs revolve; for example, they serve as the benchmark for which monitoring data are compared to assess the health of waters and to list impaired waters under CWA Section 303(d) (discussed later).

[4]As authorized by the CWA (Section 402), the National Pollutant Discharge Elimination System (NPDES) Permitting Program controls water pollution by regulating point sources (e.g., discrete conveyances such as pipes or man-made ditches) that discharge pollutants into waters of the United States.

aquatic life and wildlife, recreational purposes, and the withdrawal of water for public water supply, agricultural, industrial, and other purposes. Not surprisingly, EPA conducts a wide variety of programs and activities related to the monitoring of indicators for waterborne pathogens under the CWA as summarized in Table 1-2. It is important to note, however, that many of these listed programs and activities lie outside the committee's charge.

Regarding the attainment of water quality standards, Section 305(b) of the CWA requires states and other jurisdictions (e.g., American Indian tribes, District of Columbia) to assess and submit to EPA the health of their waters and the extent to which their water quality standards are being met every two years. In 2002, EPA released the *2000 National Water Quality Inventory* (NWQI; EPA, 2002c)—the thirteenth installment in a series that began in 1975. These NWQI reports (commonly called "305(b) reports"), as the biannual culmination of the 305(b) process, are considered by EPA to be the primary vehicle for informing Congress and the public about general water quality conditions in the United States[5] (EPA, 1997). As such, the reports characterize water quality, identify widespread water quality problems of national significance, and describe various programs implemented to restore and protect U.S. waters. Notably, states use bacterial indicators—although specific indicators, methods, and sampling practices vary from state-to-state—to determine whether waters are safe for swimming and drinking (i.e., support designated beneficial uses). Table 1-3 summarizes select findings from the 2000 NWQI report (EPA, 2002c) related to the identification of surface waters impaired by pathogens (predominantly bacteria).

In addition to establishing water quality standards, and similar to Section 305(b) of the CWA, Section 303(s) of the CWA requires states to identify waters not meeting ambient water quality standards and include them on their 303(d) list of impaired waters. Section 303(d) also requires states to define the pollutants and sources responsible for the degradation of each listed water, establish total maximum daily loads (TMDLs[6]) necessary to attain those standards, and allocate responsibility to sources for reducing their pollutant releases. The CWA further requires that water quality standards be maintained once obtained and that EPA must approve or disapprove all lists of impaired waters and TMDLs established by states (NRC, 2001). If a state submission is inadequate, EPA must establish the list or the TMDL.

Consistent with the latest NWQI report (EPA, 2002c), in 2000 EPA released

---

[5]Although positive advances have been made in recent NWQI reports, groundwater data collection under 305(b) is still too undeveloped to allow comprehensive national assessments of groundwater quality (EPA, 2002c).

[6]A TMDL can be defined as the sum of the allowable loads of a single pollutant from all contributing point and nonpoint sources that includes a margin of safety to ensure the waterbody can be used for all the purposes the state has designated. The calculation must also account for seasonal variation in water quality.

TABLE 1-3 Selected Findings and Results from the 2002 National Water
Quality Inventory

Waterbody Type	Total Size[a]	Amount[b] Assessed (% of Total)
Coastal resources: Ocean shoreline waters	58,618 miles	3,221 miles (6%)
Rivers and streams	3,692,830 miles	699,946 miles (19%)
Coastal resources: Estuaries	87,369 sq. miles	31,072 sq. miles (36%)
Coastal resources: Great Lakes shoreline	5,521 miles	5,066 miles (92%)
Lakes, reservoirs, and ponds	40,603,893 acres	17,339,080 acres (43%)

[a]Units are miles for rivers and streams; acres for lakes, reservoirs, and ponds; and square miles for coastal resources (estuaries, Great Lakes shoreline, and ocean shoreline waters).
[b]Includes waterbodies assessed as not attainable for one or more designated uses (i.e., total number of waterbody units assessed as good and impaired do not necessarily add up to total assessed).

Atlas of America's Polluted Waters (EPA, 2000a), which states that about 21,000 river segments, lakes, and estuaries encompassing more than 300,000 river and shore miles and 5 million lake acres have been reported as impaired by states and that the second leading cause of impairments (behind sedimentation or siltation) is "pathogens."[7] As for the 305(b) reports, states rely primarily on bacterial indi-

---

[7]Although the reported terminology for "pathogens" varies considerably and in many cases is unspecified, it includes primarily variations in coliforms but also includes *Escherichia coli* and enteric viruses (EPA, 2000a).

Impaired[c] (% of Assessed)	Impaired by Pathogens (Bacteria; % of Impaired)	Top Five Leading Pollutants and Causes of Impairment[d]
434 miles (14%)	384 miles (88.5%)	**Pathogens (bacteria)** Oxygen-depleting substances Turbidity Suspended solids Oil and grease
269,258 miles (39%)	93,431 miles (34.7%)	**Pathogens (bacteria)** Siltation Habitat alteration Oxygen-depleting substances Nutrients
15,676 sq. miles (51%)	4,754 sq. miles (30%)	Metals Pesticides Oxygen-depleting substances **Pathogens (bacteria)** Priority toxic organic chemicals
3,955 miles (78%)	102 miles (9.3%)	Priority toxic organic chemicals Nutrients **Pathogens (bacteria)** Sedimentation or siltation
7,702,370 acres (45%)	Not reported	Nutrients Metals Siltation Total dissolved solids Oxygen-depleting substances

[c]Partially or not supporting one or more designated uses.

[d]For states and jurisdictions that report this type of information (i.e., often a subset of the total number of states and jurisdictions that assess and report on various waterbodies; see EPA, 2002e for further information).

SOURCE: Adapted from EPA, 2002c.

cators rather than specific pathogens to assess whether waters are achieving their standards and to develop TMDLs. Indeed, EPA estimates that from 3,800 to 4,000 TMDLs will have to be completed per year to meet typical 8- to 13-year deadlines imposed on the process (NRC, 2001). It is beyond the scope of this report to discuss the 303(d) TMDL process in any detail. Rather, please refer to the 2001 NRC report *Assessing the TMDL Approach to Water Quality Management*, which reviews the program at the request of Congress and provides many recommendations for its comprehensive improvement. For example, based on that report, EPA recently provided states with guidance for integrating the development and submission of 2002 305(b) water quality reports and Section 303(d) lists of impaired waters. More specifically, the guidance recommends that states, territories, and authorized tribes submit a *2002 Integrated Water Quality Monitoring and Assessment Report* that will satisfy CWA requirements for both Section 305(b) water quality reports and Section 303(d) impaired water lists.[8]

As noted previously, regulation of recreational water (both freshwater and marine waters) is the responsibility of state and local governments. As a result, local monitoring and management programs for recreational waters vary widely, resulting in different standards and levels of protection across the nation. To help address these and related issues, in 1999 EPA issued *Action Plan for Beaches and Recreational Waters* (Beach Action Plan), a multiyear strategy to improve the monitoring of recreational water quality and the communication of public health risks associated with pathogen-contaminated recreational rivers, lakes, and ocean beaches (EPA, 1999). The Beach Action Plan describes activities of EPA's Office of Water (OW) and Office of Research and Development (ORD) to accomplish two primary objectives: (1) enable consistent management of recreational water quality programs and (2) improve the science that supports recreational water monitoring programs.

Furthermore, in October 2000, the Beaches Environmental Assessment and Coastal Health Act of 2000 was signed into effect. The BEACH Act requires coastal states to monitor beach water quality and warn the public when these waters contain dangerously high levels of disease-causing microorganisms. More specifically, it amends the CWA to require ocean, bay, and Great Lakes states to adopt minimum, health-based criteria for water quality, comprehensively test recreational beach waters for indicators of waterborne pathogens, and notify the public when contamination levels make beach water unsafe for recreation. Under the BEACH Act, EPA is required to work with states to ensure that they use the latest science to sample and test beach waters to protect the public's health. Besides requiring consistency by bringing all states up to EPA criteria, the law also requires EPA to upgrade these criteria and to develop new criteria, based on the most recent scientific studies. As a starting point, and in response to the Beach Action Plan and the BEACH Act, EPA recently published for public comment *Implementation Guidance for Ambient Water Quality Criteria* (EPA, 2002b), which builds on the seminal report *Ambient Water Quality Guidance for Bacteria – 1986* (EPA, 1986; see Table 1-2 and later discussion). When finalized, this document will help guide state, territorial, and authorized tribal water quality programs in adopting and implementing bacteriological water quality criteria to protect ambient waters designated for recreation.

---

[8]See http://www.epa.gov/owow/tmdl/2002wqma.pdf for further information.

Integrating the SDWA and the CWA

On August 6, 2001, EPA released the draft *Strategy for Waterborne Microbial Disease* (EPA, 2001), which describes a multiyear strategy for reducing the adverse impacts of microbial contamination in U.S. waters through improved water quality programs, scientific advancements, and risk communication. Moreover, it is an initial effort by EPA to begin to integrate the traditionally separate microbial assessment regulations, programs, and use of differing indicators of waterborne pathogens (see Tables 1-1 and 1-2) under the SDWA and CWA. Specifically, EPA's draft strategy report identifies the "top four approaches" to water protection, which include limiting both water contamination and exposure: (1) develop an integrated, risk-based approach to setting ambient water quality criteria and related guidance (see footnote 3) based on exposure and the application of a common set of fecal indicators across various uses of water, rather than different indicators for specific uses; (2) manage contamination sources; (3) establish monitoring and treatment standards or discharge criteria for reused water and currently unregulated industrial wastes; and (4) develop an EPA-generated microbial risk assessment paradigm. Notably, several aspects of the first approach are of particular relevance to this report and are discussed in later chapters (especially Chapters 4 and 6). The strategy report is not expected to be finalized and released until sometime in 2004 (Lisa Almodovor, EPA, personal communication, 2003).

REPORTS ON PUBLIC HEALTH IMPACTS OF, AND INDICATORS FOR, WATERBORNE PATHOGENS

Although this is the first NRC study to focus specifically on indicators for waterborne pathogens, issues surrounding their use have been discussed in several recent and historical NRC reports, as summarized in chronological order in Appendix B. Many of these reports review the public health importance of waterborne pathogens, which is discussed briefly in this chapter and in much greater detail in Chapter 2. In addition to the NRC, many federal, state, local government, and nongovernmental organizations, including the water industry and academia, have addressed the issue of the microbiological quality of drinking water and recreational water and its association with various adverse human health effects such as gastroenteritis, ear and eye infections, dermatitis, and respiratory disease. Thus, Appendix B also includes summaries of some key reports that have been conducted and published addressing these concerns.

A recurring theme of many of the reports listed in Appendix B is the need for scientifically defensible, innovative, reliable, rapid, and inexpensive approaches and methods for indicating and detecting the presence of waterborne pathogens, given their clear public health importance. Indeed, identifying, assessing, and recommending ways to help EPA address this need form the core of this report.

HISTORICAL BACKGROUND

In the United States, the principal indicators for waterborne pathogens presently in use are total coliform, fecal coliform, *Escherichia coli*, and enterococci. The coliform group (described below) is used widely as an indicator of fecal contamination of drinking water, recreational waters, and shellfishing waters, and as a measure of water treatment effectiveness. Enterococci are typically used as indicators of fecal contamination of recreational waters (EPA, 1986, 2002b). Some understanding of the historical development and application of these indicators is useful to help understand their current uses and limitations and to put this report into context.

Evolution of the Use of the Coliform Group

As noted previously, for a period of approximately 100 years following the ground-breaking work of John Snow, the public health community concentrated on preventing the transmission of waterborne bacterial disease through the fecal-to-oral route, particularly the diseases caused by *Vibrio cholerae* and *Salmonella* spp. Using methods available during that time, directly monitoring for the presence or absence of these bacteria in drinking water would not provide satisfactory or reliable protection of public health. Consequently, efforts were made to develop a more sensitive way to discern fecal contaminated water (i.e., a fecal contamination index.)

The index developed was based on *Escherichia coli*, a small bacillus first discovered by professor Theodor Escherich of Germany while he was attempting to identify the cause of cholera (Escherich, 1885). This bacterium is present at extremely high levels in the feces of warm-blooded animals. Near the end of the nineteenth century, Theobold Smith of the State of New York Department of Health developed a presumptive test for *E. coli* using a lactose-based fermentation tube test (Smith, 1891). Smith's classic fermentation tube test responds to a group of bacteria called the coliform group, of which *E. coli* is the most important member. As a result, it is commonly referred to as the coliform test. Shortly after Smith's work, the State of New York employed the coliform test to demonstrate that sewage contamination of the Mohawk River, a tributary of the Hudson River, had caused typhoid fever in persons drinking water from the Hudson downstream of the confluence of the two rivers (Mason, 1891).

In 1897, the American Public Health Association (APHA) adopted standard procedures for the coliform test, and in 1909 these procedures were published in the first edition of *Standard Methods for the Examination of Water and Wastewater* (Wolfe, 1972). In 1914, the U.S. Public Health Service (USPHS) set a standard requiring that drinking waters not show evidence of the coliform organism (U.S. Treasury Department, 1914). Technically speaking, as noted earlier, the USPHS standard applied only to waters transported across state boundaries, but it was not long before the test became a standard across the United States (APHA,

1965). Around the beginning of World War II, work done at the USPHS research center in Cincinnati demonstrated that *E. coli* measurements could be used to estimate the concentration of *Salmonella typhi* (the cause of typhoid fever) in sewage (Kerr and Butterfield, 1943) and that *E. coli* are more resistant to disinfection and environmental exposure than several other important bacterial pathogens (Wattie and Butterfield, 1944). This work historically solidified coliform measurement (1) as a means of confirming that a drinking water source was microbiologically safe and (2) for determining whether water treatment had been successful.

Method Refinement

As soon as the coliform test came into widespread acceptance, complications with its use and interpretation began to emerge. One concern was the discovery that a variety of microorganisms that read positive in the coliform test were not of fecal origin. As a result, the test method has evolved continually to become more specific. Some of the more significant developments were the so-called fecal coliform test[9] (Geldreich, 1966)—which selects for coliforms of fecal origin by using a higher incubation temperature—and, later on, the MUG test specifically identifies *E. coli* based on the action of β-glucuronidase[10] (Edberg et al., 1988).

Questions have also surfaced concerning the suitability of *E. coli* as a bacterial indicator of sewage-contaminated water in certain climates, especially tropical and subtropical climates where *E. coli* is sometimes indigenous (Bermudez and Hazen, 1988; Fujioka, 2001; Fujioka et al., 1999, Hardina and Fujioka, 1991; Hazen et al., 1987; Rivera et al., 1988). As a result, studies using a number of alternative microorganisms to coliforms or *E. coli* began to appear, particularly enterococci (Slanetz et al., 1955) and *Clostridium perfringens* (Fujioka and Shizumura, 1985; Fujioka et al., 1997). Enterococci and their taxonomically broader predecessor group, the so-called fecal streptococci, as well as *Clostridium perfringens* and its broader predecessor group the sulfite-reducing clostridia, have a long history of use and refinement as bacterial indicators of fecal contamination, as summarized below. The development and use of these alternative indicators has continued since the late 1800s because of ongoing concerns about the validity of coliforms as fecal indicators and because of certain properties that made them attractive alternatives to coliforms. Such efforts were necessary to help preserve the validity of using bacterial indicator tests as a sign of fecal contamination.

---

[9]Also referred to as thermotolerant coliform test.

[10]This test detects the presence of the enzyme β-glucuronidase, which converts the substrate methylumbelliferyl-beta-glucoronide (MUG) to a fluorogenic product if *E. coli* is present.

Fecal Streptococci and Enterococci

The prototypical enterococci, or what were previously called "fecal strepto-cocci," were first discovered and reported in the late 1880s by several research-ers. *Enterococcus faecalis* was previously named "Micrococcus ovalis" by Escherich (1887), before being called *Enterococcus faecalis* by Andrewes and Horder (1906). The genus name Enterococcus was first used by Thiercelin (1899) and Thiercelin and Jouhaud in 1903. *Enterococcus faecium* was first recognized in 1899 and further characterized by Orla-Jensen (1919). By 1900, these bacteria were recognized as being of fecal origin and were proposed as fecal indicators. Several other species of enterococci and streptococci of fecal origin were identi-fied over time. Dible (1921) proposed the name *Streptococcus faecalis* for what was previously called *Enterococcus faecalis*. Sherman (1937) proposed that the fecal enterococcal bacterial species of intestinal origin be classified in the genus *Streptococcus*, which led to widespread use of the term "fecal streptococci" for these bacteria. All of these bacteria are Gram-positive, catalase-negative, non-spore forming, facultative anaerobes with a coccoid shape and belong to the Lancefield Group D streptococci.[11] Based on molecular and immunological evi-dence however, the two species considered of likely fecal origin, *faecalis* and *faecium*, along with most other species belonging to the Lancefield Group D, such as *S. durans* and *S. avium* were reclassified from the streptococci to the enterococci in 1984. Only a few species in Lancefield Group D were left in the Streptococcus genus, notably the fecal bacteria *Streptococcus bovis* and *S. equinus*. There are currently at least 26 documented species of enterococci (Klein, 2003). The streptococci are phenotypically distinguishable from the enterococci by their inability to grow in 6.5 percent NaCl and at 10°C.

Fecal streptococci and enterococci were further evaluated as fecal indicators of water quality in the 1940s and 1950s by several groups. Efforts were made to improve the medium for their detection, with the goal of detecting those microor-ganisms of primarily fecal origin (Burman, 1961; Kenner et al., 1961; Litsky et al., 1955; Mallmann and Seligman, 1950; Slanetz and Bartley, 1957). Initial ef-forts also were made to compare these bacteria to coliforms as water quality indicators (Burton, 1949; Ostrolenk et al., 1947). In the 1970s, Cabelli and col-leagues of the EPA developed a membrane filter method to detect enterococci, specifically *E. faecalis* and *E. faecium*, in water and found that these bacteria were reliable predictors of gastrointestinal illness from primary contact recre-ation in marine waters (Levin et al., 1975). Subsequent studies have shown that the medium of this method detects at least some other enterococci species that were formerly considered fecal streptococci (Hagedorn et al., 2003).

---

[11]The genus *Streptococcus* is defined by a combination of antigenic, hemolytic, and physiological characteristics into Groups A, B, C, D, F, and G.

Clostridium perfringens and Sulfite-Reducing Clostridia

Clostridium perfringens, an anaerobic, Gram-positive, spore-forming, rod-shaped bacterium, was first reported by Welch and Nuttall in 1892 as the cause of the disease gas gangrene. It was first called *Clostridium welchii* and later *C. perfringens* by Veillon and Zuber (1898). The potential value of *C. perfringens* as a fecal indicator of water contamination was reported as early as 1899 by the city of London (Klein and Houston, 1899; as cited by Bonde, 1963). Wilson and Blair (1925) also supported the use of sulfite reducing clostridia (primarily *C. perfringens*) as fecal indicators for water. In Europe, *C. perfringens* has been used in conjunction with other sulfite reducing clostridia to detect fecal contamination in water since the 1960s (Ashbolt et al., 2001; Bonde, 1963; HMSO, 1969).

The source of *C. perfringens* and especially the spores in environmental samples has been a disputed topic. Some consider the organism to be of exclusively fecal origin but others consider the spores to be ubiquitous in soils, sediments, and other environmental media. Because the spores can survive for decades, their presence in environmental media can be difficult to interpret in the absence of a known source of fecal contamination. That is, they could either be natural environmental inhabitants or represent an historical source of fecal contamination. Uncertainties about the feces specificity of *C. perfringens* and sulfite-reducing clostridia and their extraordinary persistence in the environment are considered deficiencies in their use as fecal indicator organisms (Ashbolt et al., 2001; Cabelli, 1978). Despite these uncertainties, identification of *C. perfringens* spores became of increased interest in the 1980s and 1990s due to growing concerns about the pathogenic protozoa *Giardia lamblia* and *Cryptosporidium parvum* in water. The persistence of *C. perfringens* spores in the environment and their relative resistance to conventional water treatment suggest that they are potentially useful indicators of these highly resistant protozoa in drinking water systems and estuarine waters (Ferguson et al., 1996; Payment and Franco, 1993; Venczel et al., 1997). In addition, their apparent absence in unpolluted environmental waters in Hawaii supported their use as a fecal indicator of water quality in this tropical region (Fujioka, 2001).

Extension of the Coliform Indicator to Recreational and Shellfish Waters

Once it was widely understood that water could play an important role in the transmission of disease, it was only natural that this concern would extend from drinking water to recreational and shellfish waters. Shortly after the development of the first drinking water standards, the USPHS pursued concurrent investigations of the role of recreational and shellfish waters in enteric disease transmission by compiling data on outbreaks (Frost, 1925; Stokes, 1927a,b). Stokes (1927a,b) reported that a 1921 epidemic of typhoid fever at a boy's camp was "unquestionably attributed to bathing in polluted waters."

Around 1950, several proposed guidelines for recreational waters appeared (APHA, 1949; Cox, 1951; Scott, 1951; Streeter, 1951). In the late 1940s and early 1950s, the USPHS conducted a series of studies at bathing beaches on Lake Michigan, along the Ohio River, and on Long Island Sound (Stevenson, 1953). By the mid 1950s, a variety of bacterial indicator standards had appeared. In 1956, the City of Los Angeles conducted a survey of standards for recreational waters in 13 jurisdictions (Garber, 1956).[12] The State of Illinois used enterococci,[13] while the remaining jurisdictions used total coliforms. Statistical reporting varied over a wide range: eight jurisdictions used either a geometric mean or a median for total coliforms, three used an arithmetic mean, four used a percentage that could not be exceeded, and three used absolute maximums. The most common standard was a requirement that total coliforms not exceed 1,000 per 100 mL (see Chapter 2 for further information).

A recent review summarized current standards and proposed recreational criteria in all 50 states and various territories and tribes within the 10 EPA Regions (EPA, 2003a) and is compared to Garber's (1956) survey in Table 1-4. The EPA survey revealed the use of four different bacterial indicators (total coliform, fecal coliform, E. coli, and enteroccoci) either alone or in various combinations. As in the 1950s, this recent survey showed a wide variety of approaches being used. The most common of these was a fecal coliform limit of 200 per 100 mL (geometric mean of 5 samples taken over 30 days). Particularly notable was California, where a total of 10 different standards were being used in various jurisdictions.

Introduction of Fecal Coliform to Recreational Water Criteria

In 1968, the National Technical Advisory Committee (NTAC) convened by the U.S. Federal Water Pollution Control Administration (predecessor to the EPA) was charged with proposing microbiological criteria for recreational waters. It was NTAC's opinion that a fecal coliform measurement should be used rather than a total coliform measurement because the fecal coliform measurement is more specific.[14] Using studies showing that about 18 percent of the total coliforms detected at the Ohio River sample locations used in earlier studies were also positive in the fecal coliform test (Cabelli, 1983), the NTAC converted the total

---

[12]The States of Arkansas, California, Florida, Hawaii, Illinois, Michigan, Oregon, Virginia, West Virginia, and Wyoming; the cities of Detroit and New York; and the Tennessee Valley Authority.

[13]At that time, and as noted previously, the terms fecal streptococci and enterococci were used interchangeably. Garber refers to the work of Slanetz et al. (1955) but does not specifically identify that as the test used in Illinois.

[14]The fecal coliform measurement is similar to the total coliform measurement except that it uses an enriched lactose medium at an elevated temperature (44.5°C), so it is more selective for microorganisms found in the feces of warm-blooded animals (APHA, 1998).

TABLE 1-4 Number of Jurisdictions[a] Using an Indicator or Combination of Indicators for Recreational Waters: Mid-1950s Versus 2002

Indicators Used	1950s Bathing Standards	2002 Recreational Freshwaters	2002 Recreational Marine Waters
TC	12	0	0
FC	—	36	14
EC	—	11	0
EN	1	2	7
EN and EC	—	3	0
FC and TC	—	6	2
FC and EN	—	1	5
EC and FC	—	12	0
EN, FC, and TC	—	1	2
EC, EN, and TC	—	0	0
EC, EN, and FC	—	5	0
EC, EN, FC, and TC	—	1	0
No Standard	—	1	49

NOTE: TC = total coliform; FC = fecal coliform; EC = *Escherichia coli*; EN = enterococci

[a]2002 standards are for states and various tribes and territories within EPA Regions as of fall 2002.

SOURCES: Adapted from Garber, 1956; EPA, 2003a.

coliform density of 1,000 per 100 mL, already a common standard, to an estimated fecal coliform density of 200 per 100 mL (i.e., used a total-to-fecal coliform ratio of five-to-one). Thus, the NTAC (1968) recommended the following criteria for recreational waters: geometric mean <200 fecal coliforms per 100 mL; 90 percent of samples <400 fecal coliforms per 100 mL. These criteria were later adopted officially by EPA (1976).

Introduction of *E. coli* and Enterococci to Recreational Water Criteria

The recreational water criteria proposed by the NTAC in 1968 immediately became the subject of significant criticism (see discussion in Chapter 2). Beginning in 1972, EPA launched several epidemiologic studies designed to address the weaknesses of the previous ones. These included studies at freshwater recreational sites in Lake Erie, near Erie, Pennsylvania and Keystone Lake, near Tulsa, Oklahoma (Dufour, 1984). Similarly, marine studies were conducted at beaches in New York, Boston, and Lake Pontchartrain, Louisiana (Cabelli, 1983). These studies showed poor correlation between both total coliform and fecal coliform and reports of gastrointestinal disease. In contrast, the freshwater studies showed that both *E. coli* and enterococci were strongly correlated with gastroenteritis among swimmers, while the marine studies showed that enterococci were strongly correlated in the same way. Following these studies, EPA promulgated new recreational water criteria designed to correspond to the health risks implicated in the earlier rules (EPA, 1986). For freshwater, these criteria specified geometric means of 126 per 100 mL and 33 per 100 mL for *E. coli* and enterococci, respectively. For marine recreational waters, a geometric mean of 35 per 100 mL was specified for enterococci. Interestingly, enterococci had been proposed for this purpose much earlier (Slanetz et al., 1955) but methods available at that time were problematic.

Detection methods have continued to evolve over the last several decades for the bacteria variously referred to as fecal streptococci, enterococci, and intestinal enterococci in recreational waters (e.g., Messer and Dufour, 1998). These developments, however, have been complicated by periodic changes in the taxonomy of these related groups of bacteria and the identification of new species. Recently, efforts have been made to harmonize both the terminology used to define these bacteria as well as the methods to detect them. In this regard, the European Union and the World Health Organization have come to use the term "intestinal enterococci" to define the group of bacteria that was previously called fecal streptococci and that are now collectively called enterococci in the United States. However, it is generally believed that the different terminologies such as fecal streptococci, enterococci, and intestinal enterococci, all refer to the same group of related bacteria species and that the various methods available to detect them provide generally comparable detection (Bartram and Rees, 2000).

Appearance of Viral Pathogens

Soon after the germ theory of disease became widely accepted and scientists began to use light microscopes routinely to identify disease-causing bacteria, it became evident that certain diseases were caused by microorganisms that were not visible with the best light microscope (i.e., viruses). For example, it had long been recognized that poliomyelitis was transmitted via the fecal-oral route. In the 1940s, several investigators confirmed that the virus responsible for this disease could be found in sewage (Melnick, 1947; Paul et al., 1940; Trask and Paul, 1942). As early as 1945, an epidemic of infectious hepatitis was connected to contaminated drinking water (Neefe and Stokes, 1945). The epidemiologic evidence for a connection between infectious hepatitis and drinking water was further substantiated by a massive outbreak in New Delhi, India in 1954 that was subsequently shown to be caused by hepatitis E virus instead of hepatitis A virus (Melnick, 1957; Viswanathan, 1957; Wong et al., 1980).

Although the proposition of waterborne viral diseases was not widely accepted at first, it gradually gained acceptance, and by the mid-1960s a review cited 50 outbreaks of infectious hepatitis and 8 outbreaks of polio that were waterborne from 1946 through 1960 (Weibel et al., 1964). At about that same time, it also became evident that outside of the host, viruses are not free-living organisms and do not have metabolic requirements. This important finding supported earlier observations that under certain conditions, viruses can survive in the environment much longer than the coliform indicators (Neefe and Stokes, 1945). The superior environmental survival of important waterborne viral agents raised serious questions about the suitability of the coliform group as an indicator. Thus, while the presence of coliforms could still be taken as a sign of fecal contamination, the absence of coliforms could no longer be taken as assurance that water was uncontaminated.

Expansion Beyond Indicators for Human Fecal Contamination: Zoonoses and Protozoan Pathogens

Most, but not all, pathogens of concern in drinking water are spread by the fecal-to-oral route. In 1854, John Snow demonstrated that cholera could be transmitted through the contamination of drinking water by human feces (Snow, 1854a,b). Two years later, William Budd demonstrated that typhoid fever can be spread through the same route (Budd, 1856). Approximately 30 years later, Robert Koch and Karl Eberth isolated the specific microorganisms responsible for both of these diseases, further demonstrating the connection between disease and drinking water contaminated with human feces (Koch, 1883). As a result of these and other discoveries, by the middle of the nineteenth century, public health practitioners and researchers began to focus almost exclusively on preventing the contamination of water supplies by sewage.

This focus is particularly appropriate because a sewage-to-water connection contributes to facilitating the transmission and distribution of a waterborne pathogen by the fecal-to-oral route. Such widespread transmission can also affect the evolution of the disease organism itself in that unfettered waterborne transmission enables the continued evolution of a disease that might fatally immobilize the victim. Ordinarily, when a disease evolves to the point at which it causes such dire health effects, it loses access to future hosts because the infected host is too immobilized to expose others. On the other hand, with ready access to a public water supply, the victim's caregiver can effectively spread the disease by merely washing out the bedpan of a bedridden person (Ewald et al., 1998).

Nevertheless, for a long time it has been understood that certain diseases are zoonoses, that is, they are common to both animals and humans. Some well-known zoonotic diseases include rabies, ringworm, and plague. In fact, some researchers have suggested that many of the epidemic diseases associated with early civilization may be of zoonotic origin (Diamond, 1999). It was not a big step, then, to presume that some of the microorganisms present in animal feces could be an important source of waterborne zoonoses as well. In the 1980s, giardiasis and cryptosporiodosis became widely recognized as zoonoses transmitted by the fecal-to-oral route. Furthermore, a large-scale and well-publicized outbreak of cryptosporiodosis occurred in Milwaukee, Wisconsin in 1993 in a public water system that met SDWA microbiological water quality standards (Edwards, 1993). It is likely that many other microorganisms found in the feces of animals are the etiological agents for zoonoses as well. For example, in June 2000, the bacteria *E. coli* O157:H7 and *Campylobacter jejuni* originating from the feces of agricultural animals were found to be the cause of a well-documented waterborne disease outbreak in Walkerton, Ontario (Bruce-Grey-Owen Sound Health Unit, 2000). Finally, the intensive use of antibiotics in animal feedlots raises the specter that feces from such sources could be the source of zoonotic pathogens that possess significant antibiotic resistance (NRC, 1999b; see also Chapter 3 for further discussion).

However, human and animal feces are not the only source of enteric pathogens; microorganisms responsible for waterborne disease have also been connected to other environments (see Chapter 3 for further information). For example, Huq et al. (1983) isolated *Vibrio cholerae* from copepods found in marine waters. Fortunately, simple filtration to remove these copepods from freshwater has proven to be an effective treatment to reduce infection from drinking water (Huq et al., 1996).

As a result, whereas it made sense in the late nineteenth century to develop and use bacterial indicators to identify the presence of sewage of human origin, modern indices of microbial contamination face a much more complex and challenging task.

CURRENT STATUS OF WATERBORNE OUTBREAKS AND ENDEMIC DISEASE

Even though the association between water quality and disease has been recognized for more than a hundred years, the transmission of waterborne diseases is still a major public health concern in developed nations. In the United States, national statistics on outbreaks associated with drinking water have been collected since 1920 (Craun, 1986). Since 1971, the Centers for Disease Control and Prevention (CDC), EPA, and the Council of State and Territorial Epidemiologists (CSTE) have maintained a collaborative surveillance system of waterborne disease outbreaks (WBDOs). Currently, CDC publishes summary information on the occurrence and causes of WBDOs every two years.

In contrast, the occurrence of endemic waterborne disease has only recently become a focus of the federal government. The 1996 amendments to the SDWA (Section 1458(d)(1)) require CDC's director and EPA's administrator to jointly conduct pilot waterborne disease occurrence studies for at least five major communities or public water systems, prepare a report on the findings, and develop a national estimate of endemic waterborne disease occurrence. Furthermore, as noted previously (see also Table 1-2), the federal government has also turned its attention to recreational waters and public health by enacting the BEACH Act of 2000 to improve the quality of coastal recreational waters.

An overview of the current status of waterborne disease outbreaks and endemic disease in U.S. drinking and recreational waters is provided in the following sections. A more detailed description of the National Waterborne Diseases Outbreak Surveillance System (WBDOSS) and of the various epidemiologic studies of endemic disease associated with drinking and recreational waters is provided in Chapter 2.

Waterborne Disease Outbreaks

The United States is one of a few countries to have a national surveillance system for waterborne disease outbreaks. Surprisingly, most European countries do not yet have an adequate surveillance system for waterborne disease (the United Kingdom being a notable exception), while the situation in developing countries is even worse.

The WBDO and the foodborne disease outbreak surveillance systems at CDC are unique in that the unit of analysis is an outbreak rather than an individual case of disease, and these cases are linked to an identified exposure. In contrast, other surveillance systems typically focus on a specific disease (e.g., cryptosporidiosis, legionellosis) and collect standard epidemiologic data (e.g., person, place, and time), but do not obtain any information regarding the mode of transmission such as person-to-person contact versus waterborne or foodborne transmission for cases of cryptosporidiosis.

State, territorial, and local public health agencies are responsible for detecting and investigating WBDOs while CDC, as a federal agency, participates in outbreak investigations only by invitation or when an outbreak involves multiple states. States and territories report their outbreaks on a standard form annually and CDC then compiles, analyzes, and publishes the data every two years as a *Morbidity and Mortality Weekly Report* (MMWR) Surveillance Summary (in chronological order: Levine and Craun, 1990; Herwaldt et al., 1991; Moore et al., 1993; Kramer et al., 1996; Levy et al., 1998; Barwick et al., 2000; Lee et al., 2002). Reporting is voluntary and passive, and varies by state and territory.

Current data for drinking water and recreational water outbreaks (excluding recreational water outbreaks associated with treated waterbodies such as public swimming and wading pools) are summarized in Figures 1-1 through 1-3. Figure 1-1 provides an overview of the total number of reported WBDOs associated with drinking water by etiologic agent from the time the surveillance system was developed at CDC in 1971 through 2000. Outbreaks peaked in the early 1980s with a steady decline in numbers, except for 1992, until the 1997-1998 reporting period where the trend reversed and continued to increase through 2000. The decline in numbers of outbreaks through the 1980s and 1990s could be a result of implementation of water treatment regulations such as the SWTR; increased efforts by water utilities to produce drinking water that exceeds EPA standards; and widespread efforts by federal, state, and local public health officials to improve

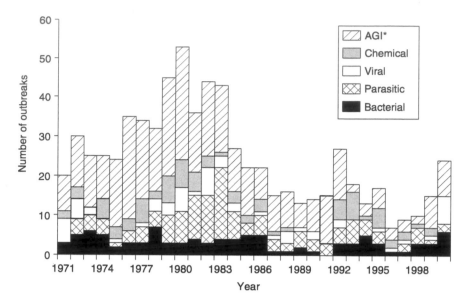

FIGURE 1-1 Number of waterborne disease outbreaks associated with drinking water by year and etiologic agent for the United States: 1971-2000 (n = 730). *AGI denotes acute gastrointestinal illness of unknown etiology. SOURCE: Adapted from Lee et al., 2002.

drinking water quality. Although the significance of the recent increase in reported outbreaks is not clear at this time, as noted previously, not all outbreaks are recognized and investigated and multiple factors can affect whether an outbreak is identified and investigated (see Chapter 2 for further information).

While the number of reported outbreaks increased from 1997 through 2000, the number of persons affected remained comparable to previous years. In this regard, changes in surveillance and reporting of WBDOs might have led to improvements in the detection of outbreaks in small systems, which tend to affect smaller numbers of individuals. In addition, better detection methods in clinical specimens as well as in water samples have increased the identification of outbreaks associated with viral pathogens. This improvement can be seen in the increase in reported viral outbreaks in 2000. Despite these improvements, however, the etiologic agent remains unknown for a large percentage of outbreaks, making the development, selection, and use of indicators for waterborne pathogens very complex.

Figure 1-2 provides an overview of the number of outbreaks by type of water, including waters that are specifically within this committee's charge; that is, surface and groundwater sources for drinking water and ambient recreational

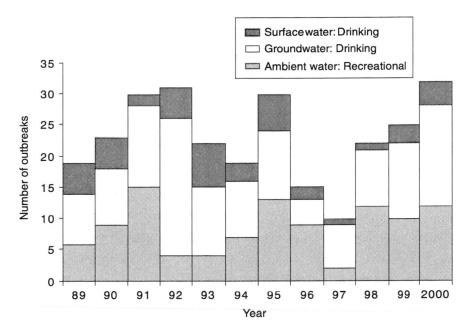

FIGURE 1-2 Number of waterborne disease outbreaks by year and water type for the United States: 1989-2000 (*n* = 278). SOURCE: Outbreak data through 2000 from the CDC's National Waterborne Diseases Outbreak Surveillance System.

waters (freshwater, and coastal marine and marine-estuarine waters). It is important to note that the reported time frame for Figure 1-2 begins in 1989 (rather than 1971 as for Figure 1-1) for consistency in classification of water type because the responsibility for the surveillance system moved within CDC divisions at that time.

No clear trend emerges in the total number of reported outbreaks during the period; however, the histogram clearly shows that groundwater outbreaks are most common (7 out of 12 years for 58 percent) followed by ambient recreational water outbreaks (4 out of 12 years for 33 percent). The two categories had the same number of outbreaks in 1990. Figure 1-2 also emphasizes the relatively few number of reported surface water outbreaks in this period, which again might be a reflection of EPA's regulations targeted at surface water systems and the ability of large water utilities to meet or exceed EPA standards. Small water utilities, often using groundwater sources, have fewer resources and are more likely to have difficulties meeting increasingly stringent water quality standards (NRC, 1997).

Figure 1-3 provides an overview of the number of reported WBDOs by illness and water type. As might be expected, acute gastrointestinal illnesses of

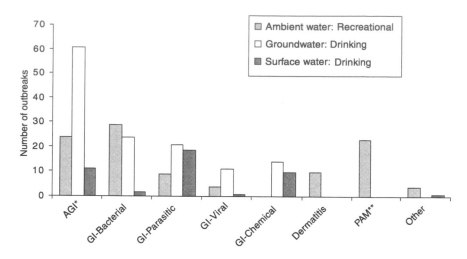

FIGURE 1-3 Number of waterborne disease outbreaks by illness and water type in the United States: 1989-2000 ($n = 278$). *AGI denotes acute gastrointestinal illness of unknown etiology. **PAM denotes primary amoebic meningoencephalitis (see Chapter 3 for further information). SOURCE: Outbreak data through 2000 from the CDC's National Waterborne Diseases Outbreak Surveillance System.

unknown etiology are most likely to be associated with groundwater sources because these outbreaks are widely thought to be caused predominantly by viruses, which are difficult to identify in clinical specimens and even more so in water samples (NRC, 1999a). Gastrointestinal outbreaks of known viral etiology were also more likely to occur in groundwater systems. Surprisingly, gastrointestinal illnesses of parasitic origin also were somewhat more likely to be associated with groundwater systems, probably because some of these systems are actually groundwater under the influence of surface water. Gastrointestinal outbreaks of drinking water supplies from bacteria and viruses were rare in surface water systems since these microorganisms are readily killed by conventional treatment practices. While gastrointestinal outbreaks associated with bacteria were most commonly reported in ambient recreational waters, outbreaks of dermatitis and cases of primary amoebic meningoencephalitis were associated only with recreational water. The "other" illness category includes one outbreak each of leptospirosis, legionellosis, and keratitis in ambient waters and one outbreak associated with algae in a surface water system.

The outbreak data summarized in Figures 1-1 through 1-3 demonstrate that the association between various pathogens and human health effects differs depending on the type of water system involved. If indicators for waterborne pathogens are to be used to predict the likelihood of water contamination with potential ensuing health effects, it is unlikely that a single indicator will suffice for these different routes of exposure. For these and other reasons, epidemiologic studies are needed to establish the causal link between the presence and density of an indicator and the associated health effects under a variety of environmental conditions. Surveillance for waterborne disease outbreaks and epidemiologic study designs are discussed further in Chapter 2.

Endemic Waterborne Disease

In 1991, Payment and colleagues reported the results of the first randomized intervention trial (i.e., in which investigators control the conditions of exposure) to evaluate whether the consumption of tap water that met current Canadian microbiological standards was associated with an increased risk of gastrointestinal disease. They compared illness rates in households drinking tap water and households drinking reverse osmosis-filtered water, which were considered pathogen free (Payment et al., 1991). The trial estimated that 35 percent of the reported gastrointestinal illness among persons drinking tap water was associated with its consumption. In 1997, Payment and colleagues conducted a follow-up intervention trial to confirm the previous results and to attempt to determine the source(s) of the illnesses (Payment et al., 1997). This second study attributed 14-40 percent of gastrointestinal illness to consumption of tap water meeting current Canadian water treatment standards. These two studies are described in more detail in Chapter 2.

Researchers' interest in the possible contribution of drinking water that met current treatment standards to the incidence of gastrointestinal illness was heightened as a result of the Payment studies and the continuing occurrence of waterborne disease outbreaks in the United States and elsewhere. As noted previously, Congress responded with new mandates in the 1996 SDWA amendments, and CDC and EPA entered into an interagency agreement in 1997 in response to the congressional mandate to conduct studies and develop a national estimate of endemic waterborne disease. The SDWA amendments were interpreted to mean that the focus of efforts should be directed at municipal drinking water. The amendments did not specify which waterborne diseases were to be studied, and after conducting several workshops, the two agencies determined that the health outcome that would be studied in this initial effort would be gastrointestinal disease.

Based on this interagency agreement, CDC has funded cooperative agreements with academic institutions to conduct two pilot intervention trials of home water treatment in households and one full-scale intervention trial along with several related "nested" epidemiology studies to help maximize the benefit of the large-scale trial. Each of these studies is reviewed in Chapter 2 along with related (including three community intervention trials conducted by EPA) studies of endemic waterborne disease. In addition, questions regarding water consumption patterns and usage behavior were added to CDC's yearly cross-sectional survey for FoodNet (CDC, 1996; see Chapter 2 for further information) beginning with the 1998-1999 cycle.

As part of the BEACH Program, EPA will be conducting new epidemiologic studies intended to correlate water quality with human health effects. The health outcomes will include gastrointestinal disease as well nongastrointestinal health outcomes in eyes, ears, skin, and the respiratory system (Rebecca Calderon, EPA, personal communication, 2002). Beach site selection criteria will include point source contamination, range of exposures, population size, geographic variety, and historical microbial testing. A pilot study was conducted during summer 2002 at one freshwater recreational beach (Indiana Dunes National Lakeshore). The current projected time line includes full-scale studies at three beaches each summer from 2003 to 2005 for a total of nine beaches. Data analysis and report preparation are expected to be completed in 2006 (see Chapter 2 for further information).

All of the aforementioned studies focus on health effects associated with water of varying quality. This water quality is measured using many methods and parameters. Because pathogens are difficult to detect in water, surrogates or indicators are often used in their stead. The following section describes the indicators for waterborne pathogens that are currently in use and provides an overview of the issues associated with the selection of appropriate indicators for waterborne pathogens.

CURRENT INDICATORS FOR WATERBORNE PATHOGENS

In the United States, predominantly bacterial indicators are used to determine (1) if drinking water sources are microbiologically safe, (2) if treatment of drinking water has been adequate, (3) if drinking water in the distribution system continues to be protected, and (4) if recreational and shellfish waters are microbiologically safe (see also Tables 1-1 and 1-2). Each of these test objectives has different requirements, and it is not likely that any one indicator or system of indicators can adequately meet all of these needs. For example, for recreational waters, shellfish waters, and source waters for drinking water, the question being addressed is the same, namely: Has this water been exposed to significant microbiological contamination? However, complexities of several kinds come into play.

First, it is important to know something about the type and source of contamination, particularly if the contamination is of fecal or nonfecal origin. At the time the coliform index was conceived, contamination by human feces was clearly the central public health issue to be addressed. Since that time it has become clear that although contamination with human fecal matter is clearly of profound and continuing public health significance, human pathogens occur in other environments as well. For example, based on the preceding discussion it is clear that animal fecal matter can be of particular significance and that the widespread use of antibiotics for animal growth promotion, as well as for control of animal diseases, may constitute an important source of antibiotic-resistant pathogens (NRC, 1999b). Finally, some enteric waterborne pathogens have natural reservoirs in the environment where they can proliferate (see Chapter 3 for further discussion of these issues). Moreover, some nonenteric waterborne pathogens are capable of proliferating in waters under the right conditions, and human exposure to the high concentrations resulting from this proliferation can create human health risks. An example is *Legionella pneumophila*, which proliferates in warm waters containing sediments and nutrients (e.g., institutional hot water systems) and causes respiratory disease through inhalation of aerosolized water (Kaufman et al., 1981).

The specific application and geographic location can also have an impact on selection of the most important indicator candidates. For example, some indicators will be more useful in temperate zones than in the subtropics, some will be more effective in surface water than in groundwater sources of drinking water, and some will be more useful in freshwaters than in marine recreational waters. Finally, a different set of indicator attributes will come into play when the effectiveness of treatment or the integrity of a drinking water distribution system is at issue (though the latter are excluded from explicit consideration in the committee's charge). These and related issues are discussed and illustrated at length in this report, especially in Chapters 4 and 6.

The timeliness of the indicator system is also more important in some applications than in others. For example, in the case of recreational waters, the results of current bacterial indicator tests are often tied directly and immediately to a decision to allow or restrict public access (see Chapter 4 for further information). It is essential that indicator systems used in such applications provide timely results because swimmers may be exposed to unacceptable levels of pathogens while the analysis is being conducted. Furthermore, beach contamination is often episodic and of short duration and a long turnaround on an indicator test runs the risk that the public is allowed access to unsafe waters, but denied access when the episode has already past (Boehm et al., 2002).

Drinking water supplies generally face different requirements for indicators for waterborne pathogens. It is sometimes possible to use indicator measurements alone to divert or avoid a water supply during a contamination episode. However, it is more common to use indicators in conjunction with other measures (e.g., sanitary surveys) to assess the overall microbiological risk associated with a given water supply source and to address that contamination by removing its source and/or installing (additional) treatment systems to serve as a protective barrier. For assessments of this sort, the accuracy and specificity of the indicator system are more important than the timeliness of the result.

SUMMARY AND CONCLUSIONS

To protect public health, and as mandated in the SDWA and CWA, it is important to have accurate, reliable, and scientifically defensible methods for determining whether source waters for drinking water and recreational waters are contaminated by pathogens and to what extent. In this regard, the development and use of bacterial indicators for waterborne pathogens began more than a century ago, when contamination of drinking waters by enteric bacterial pathogens originating from human waste constituted a major public health threat. The use of bacterial indicators (predominantly coliforms) was later expanded and adopted for use in ambient, recreational, and shellfish waters and continues to focus on identification of fecal contamination, principally of human origin. As such, the current indicator approaches have become standardized; are relatively easy and inexpensive to use; and constitute a cornerstone of local, state, and federal monitoring and regulatory programs. Although these approaches have been extremely effective in reducing waterborne disease outbreaks caused by human enteric bacteria, it is now widely understood that bacteria are not the only pathogens of public health concern; fecal contamination is not the only significant potential source of waterborne microbial pathogens; and many human pathogens and indicator organisms occur in other environments.

The number of reported disease outbreaks associated with drinking water peaked in the early 1980s, declined for more than 10 years, and increased from 1997 through 2000, although the number of persons affected has remained comparable to previous years. Recreational water outbreaks associated with ambient water did not show a specific trend. Better detection methods in clinical specimens as well as in water samples have increased the identification of pathogens, most notably viruses. Despite these improvements, the etiologic agent remains unknown for a large percentage of drinking water and recreational water outbreaks, making the selection and use of indicators for waterborne pathogens very complex.

An increased understanding of the diversity of waterborne pathogens, their sources, physiology, and ecology has resulted in a growing understanding that the use of bacterial indicators may not be as universally protective as once thought. For example, the superior environmental survival of important waterborne viruses and protozoa raised serious questions about the suitability of relying on relatively short-lived coliforms as indicators of the microbiological quality of water. That is, while the presence of coliforms could still be taken as a sign of fecal contamination, the absence of coliforms could no longer be taken as assurance that water was uncontaminated. Thus, existing bacterial indicators and indicator approaches do not in all circumstances identify all potential waterborne pathogens. Indeed, the committee concludes that no single indicator organism or small set of indicators can successfully identify or predict the presence, let alone the source, of all classes of potential pathogens—especially emerging microorganisms. Furthermore, recent and forecasted advances in microbiology, molecular biology, and analytical chemistry make it timely to assess the current paradigm of relying predominantly or exclusively on traditional bacterial indicators for waterborne pathogens to make judgments concerning the microbiological quality of source waters for drinking water and recreational waters. Nonetheless, indicator approaches will still be required for the foreseeable future since it is not practical or feasible to monitor for the complete spectrum of microorganisms that may occur in source waters for drinking water and recreational waters, and many known pathogens are difficult to detect directly and reliably in water samples.

Lastly, improvements in the timeliness of indicator analysis (i.e., rapidity of results) are needed if exposure to pathogen-contaminated water is to be prevented or controlled in a timely manner that protects public health.

REFERENCES

Andrewes, F.W., and T.J. Horder. 1906. A study of the streptococci pathogenic for man. Lancet 2: 708-713.

APHA (American Public Health Association). 1949. Standard Methods for the Examination of Water and Wastewater, 10th Edition. Washington, D.C.

APHA. 1965. Standard Methods for the Examination of Water and Wastewater, 12th Edition. Washington, D.C.

APHA. 1998. Method 9221(3) or Method 9222 (d) in Standard Methods for the Examination of Water and Wastewater, L. Clesceri, A. Greenberg, and A. Eaton, eds., 20th Edition. Washington, D.C.

Ashbolt, N.J., W.O.K. Grabow, and M. Snozzi. 2001. Indicators of microbial water quality. Chapter 13 in Water Quality—Guidelines, Standards and Health: Assessment of Risk and Risk Management for Water-Related Infectious Disease, L. Fewtrell and J. Bartram, eds., London: World Health Organization and IWA Publishing.

Bartram, J., and G. Rees. 2000. Monitoring Bathing Waters: A Practical Guide to the Design and Implementation of Assessments and Monitoring Programmes. Geneva: World Health Organization.

Barwick, R.S., D.A. Levy, G.F. Craun, M.J. Beach, and R.L. Calderon. 2000. Surveillance for waterborne-disease outbreaks—United States, 1997-1998. MMWR 49 (No. SS-4): 1-35.

Bermudez, M., and T.C. Hazen. 1988. Phenotypic and genotypic comparison of *Escherichia coli* from pristine tropical waters. Applied and Environmental Microbiology 54: 979-983.

Boehm, A., S. Grant, J. Kim, S. Mowbray, C. McGee, C. Clark, D. Foley, and D. Wellman. 2002. Decadal and shorter period variability of surf zone water quality at Huntington Beach, CA. Environmental Science and Technology 36(18): 3885-3892.

Bonde, G.J. 1963. Bacteria indicators of water pollution. Teknisk Forlag, Copenhagen 221-226.

Bonde, G.J. 1966. Bacteriological methods for the estimation of water pollution. Health Laboratory Science 3: 124.

Bonde, G.J. 1975. Bacteria indicators of sewage pollution. Pp. 37-47 in Discharge of Sewage from Sea Outfalls: Proceedings of an International Symposium, A.L.H. Gameson, ed. New York: Pergamon Press.

Bruce-Grey-Owen Sound Health Unit. 2000. The Investigative Report of the Walkerton Outbreak of Waterborne Gastroenteritis May-June, 2000. Ontario Ministry of Health. Owen Sound, Ontario.

Budd, W. 1856. Outbreak of fever at clergy orphanage. Lancet: 15 November.

Burman, N.P. 1961. Some observations on coli-aerogenes bacteria and streptococci in water. Journal of Applied Bacteriology 24: 368.

Burton, M.C. 1949. Comparison of coliform and *Enterococcus* organisms as indices of pollution in frozen foods. Food Research 14: 434-448.

Cabelli, V. 1978. Obligate anaerobic bacterial indicators. Pp. 171-200 in: Indicators of Viruses in Water and Food, G. Berg, ed., Ann Arbor Science, Ann Arbor, MI.

Cabelli, V. 1983. Health Effects Criteria for Marine Recreational Waters. Cincinnati, Ohio. EPA-600-1-80-031.

CDC (Centers for Disease Control and Prevention). 1996. FoodNet. [On-line]. Available: http://www.cdc.gov/foodnet/. Accessed October 15, 2002.

Cox. 1951. Acceptable standards for natural waters used for bathing. Proceedings American Society of Civil Engineers 77: 74.

Craun, G.F., ed. 1986. Waterborne Diseases in the United States. Boca Raton, Florida: CRC Press, Inc.

Diamond, J. 1999. Guns, Germs and Steel: The Fates of Human Societies. New York: W. Norton & Co.

Dible, H. 1921. The *Enterococcus* and the faecal streptococci: Their properties and relations. Journal of Pathology and Bacteriology 24: 3-12.

Dufour, A. 1984. Health Effects Criteria for Fresh Recreational Waters. Cincinnati, Ohio: EPA 600/1-84-004.

Edberg, S.C., M.J. Allen, and D.B. Smith. 1988. National field evaluation of a defined substrate method for the simultaneous enumeration of total coliforms and *Escherichia coli* from drinking water: Comparison with the standard multiple tube fermentation method. Applied and Environmental Microbiology 54(6): 1595-1601.

Edwards, D. 1993. Troubled waters in Milwaukee. ASM News 59(7): 342-345.

EPA (U.S. Environmental Protection Agency). 1976. Quality Criteria for Water. Washington, D.C.: EPA 440-9-76-023.

EPA. 1986. Ambient Water Quality Criteria for Bacteria - 1986. Washington, D.C.: Office of Water. EPA 440-5-84-002.

EPA. 1989a. Drinking Water; National Primary Drinking Water Regulations; Filtration, Disinfection; Turbidity; *Giardia lamblia*, Viruses, *Legionella*, and Heterotrophic Bacteria; Final Rule. Federal Register 54: 27486-27541.

EPA. 1989b. Drinking Water; National Primary Drinking Water Regulations; Total Coliforms (Including Fecal Coliforms and *E. coli*); Final Rule. Federal Register 54: 27544-27568.

EPA. 1990. Drinking Water; National Primary Drinking Water Regulations; Total Coliforms; Corrections and Technical Amendments; Final Rule. Federal Register 55: 25064-25065.

EPA. 1997. Guidelines for Preparation of the Comprehensive State Water Quality Assessments (305(b) Reports) and Electronic Updates. Washington, D.C.: Office of Water. EPA-841-B-97-002A.

EPA. 1998a. Bacterial Water Quality Standards for Recreational Waters (Freshwater and Marine Waters). Washington, D.C.: Office of Water. EPA-823-R-98-003.

EPA. 1998b. Announcement of the Drinking Water Contaminant Candidate List; Notice. Federal Register. 64: 10274-10287.

EPA. 1998c. National Primary Drinking Water Regulations: Interim Enhanced Surface Water Treatment; Final Rule. Federal Register 64: 69477-69521.

EPA. 1999. EPA Action Plan for Beaches and Recreational Waters. Washington, D.C.: Office of Water and Office of Research and Development. EPA-600-R-98-079.

EPA. 2000a. Atlas of America's Polluted Waters. Washington, D.C.: Office of Water. EPA-840-B00-002.

EPA. 2000b. National Primary Drinking Water Regulations: Ground Water Rule; Proposed Rule. Federal Register 65(91): 30193-30274.

EPA. 2000c. Stage 2 M-DBP Agreement in Principle. Microbial/Disinfection Byproducts (M-DBP) Federal Advisory Committee. Washington, D.C.: Office of Water.

EPA. 2000d. National Primary Drinking Water Regulations: Long Term 1 Enhanced Surface Water Treatment and Filter Backwash Rule; Proposed Rule. Federal Register 65: 19046-19150.

EPA. 2000e. Beaches Environmental Assessment and Coastal Health Act of 2000. Washington, D.C.: Office of Science and Technology. [On-line]. Available: http://www.epa.gov/ost/beaches/beachbill.pdf.

EPA. 2001. Strategy for Waterborne Microbial Disease (Discussion Draft). Washington, D.C.: Office of Science and Technology and Office of Water.

EPA. 2002a. National Primary Drinking Water Regulations: Long Term 1 Enhanced Surface Water Treatment Rule; Final Rule. Federal Register 67: 1811-1844.

EPA. 2002b. Implementation Guidance for Ambient Water Quality Criteria for Bacteria (Draft). Washington, D.C.: Office of Water. EPA-823-B-02-003.

EPA. 2002c. 2000 National Water Quality Inventory. Washington, D.C.: Office of Water. EPA-841-R-2-001.

EPA. 2002d. Management Briefing. Washington, D.C.: Office of Water Microbial Indicator Cross Office Team. Presentation slides received December 19, 2002.

EPA. 2003a. Bacterial Water Quality Standards for Recreational Waters (Freshwater and Marine Waters). Status Report Washington, D.C.: Office of Water. EPA-823-R-98-003.

EPA. 2003b. National Primary Drinking Water Regulations: Longterm 2 Enhanced Surface Water Treatment Rule; Proposed Rule. Federal Register 68: 47640-47795.

Escherich, T. 1885. Die Darmbakterien des Säuglings und ihre Beziehungen zur Physiologie der Verdauung. Stuttgart, Germany.

Escherich, T. 1887. Ueber Darmbakterien im allgemeinen und diejenige der Säuglinge im Besonderen, sowie der Beziehungen der letzteren zur Aetiologie der Darmerkrankungen. Centralblatt für Bakteriologie 1: 705-713.

Ewald, P., J. Sussman, M. Distler, C. Libel, W. Chammas, V. Dirita, C. Salles, A. Vicente, I. Heitmann, and F. Cabello. 1998. Evolutionary control of infectious disease: prospects for vectorborne and waterborne pathogens. Memorial Institute Oswaldo Cruz, Rio de Janeiro 93(5): 567-576.

Ferguson, C.M., B.G. Coote, N.J. Ashbolt, and I.M. Stevenson. 1996. Relationships between indicators, pathogens and water quality in an estuarine system. Water Research 30: 2045-2054.

Frost, W.H. 1925. Report of committee on sanitary control of the shellfish industry in the United States. Public Health Reports Supplement 53: 1-17.

Fujioka, R.S., and L.K. Shizumura. 1985. Clostridium perfringens, a reliable indicator of stream water quality. Journal of Water Pollution Control Federation 57(10): 986-992.

Fujioka, R.S., B. Roll, and M. Byappanahalli. 1997. Appropriate recreational water quality standards for Hawaii and other tropical regions based on concentrations of Clostridium perfringens. Proccedings of the Water Environment Federation 4: 405-411.

Fujioka, R.S., C. Stan-Denton, M. Borja, J. Castro, and K. Morphew. 1999. Soil, the environmental source of Escherichia coli and enterococci in Guam's streams. Journal of Applied Microbiology Symposium Supplement 85: 83S-89S.

Fujioka, R.S. 2001. Monitoring coastal marine waters for spore-forming bacteria of faecal and soil origin to determine point from non-point source pollution. Water Science and Technology 44(7): 181-188.

Garber, W.F. 1956. Bacteriological standards for bathing waters. Sewage & Industrial Wastes 28(6): 795-807.

Geldreich, E. 1966. Sanitary Significance of Fecal Coliforms in the Environment. Water Pollution Control Research Series. Publ. WP-2-3, FWPCA. Cincinnati, Ohio: U.S. Department of Interior.

Hagedorn, C., J.B. Crozier, K.A. Mentz, A.M. Booth, A.K. Graves, N.J. Nelson and R.B. Reneau, Jr. 2003. Carbon source utilization profiles as a method to identify sources of faecal pollution in water. Journal of Applied Microbiology 94: 792-799.

Hardina, C.M., and R.S. Fujioka. 1991. Soil, the environmental source of E. coli and enterococci in Hawaii's streams. Environmental Toxicology 6: 185-195.

Hazen, T.C., J. Santiago-Mercado, G.A. Toranzos, and M. Bermudez. 1987. What does the presence of fecal coliforms indicate in the waters of Puerto Rico? A review. Bol. Puerto Rico Medical Association 79: 189-193.

Herwaldt, B.L., G.F. Craun, S.L. Stokes, and D.D. Juranek. 1991. Waterborne-disease outbreaks, 1989-1990. MMWR 40 (No. SS-3): 1-21.

HMSO (Her Majesty's Stationary Office). 1969. The Bacteriological Examination of Water Supplies. Reports on Public Health and Medical Subjects, 4th edition, No. 71. London.

Huq, A., E.B. Small, P.A. West, M.I. Huq, R. Rahman, and R.R. Colwell. 1983. Ecology of Vibrio cholerae with special reference to planktonic crustacean copepods. Applied and Environmental Microbiology 45: 275-283.

Huq, A., B. Xu, M.A.R. Chowdhury, M.S. Islam, R. Mantilla, and R.R. Colwell. 1996. A simple filtration method for removal of Vibrio cholerae associated with planktonic copepods. Applied and Environmental Microbiology 62: 2508-2512.

Kaufman, A.F., J.E. McDade, C.M. Patton, J.V. Bennett, P. Skaliy, J.C. Feeley, D.C. Anderson, M.E. Potter, V.F. Newhouse, M.B. Gregg, and P.S. Brachman. 1981.Pontiac fever, isolation of the etiological agent (Legionella pneumophila) and demonstration of its mode of transportation. American Journal of Epidemiology 114(3): 337-347.

Kenner, B.A., H.F. Clark, and P.W. Kabler. 1961. Cultivation and enumeration of streptococci in surface water. Applied Microbiology 9: 15.

Kerr, R., and C. Butterfield. 1943. Notes on the relations between coliform organisms and pathogens. Public Health Reports 58(15): 589-607.

Klein, E., and A.C. Houston. 1899. Further report on bacteriological evidence of recent and therefore dangerous sewage pollution of elsewise potable waters. Supplement to 28th Annual Report of the Local Government Board Containing the Report of the Medical Officer for 1898-1899, London City Council, London.

Klein, G. 2003. Taxonomy, ecology and antibiotic resistance of enterococci from food and the gastrointestinal tract. International Journal of Food Microbiology 88: 123-131.

Koch, R. 1883. Bericht uber die Thatigkeit der deutschen Cholera-komission in Aegpytten and Ostindien. Wien Med Wochenschr 33: 1548-1551.

Kramer, M.H., B.L. Herwaldt, G.F. Craun, R.L. Calderon, and D.D. Juranek. 1996. Surveillance for waterborne-disease outbreaks—United States, 1993-1994. MMWR 45 (No. SS-1): 1-33.

Lee, S.H., D.A. Levy, G.F. Craun, M.J. Beach, and R.L. Calderon. 2002. Surveillance for waterborne-disease outbreaks—United States, 1999-2000. MMWR 51 (No. SS-8): 1-47.

Levin, M.A., J.R. Fischer, and V.J. Cabelli. 1975. Membrane filter technique for enumeration of enterococci in marine waters. Applied and Environmental Microbiology 30(1): 66-71.

Levine, W.C., and G.F. Craun. 1990. Waterborne disease outbreaks, 1986-1988. MMWR 39 (No. SS-1): 1-13.

Levy, D.A., M.S. Bens, G.F. Craun, R.L. Calderon, and B.L. Herwaldt. 1998. Surveillance for waterborne-disease outbreaks - United States, 1995-1996. MMWR 47 (No. SS-5): 1-33.

Litsky, W., W.L. Mallman, and C.W. Fifield. 1955. Comparison of the most probable numbers of E. coli and enterococci in river water. American Journal of Public Health 45: 1049.

Mallmann, W.L., and E.B. Seligman. 1950. A comparison study of media for the detection of streptococci in water and sewage. American Journal of Public Health 40: 286.

Mason, W. 1891. Notes on some cases of drinking water and disease. Journal of the Franklin Institute (Nov.): 1-10.

Melnick, J.L. 1947. Poliomyelitis virus in urban sewage in epidemic and nonepidemic times. American Journal of Hygiene: 240-253.

Melnick, J.L. 1957. A waterborne urban epidemic of hepatitis. Pp. 211-225 in Hepatitis Frontiers, F.W. Hartman, ed. Boston: Little Brown and Co.

Messer, J.W. and A.P. Dufour. 1998. A rapid, specific membrane filtration procedure for enumeration of enterococci in recreational water. Applied and Environmental Microbiology 64(2): 678-680.

Moore, A.C., B.L. Herwaldt, G.F. Craun, R.L. Calderon, A.K. Highsmith, and D.D. Juranek. 1993. Surveillance for waterborne-disease outbreaks - United States, 1991-1992. MMWR 42 (No. SS-5): 1-22.

Neefe, J., and J. Stokes. 1945. An epidemic of infectious hepatitis, apparently due to a waterborne agent. Journal of the American Medical Association 128: 1063-1075.

NRC (National Research Council). 1997. Safe Water From Every Tap: Improving Water Service to Small Communities. Washington, D.C.: National Academy Press.

NRC. 1999a. Setting Priorities for Drinking Water Contaminants. Washington, D.C.: National Academy Press.

NRC. 1999b. The Use of Drugs in Food Animals. Washington, D.C.: National Academy Press.

NRC. 2001. Assessing the TMDL Approach to Water Quality Management. Washington, D.C.: National Academy Press.

NRC. 2002. Biosolids Applied to Land: Advancing Standards and Practices. Washington, D.C.: National Academy Press.

NTAC (National Technical Advisory Committee). 1968. National Technical Advisory Committee (NTAC) Report of the Committee on Water Quality Criteria. Washington, D.C.

Orla-Jensen, S. 1919. The Lactic Acid Bacteria. Copenhagen: A. F. Host and Son Co.

Ostrolenk, M., N. Kramer, and R.C. Clerverdon. 1947. Comparative studies of enterococci and *Escherichia coli* as indices of pollution. Journal of Bacteriology 53: 197-203.

Paul, J.R., J.D. Trask, and S. Gard. 1940. Poliomyelitis virus in urban sewage. Journal of Experimental Medicine 71: 765-777.

Payment, P., L. Richardson, J. Siemiatycki, R. Dewar, M. Edwardes, and E. Franco. 1991. A randomized trial to evaluate the risk of gastrointestinal disease due to consumption of drinking water meeting current microbiological standards. American Journal of Public Health 81: 703-708.

Payment, P., and E. Franco. 1993. *Clostridium perfringens* and somatic coliphages as indicators of the efficiency of drinking water treatment for viruses and protozoan cysts. Applied and Environmental Microbiology 59: 2418-2424.

Payment, P., J. Siemiatycki, L. Richardson, G. Renaud, E. Franco, and M. Prevost. 1997. A prospective epidemiological study of gastrointestinal health effects due to the consumption of drinking water. International Journal of Environmental Health Research 7: 5-31.

Pontius, F.W., and S.W. Clark. 1999. Drinking water quality standards, regulations and goals. Chapter 1 in Water Quality and Treatment: A Handbook of Community Water Supplies, 5th Edition. New York: McGraw-Hill.

Rivera, S., T.C. Hazen, and G.A. Toranzos. 1988. Isolation of fecal coliforms from pristine water in a tropical rain forest. Applied and Environmental Microbiology 54: 513-517.

Scott, W. 1951. Sanitary study of shore bathing water. Bulletin of Hygiene 33: 353.

Sherman, J.M. 1937. The streptococci. Bacteriological Reviews 1: 3-97.

Slanetz, L., D. Bent, and C. Bartley. 1955. Use of the membrane filter technique to enumerate enterococci in water. Public Health Reports 70(1): 67.

Slanetz, L.W., and C.H. Bartley. 1957. Numbers of enterococci in water, sewage and feces determined by the membrane filter technique with improved medium. Journal of Bacteriology 74: 591.

Smith, T. 1891. A New Method for Determining Quantitatively the Pollution of Water by Fecal Bacteria. 13th Annual Report of the New York State Board of Health. Albany, New York.

Snow, J. 1854a. On cholera. In The Sourcebook of Medical History, L. Clendening, ed. 1942. New York: Paul B. Hoeber, Inc.

Snow, J. 1854b. On the Mode of Communication of Cholera. Self-published by Dr. John Snow of Sackville Street, Picadilly (republished by Delta Omega, the U.S. National Honorary Public Health Society).

Stevenson, A. 1953. Studies of bathing water quality and health. American Journal of Public Health 43: 529-538.

Stokes, W.R. 1927a. Report of the committee on bathing places. American Journal of Public Health 17: 121

Stokes, W.R. 1927b. A search for pathogenic bacteria in swimming pools. American Journal of Public Health 17: 334

Streeter, H.W. 1951. Bacterial Quality Objectives for the Ohio River: A Guide for the Evaluation of Sanitary Condition of Waters Used for Potable Supplies and Recreational Uses. Cincinnati, Ohio: Ohio River Valley Water Sanitation Commission.

Thiercelin, M.E. 1899. Sur un diplocoque saprophyte de l'intestin susceptible de devenir pathogene. CR Soc. Biol. 5: 269-271.

Thiercelin, M.E., and L. Jouhaud. 1903. Reproduction de l'enterocoque; taches centrales; graduations periferiques et microblastes. CRS Soc. Biol. Fil. 55: 686-688.

Trask, J., and J. Paul. 1942. Periodic examination of sewage for the virus of poliomyelitis. Journal of Experimental Medicine 75: 2-6.

U.S. Treasury Department. 1914. Bacterial standard for drinking water. Public Health Report 29: 2959-2966.

Veillon, A., and A. Zuber. 1898. Recherches sur quelques microbes strictement anaérobies et leur rôle en pathologie Arch. Med. Exp. Anat. Pathol. 10: 517-545.

Venczel, L.V., M. Arrowood, M. Hurd, and M.D. Sobsey. 1997. Inactivation *Cryptosporidium parvum* oocysts and *Clostridium perfringens* spores by a mixed-oxidant disinfectant and by free chlorine. Applied and Environmental Microbiology 63(4): 1598-1601.

Viswanathan, R. 1957. Infectious hepatitis in Delhi (1955-56), a critical study: Epidemiology. Indian Journal of Medical Research 45(suppl.): 1-29.

Wattie, E., and C. Butterfield. 1944. Relative resistance of *Escherichia coli* and *Eberthella typhosa* to chlorine and chloramines. Public Health Reports 59(52): 1661-1671.

Weibel, S.R., F.R. Dixon, B.B. Weidner, and L.J. McCabe. 1964. Waterborne-disease outbreaks, 1946-1960. Journal of the American Water Works Association 56: 947-958.

Welch, W.H., and G.H.F. Nuttall. 1892. A gas-producing bacillus (*Bacillus aeorgenes capsulatus*, Nov. Spec.) capable of rapid development in the body after death. Johns Hopkins Hospital Bulletin 3: 81.

Wilson, W.J., and E.M.M. Blair. 1925. Correlation of the sulfite reduction test with other tests in the bacteriological examination of water. Journal of Hygiene, Cambridge 24: 111.

Wolfe, H. 1972. The coliform count as a measure of water quality. Pp. 333-346 in Water Pollution Microbiology, R. Mitchell, ed. New York: Wiley-Interscience.

Wong, D. C., R.H. Purcell, S.R. Prasad, M.A. Sreenivasan, and K.M. Pavri. 1980. Epidemic and endemic hepatitis in India: evidence for non-B hepatitis virus aetiology. Lancet 2: 876-879.

2

Health Effects Assessment

INTRODUCTION

The foremost goal of developing and using indicators for waterborne pathogens is public health protection. This chapter provides an overview of health effects assessment for waterborne pathogens and their indicators, and includes a brief review of surveillance and epidemiologic study designs, an historical review and current status of health effects assessment, and a detailed discussion of quantitative microbial risk assessment. Furthermore, health effects assessment is discussed throughout this chapter in the context of drinking water and of fresh and marine recreational waters. This chapter also includes a description of the national surveillance system for waterborne disease outbreaks and several related epidemiologic studies currently being conducted. The final section is a summary of the chapter and its conclusions and recommendations.

This chapter is not intended to serve as a comprehensive review of epidemiology as a methodologic tool or waterborne disease in humans, both of which are beyond the scope of this report. Rather, it provides some substantive background information on epidemiology and health effects assessment within the overall context of indicators for waterborne pathogens as discussed throughout this report and especially in support of a phased approach to microbial water quality monitoring that is provided in Chapter 6.

Approaches to Health Effects Assessments

Health effects assessments for waterborne pathogens can be based on a number of approaches, all of which have been used to document and quantify the health risks resulting from microorganisms in water. These approaches include (1) assessments of epidemiologic evidence for waterborne-based outbreaks; (2) human volunteer studies showing that a known or potential waterborne pathogen is infectious by the oral ingestion route and capable of causing infection and disease at particular doses (dose-response studies); (3) various types of retrospective and prospective epidemiologic studies for health effects assessments; (4) estimates of health risks by linking epidemiologic evidence for disease to measured concentrations of either pathogens or indicators in the water; (5) estimates of the ratios of pathogens to indicators in the exposure vehicle (e.g., feces, sewage, fecally contaminated water); and (6) quantitative microbial risk assessments that integrate human exposure and health effects data for quantitative risk estimations or characterizations.

Health Effects Concerns and Early Studies of Microbial Water Quality

Outbreak Investigations and Risk Estimates from
Pathogen-to-Indicator Ratios in Water

As discussed in Chapter 1, concerns about the sanitary quality of drinking water and the risks of enteric infectious diseases in the United States go back to at least the late 1800s, when enteric disease outbreaks were first recognized and linked at least tentatively to these exposure routes. Similar concerns for U.S. recreational and shellfish waters started in the 1920s. The initial recognition of and concern about infectious disease risks from these sources of exposure focused on enteric bacterial diseases, and early health effects assessments of enteric bacterial pathogens and waterborne outbreaks date back to the early 1920s. Human health risks from enteric viruses and parasites in water were first recognized and addressed during and shortly after World War II. However, civilian risks from these waterborne pathogens were not widely documented and appreciated until studies of waterborne outbreaks and waterborne pathogen occurrence were first reported in the 1960s. The recognized viral and protozoan pathogens of initial concern were infectious hepatitis viruses, polio, and other enteroviruses, and *Entamoeba histolytica* and *Giardia lamblia*, respectively.

Perhaps the first attempts at linking health effects assessments of waterborne pathogens to microbial water quality were based on ratios of *Salmonella typhi* to fecal indicator (coliform) bacteria in feces and sewage and the allowable limits of coliforms in drinking water and, later, in recreational bathing and shellfish waters (Kehr and Butterfield, 1943; Prescott et al., 1945).

Early Health Effects Assessments of Enteric Pathogens from Human Dose-
Response Studies

The first human health effects dose-response studies appear to be with the protozoan parasites *Entamoeba histolytica* and other *Entamoeba* species conducted using enemy prisoners in the Philippines by U.S. Army medical officers early in the twentieth century (Walker and Sellards, 1913). This study also showed for the first time that *E. histolytica* caused dysentery, that the cysts and trophozoites (see also Chapter 3) were different forms of the same microorganism, and that other *Entamoeba* species (notably *Entamoeba coli*) did not cause dysentery or other enteric disease. It was not until the 1950s that a researcher attempted to determine the number of enteric parasites necessary to cause infection in human dose-response studies with *Entamoeba coli* (Rentdorff, 1954a) and *Giardia lamblia* (Rentdorff, 1954b). Dose-response data on bacterial infectivity from human volunteer studies date back to at least the 1940s when different doses of *Shigella paradysenteriae* (now *S. flexneri*) were administered in vaccine trials (Shaughnessy et al., 1946).

For enteric viruses, the first human volunteer studies were with infectious hepatitis viruses during and after World War II (Cameron, 1943; Lainer, 1940; MacCallum and Bradley, 1944; Voegt, 1940). Studies by MacCallum and Bradley's group are considered the first to distinguish infectious from serum hepatitis. Some of these early studies provided the first dose-response data for infectious hepatitis, but lack of knowledge about actual virus concentrations in the inocula has hampered the use of these data and subsequent infectious hepatitis human volunteer dose-response data for health risk assessments. Studies to estimate dose-response for virus infectivity were conducted using candidate live oral poliovirus vaccines in the 1950s (Koprowski, 1956; Sabin, 1955) and 1960s (Katz and Plotkin, 1967), and infectivity for humans could be related to virus concentrations as measured by other methods. Mathematical modeling of the data sets, taking into account the number of subjects used at each dose and the sensitivity of the dose-response study, was not undertaken until the 1980s (Haas, 1983b). However, many data sets, along with the best-fit models for bacteria, protozoa, and viruses, have since been compiled (Haas et al., 1999b).

Prospective Epidemiologic Studies of Microbial Water Quality and Health Effects

Prospective epidemiologic studies have attempted to link health effects in exposed individuals to the microbial quality of water. This approach has been used primarily for recreational waters and dates back to studies by Stevenson (1953) on Lake Michigan and the Ohio River. Those studies reported epidemiologically detectable health effects in bathers from waters containing about 2,300-2,400 total coliforms per 100 mL. Based on several lines of evidence—including outbreak data, the ratios of *Salmonella typhi* to fecal indicator (coliform) bacteria in feces and sewage, and epidemiologic studies of enteric illness in bathers at beaches having different levels of fecal contamination—the U.S. Public Health Service (USPHS) and later the Federal Water Pollution Control Association

(FWPCA; predecessor to the U.S. Environmental Protection Agency) developed bacteriological quality guidelines for recreational waters, as noted in Chapter 1 (Cox, 1951; NTAC, 1968; Scott, 1951).

Public Health Risk Assessment Framework

The ultimate objective of determining the microbiological quality of water is to identify and then minimize the public health risk from consuming water intended for drinking and from exposure to recreational water. Data are used to develop approaches to remediate or control this public health risk by reducing the potential exposure to levels that are considered acceptable (e.g., by controlling contamination sources) or developing communication strategies to prevent exposure (e.g., by closing a beach).

Indicators are measured for many purposes (see Chapter 4 for a detailed discussion of indicator applications). In terms of public health protection, indicators for pathogens in water intended for drinking are measured to determine the level of microbial contamination of source water (see also Chapter 6), whether existing water treatment processes are adequate, and whether the integrity of the distribution system has been breached. In addition, indicators can be used to measure the quality of the water in unregulated private wells. The measurement of indicators in the recreational water setting is typically conducted to determine if the level of contamination of surface waters such as oceans, lakes, and rivers is sufficiently elevated to pose a human risk and, therefore, to determine whether warnings should be issued or recreational waters should be closed to the public.

In drinking water and food, philosophically a zero-tolerance approach has been taken for indicators. Thus, it is presumed that if a measured indicator concentration is zero through water protection and treatment, the health risk is also zero. However, this traditional strategy does not provide an effective framework for decision making in the context of what is currently known about indicators. All ambient waters (including groundwater) will be subject to some level of microbial indicators and contamination whether associated with fecal sources (both human and nonhuman) or with naturally occurring microorganisms. Thus, the regulatory question remains, What measurable microorganisms in water best represent a risk to human health and at what levels would they be of concern?

Such criteria and standards can be established by determining two relationships: (1) between the density of the indicators and the occurrence of adverse health outcomes, and (2) between the density of the indicators and the presence of pathogenic organisms in the water. Although the association between the occurrence of a pathogen (or its indicator) in water and a human health effect is a difficult one to determine, epidemiologic studies, surveillance, and risk assessment are useful tools to help establish this association.

Risk assessment is a process that allows for the integration of scientific data regarding an environmental hazard into a framework that addresses the risk of exposure and its potential health impacts (NRC, 1983). The process is quantitative in nature and attempts to address both the nature and the magnitude of the risk. This process has proven invaluable to the regulatory community, industry, and risk managers and has direct application to public health risk from water. The value of such a framework is that many different types of information—various indicator data, epidemiologic data, and data specific to the nature of the exposure (e.g., recreational or irrigation waters versus source of potable water supply)— can be used to define public health safety goals.

In ambient recreational waters, there is a need to understand the nature and level of the risk and, therefore, to take a risk assessment approach. In the risk assessment process, hazard identification has traditionally been separated from exposure analysis. In this case, the nature of the microbial hazard and its identification are closely tied to the sources and fate of the pathogenic microorganisms and, thus, the exposure. For example, enteric viruses detected at beaches can be tied to human fecal inputs and the ability of the virus to survive and cause illness at low doses and concentrations.

Many attributes of indicators (see Chapter 4 for further information) and indicator methods that may lend themselves to the risk assessment process are currently available (e.g., identifying sources of microbial contamination), whereas other attributes will be difficult to determine and may not prove feasible (e.g., establishing a quantitative relationship between concentrations of indicators and the degree of public health risk). As in most science-based evaluations, uncertainty will have to be described, and quantifying uncertainty is most problematic in the exposure portion of the analysis where indicators are used to estimate the potential for exposure to actual pathogens. A microbial risk framework can be developed and used to understand the basic principles and data gaps in the study of public health risks associated with the characterization of recreational water quality using a variety of methodologies. Such an approach will lead to a decision support system for data gathering (types of data and methods) and for response and mitigation efforts.

SURVEILLANCE AND EPIDEMIOLOGIC STUDY DESIGNS

As noted previously, health outcomes can be linked to exposure data by various epidemiologic methods. A brief overview of these methods follows. Various introductory and advanced textbooks, as well as review articles on epidemiologic methods, can be consulted for more comprehensive coverage of this topic and for detailed definitions of various epidemiologic terms used in this chapter and report (e.g., Gordis, 2000; Last, 2001; Lavori and Kelsey, 2002; Matthews, 2000; Meinert, 1986; Rothman and Greenland, 1998; Rothman, 2002; Schlesselman, 1982).

Surveillance

Modern public health surveillance of disease was defined by Langmuir (1963) as "the continued watchfulness over the distribution and trends of incidence through the systematic collection, consolidation and evaluation of morbidity and mortality reports and other relevant data." It is now standard practice to add to this definition the concept of applying these data to prevention and control of disease.

The steps in surveillance include the systematic collection of data, analyses to produce statistics, interpretation to provide information in a timely manner, actions taken as a result of the data, and continued surveillance to evaluate the success of the actions taken. Guidelines for evaluating surveillance systems have been proposed by the Centers for Disease Control and Prevention (CDC, 2001).

Epidemiologic Study Designs

Epidemiologic studies fall into two general categories: (1) experimental studies (e.g., randomized controlled intervention or clinical trials) in which investigators control the conditions of exposure in the study and (2) observational studies (e.g., cohort, case-control, cross-sectional, and ecologic studies) in which investigators do not control the exposure or most other aspects of the process being studied.

Of the epidemiologic studies, randomized controlled trials provide the strongest epidemiologic evidence of an etiologic association between exposure and outcome, followed in decreasing order by cohort studies, case-control studies, cross-sectional studies, and ecologic studies.

Randomized Controlled Trial

This epidemiologic experimental design is regarded as the most scientifically rigorous method of hypothesis testing available. Subjects are randomly allocated into two groups, one that will receive an experimental treatment or intervention and the other that will not. Randomization tends to produce comparability between the two groups with respect to factors that might affect the health outcome being studied and, thus, to minimize the potential for confounding variables.[1] Additional objectivity is provided when subjects, investigators, and statisticians analyzing the data are unaware of the subject's allocation to a particular treatment or intervention (known as randomized triple-blinded trials). The scien-

---

[1]Confounding variables ("confounders") are variables that can alternatively cause or prevent the outcome of interest in an epidemiologic study and are associated with the factor under investigation. As such, confounding variables may be due to chance or bias, and unless adjusted for, their effects cannot be distinguished from those factors being studied.

tific rigor of this study design is its chief advantage, while its cost, often in the millions of dollars, is its greatest disadvantage. Therefore, randomized controlled trials are generally used only when a well-defined hypothesis is being tested.

Randomized controlled trials have additional benefits. They provide a temporal association between the exposure and the health outcome, which is one measure of causality, because the exposure precedes the outcome. They also allow for the calculation of incidence rates of disease in each group (i.e., the occurrence of a certain disease or health outcome in a group or population over a specified period of time) and their relative risk rather than being limited to calculations of odds ratios. As defined in *A Dictionary of Epidemiology*, edited by Last (2001), a relative risk (RR) or risk ratio is "the ratio of two risks, usually exposed/ not exposed." The odds ratio (OR) is defined differently according to the situation (e.g., calculation of odds of exposure or odds of disease). As defined in Last (2001), "The exposure-odds ratio for a set of case-control data is the ratio of the odds in favor of exposure among the cases to the odds in favor of exposure among non-cases" and "the disease-odds ratio for a cohort or cross-sectional study is the ratio of the odds in favor of disease among the exposed to the odds in favor of disease among the unexposed." Problems associated with randomized control trials include noncompliance, participant dropout, and generalizability of results.

A related type of experimental design is a community intervention trial in which the exposure is assigned to groups of people rather than singly. This type of experiment is often used to study environmental exposures. Most community intervention trials do not employ random assignment for the experimental treatment; rather, they use a cross-over design (i.e., before and after treatment) where the community serves as its own control.

Cohort Study

A cohort is defined as a group of persons who are followed over a period of time and usually includes individuals with a common exposure. A cohort study involves measuring the occurrence of disease within one or more cohorts that have differing exposures during a certain period of follow-up. John Snow's study of the cholera outbreak in 1854 (see Chapter 1 for a brief description; Snow 1854) is an elegant example of a cohort study. Cohort studies can be prospective (exposure information is recorded at the beginning of the follow-up and the period of time at risk is forward in time) or retrospective (cohorts are identified from recorded information and the follow-up time occurred before the beginning of the study). Cohort studies have several advantages. They allow for the association of multiple health outcomes or diseases, or multiple endpoints within the progression of one disease, along with the exposure of interest. Like randomized controlled trials, cohort studies provide a temporal association between exposure and health outcome. Finally, incidence rates of disease in the cohorts being assessed and their relative risk can also be calculated.

Cohort studies are subject to several types of potential bias, including bias in the selection of the cohorts' exposure, bias in assessment of the health outcome, and bias if the two cohorts have differing response rates. A cohort design is generally selected when there is good evidence of an association of a health outcome with a certain exposure and when the exposure is relatively rare but the incidence of disease among the exposed group is high. Attrition of the study population is minimized when the time between exposure and disease is short. Although not as costly as a randomized experiment, cohort studies are generally more expensive that other types of epidemiologic designs. As with most epidemiologic studies, cohort studies are subject to confounding. Known confounding factors can be controlled for in the analysis of the data, but unknown confounders are by their nature impossible to adjust for in the analysis.

Case-Control Study

This type of study aims to achieve the same goals as the cohort study while minimizing the need to obtain information on exposure and outcome from large populations. Samples are taken from the source population to reduce the number of study participants. Properly designed and conducted, case-control studies provide information that is similar to what could be collected from a cohort study but at considerably less cost and time. In this study design, the investigator selects individuals with the health outcome of interest (cases) and appropriate individuals without the health outcome (controls), collects information regarding their past exposure, and then compares the rates of exposure of the two groups. Issues to be considered when using this study design include the ascertainment of cases of disease (e.g., diagnostic criteria, population source, incident or prevalent cases) and the selection of appropriate controls (i.e., should controls be comparable to cases in all respects other than having the disease, and how many variables between cases and controls should be matched?). A case-control study design is often used when investigators want to determine the association of a health outcome, especially a rare one, with multiple rather than single exposure factors.

Case-control studies are subject to several biases, including recall bias (i.e., cases might be more likely to remember their past exposure than controls), selection bias, and nonresponse bias. Case-control studies suffer from the same problems with confounding factors as cohort studies. Another disadvantage of case-control studies is the inability to calculate incidence rates and their differences and ratios (e.g., relative risks); investigators can calculate only the ratio of incidence or prevalence rates or risks (e.g., odds ratios; see previous discussion of relative risk and odds ratio).

Cross-Sectional Study

This type of study provides a snapshot of the status of a target population with regard to exposure status, health outcome, or both at a specific point in time. Cross-sectional studies attempt to enumerate the population and assess the prevalence of various characteristics. This design is characterized by the fact that only one set of observations is taken from each person. Although it cannot measure disease incidence because information across time is not available, disease prevalence can be assessed. Cross-sectional studies often are the first type of epidemiologic study conducted to determine the association between a health outcome and several possible exposure variables. As the hypothesis to be tested is refined, investigators typically progress to one of the other study designs. In some instances, however, cross-sectional surveys are conducted repeatedly over long periods as a form of disease surveillance system.

Ecologic Study

An ecologic study, also known as an aggregate study, compares groups rather than individuals. This design is most often used when individual-level data are missing. Although ecologic studies are relatively easy and inexpensive to conduct, their results are often difficult to interpret. Ecologic studies are used to study environmental exposures because it is difficult to accurately measure relevant exposures or doses at the individual level for large numbers of persons. In addition, exposure levels may vary little within a study area. The major limitation of ecologic analyses is the so-called ecologic bias in which the expected ecologic effect fails to accurately represent the biologic effect at the individual level. Robinson (1950) was the first to describe mathematically how ecologic associations could differ from the corresponding associations at the individual level. This phenomenon has become widely known as the ecologic fallacy.

Other Types of Studies

Some epidemiologic studies can be considered a specific type of the aforementioned study designs or use methods that incorporate multiple elements of these basic study designs. Longitudinal time series studies and seroprevalence studies are two such examples. Longitudinal time series are cohort-type epidemiologic studies that correlate exposure variables of interest (e.g., an environmental indicator such as water turbidity or the presence of a waterborne pathogen like *Cryptosporidium parvum* in water) with health outcomes (e.g., a clinical measurement such as an immunologic marker of exposure in a serologic specimen or a direct measurement such as occurrence of diarrhea) over a specified period of time. These studies incorporate temporal factors in their analyses with exposure occurring before the health outcome. The need to include the (most) appropriate time lag between measurement of the exposure and health outcome make these analyses complex and difficult to successfully accomplish as well as making interpretation of the results difficult at times. However, these studies can provide good epidemiologic associations between exposures and adverse health outcomes.

Seroprevalence studies are a specific type of cross-sectional study design. These studies measure the prevalence of a serologic marker in study participants as the health outcome of interest. Serologic markers can be difficult to interpret, however, because their measurements represent historical exposure and it is not always clear when and for how long the selected marker is present after exposure to a pathogen of concern. In addition, there may be multiple markers to choose from (e.g., circulating antibodies to several antigens), which further complicates interpretation of results. Nevertheless, seroprevalence studies are useful in determining population exposure to a pathogen even when the pathogen itself cannot be detected. Lastly, measuring seroprevalence is especially useful in investigations of waterborne disease outbreaks because it can establish that individuals were previously exposed and infected.

HISTORICAL REVIEW AND CURRENT STATUS OF HEALTH EFFECTS ASSESSMENT

The following section provides an overview (historical and current) of the most salient epidemiologic assessments of health effects associated with drinking water and recreational water exposure. Please refer to the previous section for a review of the surveillance and epidemiologic terms and methods used in the health assessments described below.

U.S. National Waterborne Diseases Outbreak Surveillance System

Surveillance for outbreaks associated with drinking water and recreational water has been going on since 1920 (Craun, 1986). The CDC, the U.S. Environmental Protection Agency (EPA), and the Council of State and Territorial Epidemiologists (CSTE) have maintained a collaborative surveillance system of waterborne disease outbreaks (WBDOs) since 1971 (see also Chapter 1). The National Waterborne Diseases Outbreak Surveillance System (WBDOSS), located at CDC, collects data regarding outbreaks associated with drinking water and recreational water. Moreover, in recent years (1999-2000), the WBDO surveillance system has also provided data on outbreaks that occurred as a result of occupational exposure to water. The primary objective of collecting outbreak data is ultimately to reduce the occurrence of WBDOs by characterizing the epidemiology of the outbreaks, identifying the etiologic agents, and determining the reasons for the occurrence. Results from these efforts provide the opportunity to issue public health prevention and control messages.

Characteristics of the Surveillance System

State, territorial, and local public health agencies are responsible for detecting and investigating WBDOs. As a federal agency, CDC participates in outbreak investigations only by invitation from a state or territorial epidemiologist or if an outbreak involves multiple states. Reporting is voluntary and passive, and varies by state. States and territories report their outbreaks annually on a standard (hard copy) form (CDC Form 52.12), and CDC compiles, analyzes, and publishes the data. Since 1989, when responsibility for the surveillance system was moved to CDC's Division of Parasitic Diseases, the data have been published every two years as a *Morbidity and Mortality Weekly Report* (MMWR) Surveillance Summary (Barwick et al., 2000; Herwaldt et al., 1991; Kramer et al., 1996; Lee et al., 2002; Levy et al., 1998; Moore et al., 1993). Both the surveillance system's submitted hard copy report forms and the electronic database reside at CDC.

Two major categories of data are reported on the forms: (1) epidemiologic data such as type of exposure; number of persons exposed, ill, and hospitalized; number of fatalities; symptoms; etiologic agent; and results from clinical laboratory data; and (2) environmental data such as the type of water system involved, results from water testing, and factors that contributed to contamination of the water. CDC contacts the state's environmental agency if additional information regarding source water, treatment, or supply is needed to flesh out the investigation and to work through the issues that led to the contamination. Completion rates for the report forms vary tremendously by outbreak investigation, as well as by the intensity and scope of the investigation.

Unlike most surveillance systems, the unit of analysis in WBDOSS is an outbreak rather than an individual case of a specific disease. Two major criteria must be met for an event to be classified as an outbreak: (1) at least two persons must have experienced a similar illness after consumption of a common source of drinking water or after exposure to water used for recreational purposes and (2) epidemiologic data must implicate water as the probable source of the illness. However, the stipulation that at least two persons be ill is waived for single cases of laboratory-confirmed primary amoebic meningoencephalitis (PAM; see also Chapter 3) and for single cases of chemical poisoning if water quality data indicate contamination by the chemical. An outbreak that meets both criteria will be included in the surveillance system whether the etiologic agent is infectious, chemical, or unidentified.

WBDOs are classified (Class I-IV) according to the strength of the evidence implicating water. Epidemiologic data are weighted more heavily than water quality data, and outbreaks that are reported without supporting epidemiologic data are excluded from the surveillance system (see Table 2-1).

In addition, each drinking water system associated with an outbreak is classified by the following types of problems: untreated surface water, untreated groundwater, treatment deficiency (e.g., inadequate disinfection), distribution

TABLE 2-1 Classification of Investigations of Waterborne Disease Outbreaks in the United States[a]

Class[b]	Epidemiologic Data	Water Quality Data
I	Adequate[c] a) Data were provided about exposed and unexposed persons; and b) Relative risk or odds ratio was ≤ 2, or the *P*-value was <.05	Provided and adequate historical information or laboratory data such as the history that a chlorinator malfunctioned or a water main broke, no detectable free-chlorine residual, or the presence of coliforms in the water
II	Adequate	Not provided or inadequate (e.g., stating that a lake was crowded)
III	Provided, but limited a) Epidemiologic data provided did not meet the criteria for Class I; or b) The claim was made that ill persons had no exposures in common besides water, but no data were provided	Provided and adequate
IV	Provided, but limited	Not provided or inadequate

[a]Outbreaks of *Pseudomonas* and other water-related dermatitis and single cases of primary amoebic meningoencephalitis or of illness resulting from chemical poisoning are not classified according to this scheme.

[b]Based on the epidemiologic and water-quality data provided on CDC Form 52.12.

[c]Adequate data were provided to implicate water as the source of the outbreak.

SOURCE: Adapted from Lee et al., 2002.

system deficiency (e.g., cross-connection), and unknown or miscellaneous deficiency (e.g., contamination of bottled water).

Usefulness of the Surveillance System

Outbreak data gathered through this surveillance system are useful for identifying deficiencies in providing safe drinking water and recreational water, evaluating the adequacy of current regulations for water treatment, and monitoring water quality. In addition, outbreak data are used to determine or update the biology and epidemiology of etiologic agents and to influence research priorities.

However, the WBDO surveillance system does not consistently provide information that would help researchers link indicator and pathogen data with health outcome data because the collection of water quality data is not required for inclusion of an outbreak in the system. In addition, the utility of coliform data as an indicator of outbreak vulnerability of public water systems has come into question in a recent study by Nwachuku et al. (2002). That study compared Total

Coliform Rule (TCR; see also Table 1-1) violations for water systems that had and had not reported an outbreak from 1991 to 1998. Their findings suggested that the TCR is not able to identify those water systems that are vulnerable to an outbreak. The authors of that study suggest that source water be examined using a wide variety of indicators because simply monitoring treated drinking water for one indicator, most often coliforms, will not provide a useful measure of the water's overall microbial quality. The difficulty of collecting water samples in a timely manner is a complicating factor in attempts to epidemiologically link contaminated water with adverse health outcomes.

Adding to the difficulty of linking water quality data to health outcome data is the fact that health agencies and environmental agencies are separate in approximately 70 percent of states, and this probably holds true at county and local levels as well (Lynn Bradley, Association of Public Health Laboratories, personal communication, 2003). Thus, responsibility for investigating waterborne outbreaks will rest with either the health staff or the environmental staff. Therefore, the thoroughness of the epidemiologic versus the environmental components of any outbreak investigation will vary with each local or state agency. Better coordination between the two components of a waterborne disease investigation would increase the completeness of investigations. In this regard, participants in a recent colloquium on the burden of gastrointestinal illness recommended a team approach to outbreak investigations and standardized protocols and case definitions (Payment and Riley, 2002).

The importance of, and approaches to, waterborne disease surveillance and other epidemiologic methods of estimating waterborne disease burdens, etiologies, and causes have been addressed by the World Health Organization (WHO; Fewtrell and Bartram, 2001). Several chapters of that WHO report describe approaches and limitations of acquiring, interpreting, and applying epidemiologic information on waterborne disease to prevention and control measures. The report also discussed the development of water quality criteria, guidelines, and standards. Standardized protocols for waterborne outbreak investigations, especially for large outbreaks, could help ensure comprehensive investigations that include the collection of epidemiologic, clinical, laboratory, and environmental data. Such investigations would allow researchers to maximize the information obtained and provide opportunities to associate various exposure factors with health outcomes.

Representativeness of Outbreaks Reported to the WBDO Surveillance System

Many factors affect the likelihood that a waterborne outbreak will be recognized and investigated; such factors lead to concerns about the representativeness of WBDOs that are reported, including the following (Lee et al., 2002):

• The larger the outbreak, the more likely it is to be detected over background incidence of illness or related symptoms.

• The more severe the illness (e.g., bloody diarrhea caused by *Escherichia coli* O157:H7), the more likely the outbreak will be detected.

• The public's awareness that an outbreak is occurring will more readily lead people to call their local health department to report an illness.

• A clinician's specific interest in an agent (e.g., *Cryptosporidium parvum* at a time when this protozoan was not well known) will help ensure that a request for laboratory testing to identify the agent is more likely.

• State and private laboratories vary in their routine testing practices for pathogens in stool specimens, making the detection of any pathogen more or less likely (e.g., *Cryptosporidium parvum* may not be included automatically in a request for routine ova and parasite testing of a fecal specimen).

• Local and state health departments have limited budgets and allocate resources according to the perceived health risks to their communities.

In general, outbreaks with high attack rates (i.e., the cumulative incidence of infection in a group observed during an epidemic) or a large number of cases of illness associated with severe symptoms in a state that has had previous waterborne outbreaks are likely to be recognized. Outbreaks that are more likely to be missed include those that have low attack rates, are associated with mild symptoms, and are caused by an etiologic agent that is not easily identified such as a virus. It is important to note that a large proportion of identified outbreaks to date have occurred in small communities. This indicates that increases in outbreak-related cases of illness in larger communities and cities might be missed because such cases are typically reported to a much wider variety of physicians and laboratories.

Sensitivity of the Surveillance System and Underreporting

The sensitivity of the WBDO surveillance system (i.e., the probability that an actual outbreak will be identified correctly, reported to CDC, and recorded into the surveillance database) is unknown because the actual (total) number of WBDOs cannot be determined. However, the sensitivity of the system is probably low because of underreporting of WBDOs, likely caused by lack of recognition that an outbreak is occurring or has already occurred. The multiple sequential barriers that can exist to reporting cases of outbreak-related illness are listed below:

• person gets infected when exposed to agent;
• person becomes symptomatic;
• person seeks medical care;
• health care worker orders a laboratory test;
• person provides the requested specimen (e.g., stool);
• laboratory tests for the specified agent;

- test result is positive;
- positive result is reported to the health department;
- health department reviews and analyzes the reports in a timely manner;
- health department concludes that an outbreak might be occurring; and
- health department investigates the potential outbreak.

Enough nonoccurrences or failures at any of these steps could result in a missed WBDO.

To complicate matters, standardized clinical and environmental laboratory methods that are both sensitive and specific are lacking for many viruses, and routine testing for parasites in fecal samples is not always done. For example, the incubation period for parasitic diseases such as cryptosporidiosis averages 7 days and can be as long as 14 days, making the association between illness and water exposure much more difficult.

Planned Improvements to the Surveillance System

The hard copy form used to report WBDOs to the WBDOSS was revised recently to allow for more specific reporting of water quality data (e.g., turbidity, total and fecal coliforms, other indicators of waterborne pathogens as needed). While states sometimes report finished water quality data, inclusion of source water quality monitoring data for a variety of indicators would likely contribute important information. More options are provided on the outbreak reporting form for listing the types of problems and deficiencies encountered in drinking water systems (e.g., lack of filtration, lack of backflow prevention, cross-connection, negative pressure) and in recreational water settings (e.g., heavy bather density, animal or human fecal contamination). The system will expand the types of outbreaks that it includes, such as outbreaks of legionellosis and wound infections associated with exposure to recreational water.

As a result of these recent and planned changes, CDC will interact more actively with state waterborne disease coordinators to make them aware of the report revisions. CDC also plans to use this opportunity to review problems with outbreak reporting and to emphasize the need for collecting water quality data in addition to the epidemiologic data that are routinely collected.

Epidemiologic Studies of Diseases Attributed to Drinking Water

For drinking water, experimental and observational epidemiologic studies have focused on determining if there is an association between water consumption and adverse health outcomes, especially gastrointestinal illness. There has been an ongoing debate in the United States about the extent to which infectious diseases may be transmitted to humans through drinking water that meets federal standards for water quality. These concerns have been heightened by: (1) the continuing occurrence of waterborne disease outbreaks (see Figures 1-1, 1-2, and 1-3); (2) two outbreaks associated with public water systems that met Safe Drinking Water Act (SDWA) standards (Goldstein, 1996; Mac Kenzie et al., 1994) but resulted in a number of deaths; (3) a 2002 outbreak of primary amoebic meningoencephalitis caused by *Naegleria fowleri* associated with untreated groundwater;[2] and (4) the findings of Payment et al. (1991, 1997; see more below and Chapter 1), which suggested that between 14 and 40 percent (depending on population group and water exposure) of gastrointestinal illness in a community might be attributable to waterborne pathogens.

Various methods can be used to estimate the strength of an association between an exposure variable and a health outcome variable, and EPA and CDC are in the process of developing a congressionally mandated (i.e., SDWA Amendments of 1996) national estimate of waterborne disease occurrence. As noted in Chapter 1, the congressional language "waterborne disease occurrence" has been interpreted to mean that the focus of the study should be on "gastrointestinal disease attributable to municipal drinking water." A review of the epidemiologic studies that have and will provide data components (attributable fractions[3] from select water systems and gastrointestinal rates in the general population) to be used in developing the national estimate are summarized below.

The Canadian Intervention Trials

As discussed in Chapter 1, at the time that Payment and colleagues (1991) designed their first intervention trial, it was generally thought that drinking water meeting Canadian water quality standards posed a minimal health risk to consumers. However, the advent of more sensitive detection methods for waterborne viruses and parasites, coupled with the continued occurrence of waterborne outbreaks, raised some doubts as to whether municipal waters were free of microorganisms that could be pathogenic to humans. In addition, there was concern that endemic waterborne disease might be occurring without being recognized. For these reasons, Payment and colleagues designed a randomized trial to measure the rates of gastrointestinal illness related to the consumption of tap water that met then current water quality standards. However, the source of the tap water used in the study was surface water contaminated with human sewage. It was estimated that 35 percent of the reported gastrointestinal illness among tap water drinkers was related to consumption of the drinking water and was thus prevent-

---

[2]See http://www.uswaternews.com/archives/arcrights/3attfil7.html and http://www.maricopa.gov/public_health/docs/alerts/newsrelease-nov1102f.pdf for further information.

[3]For the purpose of this report, the term "attributable fraction" refers to the proportion of all cases that can be explained by (attributed to) a particular exposure (see Last, 2001, for further information on how this term is defined and used in epidemiology).

able. The study also included testing of raw source water and finished treated water for several indicators of water quality, including turbidity, chlorine, total bacteria, total and fecal coliforms, *Pseudomonas aeruginosa*, *Aeromonas hydrophila*, *Clostridium perfringens*, and human enteric viruses and bacteriophages. Despite the association of tap water with gastrointestinal illness, researchers did not find any correlations between illness and any of the physical, chemical, and microbial indicators measured.

As the debate continued about whether coliform-free drinking water was also pathogen-free, Payment and colleagues followed up their first intervention trial with a second randomized trial in the same study area (Payment et al., 1997). Large-volume analyses of raw source water were conducted for human enteric viruses, *Giardia lamblia*, *Cryptosporidium parvum*, *Clostridium perfringens*, and coliphages. Analyses were also conducted in post-filtration, pre-disinfection, and finished treated water. The tap water met or exceeded then-current water quality standards, and the distribution system was in compliance with regulations for coliforms and chlorine residual. The results of this trial indicated that 14 to 40 percent (depending on the age group and study assignment group) of gastrointestinal illness was attributable to tap water that met water quality standards. Furthermore, the results indicated that the water in the distribution system appeared to be partly responsible for the illnesses.

The Australian Intervention Trial

As a result of the two Payment studies and based on modeling data of *Cryptosporidium* oocysts obtained from quantitative microbial risk assessment (QMRA; discussed later) (Haas et al., 1996), new public health concerns were raised about the microbiological safety of drinking water in developed countries and the possibility that substantial endemic waterborne disease was going unrecognized. These concerns led to a randomized controlled intervention trial in Melbourne, Australia (Hellard et al., 2001). These researchers sought to improve on the previous trials by Payment and colleagues by incorporating double blinding into the study design (i.e., participants and investigators were unaware of the participants group assignment). Source water for the study area was obtained from uninhabited and protected forest catchments without farming or recreational water activity.

Six hundred households were randomly allocated to receive either a treatment device or a sham device and were followed for 68 weeks. This trial did not find a statistically significant difference between the rates of gastrointestinal illness in each group. Pathogens in fecal specimens were not found to be more common in the group that received the sham device. Routine water quality monitoring was performed by the water utility at customer properties. Water samples were tested for total and fecal coliforms, heterotrophic plate count bacteria (HPCs), and free and total chlorine. A composite sample from the four water mains that supplied the study area was collected weekly and analyzed for *Giardia lamblia*, *Cryptosporidium parvum*, *Aeromonas*, *Clostridium perfringens*, and *Campylobacter*. One possible explanation for the differing results between the Hellard and Payment trials is the quality of the source water. This raises the issue of generalizability of results from studies conducted in specific populations and geographic locations using specific types of water systems.

The U.S. Intervention Trials and Observational Studies

As mentioned previously, EPA and CDC decided that one method of developing a national estimate of endemic waterborne disease would be to obtain attributable fractions from controlled experiments and apply them to incident gastrointestinal illness rates in general populations determined from observational studies, such as cross-sectional surveys.

Pilot Water Evaluation Trials (Pilot WET and HIV WET) Two trials were designed in different populations to minimize the problems encountered in the previous intervention studies. The objectives were to assess (1) the effectiveness of triple blinding to group assignment (i.e., participants, investigators, and analysis team); (2) the effectiveness of the treatment device; (3) the logistical obstacles that could be encountered in conducting the trial; (4) the effectiveness of data collection tools (e.g., health diary); and (5) the ability to collect fecal and serologic specimens and to test for various enteric bacteria, viruses, and parasites.

In the first pilot trial, the study population was composed of healthy adults and children living in a northern California community supplied by surface water and included 77 households (Colford et al., 2002). The water treatment plant serving the study area used conventional treatment with chloramination, and the finished water met all federal and state drinking water standards. Household tap water was tested in a subsample of the households for HPCs, total coliforms, copper, lead, and sulfites. Study results showed that participants could successfully be blinded to their treatment assignment. While not a primary objective of the study because of its small sample size, incidence rates of gastrointestinal illness were compared between the two groups, and the attributable fraction—although not statistically significant (most likely because of a lack of statistical power)—was unexpectedly found to be 24 percent. No significant differences were found in water consumption patterns between the two groups.

The second pilot trial was designed with the same objectives as the first except for the target population, which was an HIV-infected cohort. Immunocompromised persons such as those with HIV are at increased risk of infection and severity of illness from pathogenic organisms because their immune systems are less able to protect them. This study population (50 HIV-infected patients) was from a northern California community that receives approximately 80 percent of its water from the largest unfiltered surface water supply on the

West Coast. Household tap water was analyzed for HPCs, copper, lead, sulfate, and residual chlorine. This study has been completed and an attributable fraction has been calculated that could be applied to incidence rates of gastrointestinal illness in immunocompromised populations when calculating estimates of water-borne diseases (the manuscript has been submitted for publication and is currently under review [Deborah Levy, CDC, personal communication, 2004]). In addition, in preparation for this trial, a cross-sectional survey was conducted to determine the prevalence of gastrointestinal illness and drinking water consumption patterns in the HIV-infected study population. Forty-seven percent of respondents reported a gastrointestinal illness in the seven days before being surveyed (Eisenberg, 2002a). While drinking boiled or filtered water was not associated with diarrhea, those who drank bottled water were at significantly increased risk. Extending their work from one HIV clinic to two additional clinics, the researchers found that the risk of diarrhea was lower among those consuming boiled water, although this finding was not statistically significant and the relative risk of diarrhea for "always" versus "never" drinking bottled water was also nonsignificantly elevated (Eisenberg, 2002b).

Big WET and Nested Epidemiologic and Water Quality Studies Big WET is a full-scale, randomized, triple-blinded, controlled intervention of drinking water treatment. The study was conducted from October 2000 through June 2002 in Davenport, Iowa, a community that derives its drinking water solely from surface water. The water treatment plant serving the study area used conventional treatment with chloramination, and the finished water met all federal and state drinking water standards. The primary objective of this trial was to determine the incidence of gastrointestinal illness in groups randomly assigned to receive either a water treatment or a sham device and to calculate the fraction of the gastrointestinal illness attributable to consumption of the drinking water. Four hundred households, including adults and children, were monitored for one year.

To maximize the benefit of a large-scale trial, several nested studies were included:

1. Because a secondary objective of the trial was to gather data that would aid in the formulation of a national estimate of waterborne endemic disease, a cross-sectional telephone survey was administered to 400 randomly selected persons once a quarter for the 12-month follow-up period. The survey questionnaire, modeled after CDC's FoodNet survey (described below), collected information about water consumption patterns in and out of the home, swimming activities, symptoms of gastrointestinal illness, and burden of illness as measured by missed school, work, and recreational activities. Community surveillance data are increasingly being used in studies of gastrointestinal illness but have not been validated against data collected from individual reporting of illness. Thus, Big WET

provided a unique opportunity to compare gastrointestinal illness reports from daily health diaries and from a random digit dialing telephone survey.

2. As a result of unexpected severe flooding during the study period, participants completed an additional survey related to exposure to flood water. The objectives of this study were to determine whether rates of gastrointestinal illness were elevated during the flood and whether contact with flood water was associated with an increased risk of gastrointestinal illness. The results of this study have been accepted for publication in the American Journal of Epidemiology (Deborah Levy, CDC, personal communication, 2004).

3. A water sampling program was implemented and funded by the American Water Works Association Research Foundation (AWWARF Project #2580). This study was designed to determine whether a significant relationship exists between water quality indicators and gastrointestinal illness and, in the event of a difference in gastrointestinal rates in the intervention trial, to determine the most likely source of water contamination. Throughout the intervention trial, raw water, effluent water, and water entering the distribution system were tested for common water quality indicators and waterborne pathogens. These indicators and pathogens included total and fecal coliforms, *E. coli*, *Giardia lamblia*, *Cryptosporidium parvum*, *Clostridium perfringens*, *Bacillus subtilis*, somatic and male-specific coliphage, HPCs, culturable enteric viruses, and algae, as well as water pH, temperature, turbidity, alkalinity, chlorine, phosphate, and hardness. Water from household taps was tested for total coliforms, HPCs, pH, temperature, turbidity, and total chlorine. The AWWARF project report has been completed and is in the process of being published (Deborah Levy, CDC, personal communication, 2004).

4. EPA provided additional funding to conduct a brief plumbing survey of a subsample of households participating in the trial to ensure that frequency of cross-connections within the households could be controlled for in analyses of the association between water consumption and gastrointestinal illness.

Big WET and all of the nested studies have been completed. Primary analyses of the data from the trial have been completed, and the manuscripts are currently being written and submitted for publication (Deborah Levy, CDC, personal communication, 2004). The primary results of the Big WET study were presented at the annual conference of the International Society of Environmental Epidemiology in Perth, Australia on September 26, 2003 (Colford et al., 2003). The study did not find a reduction in gastrointestinal illness after the use of a treatment device designed to be highly effective in the removal of microorganisms from tap water. Secondary epidemiologic, clinical, and environmental analyses are being conducted and should be completed in early 2004. Analyses linking data from Big WET trial with data from the nested studies are ongoing and the flood-related manuscript has already been accepted for publication. Although this intervention trial was expensive to conduct, a variety of epidemiologic, clinical, and environmental issues were addressed because of the possibility of nesting multiple smaller studies within the larger study. As demonstrated with Big WET, and given the lack of funding often encountered, fewer but more comprehensive epidemiologic studies rather than multiple smaller studies that focus only on one or two issues should be funded and conducted when working within a fixed or constrained budget, thus providing a better cost-benefit ratio. Budgetary constraints notwithstanding, the same principles of thorough epidemiologic investigation apply to epidemiologic studies as to outbreak investigations. That is, the need to collect epidemiologic, clinical, laboratory, and environmental (both pathogens and indicators whenever feasible) data.

FoodNet Cross-Sectional Survey The Foodborne Diseases Active Surveillance Network (FoodNet) is the principal foodborne disease component of CDC's Emerging Infection Program (EIP). FoodNet is a collaborative project of CDC,[4] 10 EIP states,[5] the U.S. components, including an active laboratory-based surveillance; a survey of clinical laboratories, physicians, and a randomly selected population in the EIP catchment areas; and epidemiologic studies.

The survey of the EIP population is relevant to the development of a national estimate of waterborne disease. Before FoodNet, few studies provided reliable estimates of population rates of gastrointestinal illness (Hodges et al., 1956; Monto and Koopman, 1980). The FoodNet population survey collects information on recent gastrointestinal illness and will serve as one source of the incidence rates of gastrointestinal illness in the population that are needed to calculate the national estimate. Attributable fractions obtained in the intervention trials will be applied to these population rates. Population-based estimates of the burden of gastrointestinal illness in the United States calculated from FoodNet 1996-1997 data were published in 1999 and again in 2002 (Herikstad et al., 2002; Mead et al., 1999).

Community Intervention Trials As part of the national estimate of endemic waterborne disease, EPA has conducted three intervention trials in communities that upgraded their treatment plant operations (e.g., adding filtration). The design was a matched pre- and post-community intervention (i.e., improved water treatment). The target population is randomly selected from the community, and all participants completed health diaries and provided serum and fecal specimens. The primary objective is to compare incident rates of gastrointestinal illness before and after the intervention, while the second objective is to compare

---

[4]For further information about FoodNet, see http://www.cdc.gov/foodnet/.

[5]California, Colorado, Connecticut, Georgia, Maryland, Minnesota, New Mexico, New York, Oregon, and Tennessee.

seroprevalence rates for *Cryptosporidium parvum* for the two periods. EPA already has water quality data for these water systems collected under the Information Collection Rule (ICR; see Table 1-1). Additional information about these trials is provided in Table 2-2. Results from these community interventions will provide, like their counterparts in the experimental interventions, waterborne attributable fractions that can be applied to incident gastrointestinal rates for target populations of interest. Furthermore, several other ongoing studies that use a variety of epidemiologic designs and measure the association between water consumption and gastrointestinal illness are also summarized in Table 2-2.

Other Epidemiologic Studies

Other observational studies have been conducted to address the association between drinking water consumption and illness and they are briefly noted below. Details of these studies can be found in the original published manuscripts, which are referenced. Several time series studies by Schwartz and Levin (1999) and Schwartz et al. (1997, 2000) attempted to show an association between water turbidity and illness. These studies are difficult to interpret because of concerns about the temporal associations as well as the reliability of the turbidity measurements. Naumova et al. (2003) used time series analyses to study the association between emergency room visits and hospitalization caused by gastrointestinal illness and drinking water turbidity before and during the Milwaukee waterborne *Cryptosporidium* outbreak of 1993 (see also Chapter 1).

EPA and AWWA have conducted or funded several studies that compared seroprevalence rates of antibodies to *Cryptosporidium parvum* in paired cities with different types of source water and water systems (AWWA, 1999; Frost et al., 2001, 2002). General epidemiologic descriptions of time series studies and seroprevalence studies are provided near the beginning of this chapter.

U.S. National Estimate of Waterborne Disease Occurrence—Current Status and Future Direction

The SDWA amendments of 1996 (1458[d]) set a time line for the development of a national estimate of waterborne disease occurrence at five years after its promulgation in August 1996. However, because of variables involved in designing and conducting epidemiologic studies, the effort will take longer to complete. EPA and CDC already have developed the framework for calculating the estimate using waterborne attributable fractions and incidence rates of gastrointestinal illness, and are currently waiting for results from the trials. In the meantime, EPA work is focused on developing a model for characterizing water systems according to microbial risk while CDC efforts are focused on analyzing FoodNet water data and waterborne disease outbreak data. An additional component that the two agencies would like to incorporate into the calculation of the national estimate, if feasible, is seroprevalence of *Cryptosporidium parvum* from the fourth cycle (1999-2000) of the National Health and Nutrition Examination Survey (NHANES IV).[6] CDC conducts NHANES periodically and is the only national source of objectively measured health data that can provide estimates of both diagnosed and undiagnosed medical conditions in the population. Data are collected for biomedical research, public health, tracking of health indicators, and policy development through physical examinations, clinical and laboratory tests, and interviews. Seroprevalence rates have been calculated for target populations across the United States, and CDC will attempt to determine if there is an association between these rates and the type of water system in the community from which the population was selected (EPA will provide the water system information).

The development of a national estimate of gastrointestinal illness that can be attributed to municipal drinking water is expected to be an evolving effort. The first few studies were conducted in communities with surface water systems because these were thought to be more vulnerable to contamination and therefore more of a public health risk. As a result, information currently is lacking for groundwater systems. However, EPA recently funded a study in a groundwater community in Florida and has plans to fund more through their Science to Achieve Results (STAR) grants program.[7] The primary objective of the Florida study is to estimate the risks of endemic gastrointestinal illness associated with consumption of conventionally treated groundwater and to determine the relative contributions of source water quality, treatment efficacy, and distribution system vulnerability to endemic waterborne disease. The study design proposed was a 12-month, double-blinded, randomized intervention trial that will include 900 households and will measure rates of gastrointestinal illness in groups with drinking water that receives different levels of treatment.

The national estimate of endemic waterborne disease will be updated and refined as data become available from the ongoing as well as new studies and additional data sources are mined (Deborah Levy, CDC, personal communication, 2004). Potential sources of these data include surveillance databases, additional epidemiologic studies, additional populations and geographic sites, systematic reviews and meta-analyses, and public health and microbial risk assessment. Furthermore, the estimate will be expanded, ideally to include other waterborne illnesses (e.g., respiratory illness) and other types of water (e.g., ambient and treated recreational waters, private wells). Finally, collaboration with WHO would expand the effort and include developed and developing countries across the world.

---

[6]For further information about NHANES, see http://www.cdc.gov/nchs/nhanes.htm.

[7]For further information about EPA's STAR grants program, see http://es.epa.gov/ncer/grants/.

TABLE 2-2 Summary of Key Characteristics of Select Epidemiologic Studies Associating Drinking Water with Health Outcomes

Study Sponsor/ Primary Institution	Study Design	Geographic Location— Water System	Population	Study Size
National Institutes of Health/ University of California, Berkeley	Randomized triple-blinded intervention trial	Sonoma and Santa Rosa, California— mixed water system	Elderly	500 households
CDC/Tufts University	Longitudinal time series	Lowell and Newton, Massachusetts— surface water systems, one with filtered river water and one unfiltered from a partially protected watershed	Children	1,000+ persons per city
CDC/Tufts University	Cohort study	Lowell and Newton, Massachusetts— surface water systems, one with filtered river water and one unfiltered from a partially protected watershed	Children	400 households
EPA	Community intervention trial before and after installation of a filtration plant	Undisclosed city, Massachusetts— surface water system	General population	300 households
EPA	Community intervention trial before and after installation of a filtration plant with ozonation	Seattle, Washington —surface water system	Children and elderly	300 households

Exposure Measurements	Health Outcomes	Epidemiologic Measure of Association	Status (as of January 2004)
Consumption of tap water treated with either real or sham device	Gastrointestinal Illness	Attributable fraction	Data collection ongoing
Consumption of tap water. Water quality measured by turbidity and presence of *Cryptosporidium* oocysts in finished water	Prevalence of *Cryptosporidium* antibodies in serum and saliva specimens, and episodes of gastrointestinal illness	Correlations of exposures and outcomes over time	Data analysis ongoing
Consumption of tap water and exposure to recreational water. Water quality measured by turbidity and presence of *Cryptosporidium* oocysts in finished water	Prevalence of *Cryptosporidium* antibodies in serum and saliva specimens, and episodes of gastrointestinal illness	Relative risk and correlations of exposures and outcomes over time	Data analysis ongoing
ICR data on source water monitoring	Seroprevalence of *Cryptosporidium* antibodies and episodes of gastrointestinal illness before and after the intervention	Attributable fraction	Study completed, results not published
ICR data on source water monitoring	Seroprevalence of *Cryptosporidium* antibodies and episodes of gastrointestinal illness before and after the intervention	Attributable fraction	Study completed, AWWARF report (Project #2367) in preparation

continued

TABLE 2-2 Continued

Study Sponsor/ Primary Institution	Study Design	Geographic Location— Water System	Population	Study Size
EPA	Community intervention trial before and after installation of microfiltration	Texas groundwater system under the influence of surface water	General population	200 households
San Francisco Department of Public Health and University of California, San Francisco	Case-control study	San Francisco, California	Persons with AIDS	49 cases and 99 matched controls

In conclusion, a substantial effort to determine the potential health risks associated with the consumption of drinking water has been going on in the United States since the late-1980s, while surveillance for waterborne disease outbreaks has been continuous for several decades. However, most of these efforts have not focused on associating specific waterborne pathogens with indicators, and associating pathogens and their indicators with illness. Linking gastrointestinal illness with water consumption is not the epidemiologic equivalent of linking the illness to waterborne pathogens and indicators despite the intuitive understanding that the two hypotheses are closely related.

Data collection efforts in outbreak investigations of drinking water have concentrated mostly on identifying the epidemiologic link to water consumption and identifying a pathogen in clinical specimens rather than in water samples. When water samples are tested for bacterial indicators such as total and fecal coliforms, the samples that are collected are typically from finished water (especially at the treatment plant and not in the distribution system) rather than source water, in part because only the treated water is ingested. As mentioned previously, in many outbreak investigations the water is no longer contaminated or the indicator/pathogen is present in such low concentrations by the time water samples are collected that the resulting water analyses are negative. It is often difficult to collect water samples quickly enough that the sample is representative of the water quality that was likely responsible for the outbreak. Nevertheless, timely and thorough investigations of drinking water outbreaks can provide epidemiologic data associating poor water quality with adverse health outcomes.

To date, few data exist to correlate indicators with pathogens in drinking

Exposure Measurements	Health Outcomes	Epidemiologic Measure of Association	Status (as of January 2004)
ICR data on source water monitoring	Seroprevalence of *Cryptosporidium* antibodies and episodes of gastrointestinal illness before and after the intervention	Attributable fraction	Data collection ongoing
Tap water consumption inside and outside the home	Development of cryptosporidiosis	Odds ratio Attributable fraction	Results published (Aragon, 2003)

water and with endemic adverse health outcomes in the United States. Although Payment's study (1991) and Colford's pilot study (2002) did find an association between drinking water and gastrointestinal illness, the Payment study did not find any correlation between illness and the indicators that were measured and the Colford study was not designed to look at this association. In addition, the Big WET study did not show an association between drinking water and illness and it remains to be seen if any of the secondary analyses when linked to the water quality data collected in the nested study will provide measurable epidemiologic associations. To have adequate statistical power to address the epidemiologic association of health outcomes with specific indicators and waterborne pathogens, the study sample size must be large, and therefore the costs can become prohibitively expensive. Once again, if water quality is monitored, the choice is to test the water that is consumed rather than the water at its source and along its distribution system. Nevertheless, recent studies have begun to include tests for indicators and pathogens in water, and it remains to be seen if the results will show correlations with adverse health outcomes.

Epidemiologic Studies of Diseases Attributed to Recreational Water

As reviewed in Chapter 1, the development of microbial water quality criteria and standards based on health effects assessments from prospective epidemiologic studies did not occur until the 1950s. Of particular importance, Stevenson (1953) reported the results of a series of USPHS studies at three pairs of bathing sites on Lake Michigan, along the Ohio River, and on Long Island Sound. The

Stevenson study results indicated that swimmers showed increased rates of illness over non-swimmers per 1,000 person-days of activity, and illness rates tended to rise with increased swimming days. The symptoms reported included eye, ear, nose and throat, as well as gastrointestinal illness. Overall, there was not a consistent correlation found between levels of illness and levels of coliform bacteria measured in bathing water when the data for all study sites were examined. However, specific data from Lake Michigan freshwater beaches suggested increased illness rates at total coliform levels of about 2,300 per 100 mL. This value was subsequently lowered by a factor of approximately two in order to obtain a total coliform indicator standard of 1,000 per 100 ml. Thus, as discussed in Chapter 1, this value of total coliforms became the basis for the subsequent bacteriological water quality criteria recommended by the National Technical Advisory Committee (NTAC) of the FWPCA, after conversion to fecal coliforms concentrations based on the ratio of total to fecal coliforms (NTAC, 1968).

As noted previously, the Stevenson study and the subsequent development of the FWPCA-NTAC recreational water quality criteria derived from it in 1968 were criticized as inadequate in several ways. This criticism ultimately resulted in a National Research Council (NRC, 1972) report, *Drinking Water and Health, Volume 1* (see also Appendix B). Notably, the NRC report concluded, "No specific recommendation is made concerning the presence or concentrations of microorganisms in bathing water because of the paucity of valid epidemiological data" (NRC, 1972). The fecal coliform measurement itself was also criticized, with the report noting that thermotolerant bacteria such as *Klebsiella* spp. read positive in this test but are not necessarily fecal in origin. In addition, the NTAC fecal coliform criteria did not account for the considerable daily variability in water quality; the relatively loose definition of swimming did not require immersion of the head, which would result in greater exposure to water; and beach-going but nonswimming control participants were not included.

Because of the recognized deficiencies of previous studies (see also Chapter 1), EPA conducted prospective epidemiologic-microbiological studies in the 1970s to compare rates of gastrointestinal illness in swimmers and beach-going non-swimmers at fresh and marine beaches differing in microbial water quality and sources of fecal contamination. These studies by Cabelli and colleagues (Cabelli et al., 1982) used more rigorous definitions of gastrointestinal illness and included a number of different microbial indicators of fecal contamination in water, including enterococci and *Escherichia coli*.

From these studies it was concluded that concentrations of enterococci best correlated with gastrointestinal illness (e.g., vomiting, diarrhea, nausea, stomach ache) attributable to swimming in marine waters and that both enterococci and *E. coli* best correlated with such illness in fecal contaminated freshwaters. Log-linear relationships between mean enterococcus or *E. coli* density per 100 mL and swimming-associated rates for gastrointestinal symptoms per 1,000 persons were subsequently developed. These became the basis for current marine and fresh recreational water quality criteria and guidelines that were calculated using geometric mean values of several (generally 5 or more) equally spaced samples over a 30-day period: *E. coli* not to exceed 126 per 100 mL, or enterococci not to exceed 33 per 100 mL in freshwater; enterococci not to exceed 35 per 100 mL in marine water (Cabelli, 1983; Dufour, 1984; EPA, 1986). Single-sample maximum allowable densities were also promulgated based on beach use. These values were based on risk levels of 8 and 19 gastrointestinal illnesses per 1,000 swimmers at freshwater and marine beaches, respectively, and they were estimated to be equivalent to the risk levels for criteria of 200 fecal coliforms per 100 mL.

These pivotal studies by EPA prompted numerous epidemiologic-microbiological studies of similar and improved design in many parts of the world (see Table 2-3 and systematic review of these and other recreational water studies by Wade et al., 2003). Many of the more recent studies attempted to improve and expand upon these prospective epidemiologic studies by EPA in several respects. For example, some addressed a broader range of swimming-associated health effects such as respiratory illness, while others obtained better estimates of the microbial quality of water to which bathers were actually exposed by more intensive and extensive sampling to address spatial and temporal variability of water quality. Some studies measured concentrations of even more microbial indicators such as coliphages, and others measured concentrations of enteric pathogens such as enteric viruses and parasites.

Throughout the 1980s and 1990s, data from various recreational water quality studies around the world began to emerge. Consistent with its goal to develop a harmonized framework for science-based guidelines on water quality and health, WHO developed a uniform approach to recreational water quality, commonly referred to as the "Annapolis Protocol" (Bartram and Rees, 1999; WHO, 1999). This approach provides a harmonized risk assessment and management framework for recreational water. It was developed in response to the need to establish an effective and harmonized approach to monitoring and managing fecal contamination of recreational waters. Some of the key recommendations in the Annapolis Protocol include the following: (1) moving away from sole reliance on "guideline" values of fecal indicator bacteria toward use of a qualitative ranking of fecal loading in recreational water environments, supported by direct measurement of appropriate fecal indicators and (2) provisions to account for the impact of actions to discourage water use during periods or in areas of higher risk. The protocol has been tested in several countries, and recommendations resulting from these tests have been included in new *WHO Guidelines for Safe Recreational Water Environments* (see WHO, 2003).

The results of many of the historical and more recent prospective epidemiologic-microbiological studies were compiled and summarized in a review article by Prüss (1998) as part of the WHO effort to develop and harmonize recreational water quality criteria and guidelines (Fewtrell and Bartram, 2001; WHO, 1999).

TABLE 2-3 Selected Studies Used for Analysis of Health Effects and
Microbial Water Quality Relationships in Recreational Waters

Authors	Year	Country	Study Design	Water
Stevenson [a,b,c]	1953	United States	Randomized controlled Trial	Fresh
Mujeriego et al. [a,d]	1982	Spain	Retrospective cohort (cross-sectional study)	Marine
Cabelli et al. [a,d]	1982	United States	Prospective cohort	Fresh, Marine
Cabelli [a,b,d]	1983	Egypt	Prospective cohort	Marine
Dufour [a,d]	1984	United States	Prospective cohort	Fresh
Seyfried et al.	1985	Canada	Prospective cohort	Fresh
Fattal et al., UNEP/WHO no. 20 [a,c]	1987	Israel	Prospective cohort	Marine
Lightfoot	1989	Canada	Prospective cohort	Fresh
Ferley et al. [a,b,d]	1989	France	Retrospective cohort	Fresh
Cheung et al. [a,d]	1989	Hong Kong	Prospective cohort	Marine
UNEP/WHO no. 53 [a,c,d]	1991	Spain	Prospective cohort	Marine
UNEP/WHO no. 46 [a,c]	1991	Israel	Prospective cohort	Marine
Fewtrell et al. [c,e]	1992	United Kingdom	Prospective cohort	Fresh
Corbett et al. [c,d]	1993	Australia	Prospective cohort	Marine
Pike [a,b,d]	1994	United Kingdom	Prospective cohort (cross-sectional study)	Marine
Kay et al. [c]	1994	United Kingdom	Randomized controlled Trial	Marine
Medical Research Council[b,d]	1995	South Africa	Prospective cohort	Marine
Kueh et al. [a]	1995	Hong Kong	Prospective cohort	Marine
Bandaranayake [c]	1995	New Zealand	Prospective cohort	Marine
Van Dijk et al. [b]	1996	United Kingdom	Prospective cohort	Marine
Haile et al.	1996	United States	Prospective cohort	Marine
Fleisher et al. [c]	1996	United Kingdom	Randomized controlled Trial	Marine

[a]Control for less than three confounders reported, or not reported at all.
[b]Exposure not defined as head immersion, head splashing, or water ingestion.
[c]<1,700 bathers and 1,700 nonbathers participating in the study.
[d]Only use of seasonal mean for analysis of association with outcome reported.
[e]Exposure is white-water canoeing, considered similar to swimming, with intake likely.
SOURCE: Adapted from WHO, 1998.

Of the 37 studies evaluated by Prüss, 22 qualified for inclusion in the evaluation.
Figure 2-1 presents the relationship between indicator organism density in marine water and illness risk for bathers. A similar compilation of the results for
studies in fresh recreational waters is shown in Figure 2-2. Of the 22 studies
selected for analysis, 17 were prospective cohort studies, 2 were retrospective

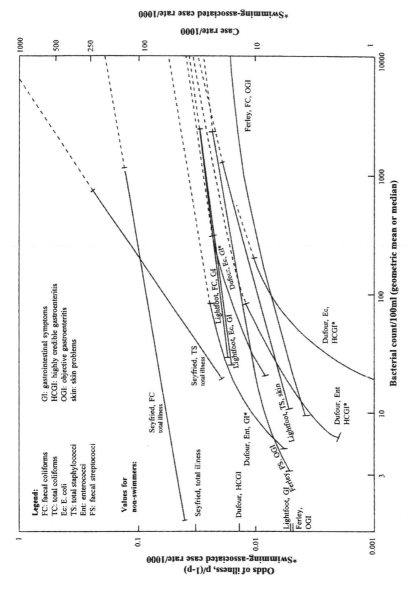

FIGURE 2-1 Relationships of risks of illness in swimmers to the microbial quality of water—marine recreational waters. Note: Solid lines indicate actual data from original studies whereas dashed lines are extrapolations of data by Prüss (1998) or Pike (1991). SOURCE: Reprinted, with permission, from Prüss, 1998 © Oxford University Press.

84

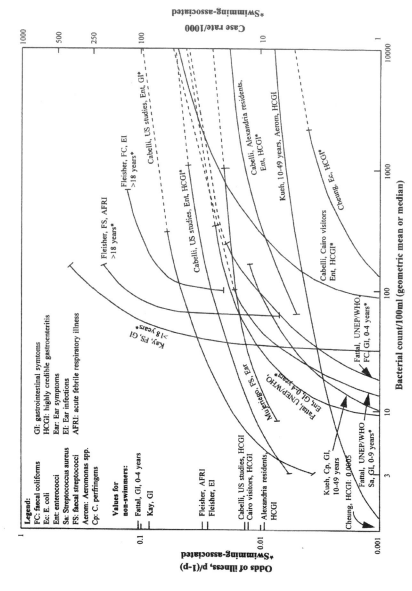

FIGURE 2-2 Relationships of risks of illness in swimmers to the microbial quality of water—fresh recreational waters. Note: Solid lines indicate actual data from original studies whereas dashed lines are extrapolations of data by Prüss (1998) or Pike (1991). SOURCE: Reprinted, with permission, from Prüss, 1998 © Oxford University Press.

cohort studies, and 3 were randomized controlled trials, as summarized in Table 2-3. Of the studies examined, the rate of certain symptoms or groups of symptoms was found to be significantly related to the count of fecal indicator bacteria in recreational water. Thus, there was a consistency across the various studies evaluated and gastrointestinal symptoms were the most frequent health outcome for which significant dose-related associations were reported. In marine waters, fecal streptococci or enterococci were the fecal indicators that best predicted gastrointestinal illness. In freshwaters, increased concentrations of fecal coliforms or *Escherichia coli* as well as fecal streptococci and enterococci were predictive of increased gastrointestinal illness risks. Staphylococci concentrations were also found to be predictive of increased risks of illness, including ear, skin, respiratory, and gastrointestinal illness. Although these latter relationships were attributed to the effects of bather density, this was not actually proven.

Based on the studies evaluated by Prüss (1998), strong and consistent associations have been reported between microbial indicators and various adverse health effects to include temporal and dose-response relationships. Furthermore, these studies have biological plausibility and analogy to clinical cases from drinking contaminated water. However, various biases commonly occur with epidemiologic studies as summarized in Table 2-4.

For marine bathing waters, randomized controlled trials in the United Kingdom (Fleisher et al., 1996; Kay et al., 1994) probably contained the least amount of bias. These studies also provide the most accurate measure of exposure, water quality, and illness compared to observational studies where an artificially low threshold and flattened dose-response curve (due to misclassification bias) were likely to have been determined. Therefore, the United Kingdom randomized controlled trials form the key studies for derivation of guideline values for the microbiological quality of recreational waters.[8] It should be recognized that these recommended guidelines values are from studies in temperate waters and are not characteristic of the tropical and subtropical waters found in many areas of the United States (e.g., the U.S. Gulf coast).

Based on analyses of data from numerous studies on the relationships between swimming-associated health effects and the microbial quality of bathing water, the WHO and other international as well as national entities have concluded that fecal streptococci and enterococci currently are the fecal indicator microorganisms that best predict health risks in recreational waters (WHO, 2001). Rather than classify recreational waters as either acceptable or unacceptable, WHO experts chose instead to establish a five-tiered classification system (i.e.,

---

[8]Guideline values are nonregulatory values for constituents in water, in this case microbial indicators, developed by the World Health Organization (see Bartram and Rees, 1999 and WHO, 2003 for further information).

TABLE 2-4 Types of Biases Potentially Encountered in Recreational Water
Quality Health Effects Studies

Type of Bias	Description
Use of indicator microbes to assess water quality	Temporal and spatial indicator variation is substantial and difficult to relate to individual bathers (Fleisher, 1990) unless study design is experimental (Kay et al., 1994; Fleisher et al., 1996); limited precision of methods for counting indicator organisms, causing measurement error (Fleisher, 1990; Fleisher et al., 1993); bacterial indicators may not be representative of viruses, which may be important etiological agents
Use of seasonal means to assess water quality	Some studies use seasonal means and not daily measurements of indicator organisms to characterize individual exposure, thus adding substantial inaccuracy
Assessment of exposure pathway	Certain studies do not account for the potential infection pathway to definite exposure (e.g., mainly head immersion or ingestion of water for gastrointestinal symptoms). Difficulties in exposure recall further increase inaccuracy of individual exposure
Non-control for confounders	Non-control for confounders (e.g., food and drink intake, age, sex, history of certain diseases, drug use, personal contact, additional bathing, sun, socioeconomic factors) may influence the observed association
Selection of unrepresentative study population	Results reported for certain study populations (e.g., limited age groups regions with certain endemicities) are a priori not directly transferable to populations with other characteristics
Self-reporting of symptoms	Most observational studies relied on self-reporting of symptoms by study populations. Validation of symptoms by medical examination (Fleisher et al., 1996; Kay et al., 1994) would reduce potential bias. External factors, such as media or publicity, may have influenced self-reporting
Response rate	Response rates were >70% in all, and >80% in most studies. Differential reporting (e.g., higher response among participants experiencing symptoms) would probably not have major consequences
Recruitment method	Recruitment methods were to approach persons on beaches in almost all observational studies and by advertisement for randomized controlled studies
Interviewer effect	Differences in methodology of data collection among interviewers may influence study results

SOURCES: Adapted from Prüss, 1998; Stavros and Langford, 2002; WHO, 2001.

very poor, poor, fair, good, or very good) based on microbial water quality (using
fecal streptococci or enterococci indicator counts) and sanitary condition (based
on sanitary inspection or survey) to identify likely health risks (WHO, 2001,
2003).

It is important to note that few studies used to establish the WHO recreational microbial water quality classification system based on fecal streptococci or enterococci included a wide range of candidate fecal indicator microbes or pathogens. Therefore, it must not be assumed that fecal streptococci or enterococci are the only or even the most predictive fecal indicators of health risks from recreational water quality exposures.

Other candidate indicators that have not been adequately studied or for which reliable methods were previously not available may eventually prove to be more predictive and reliable fecal indicators of human health risks than are fecal streptococci or enterococci.

Systematic Review of Recreational Water Studies

At the request of this committee, Colford (2002) and colleagues (Wade et al., 2003) were asked to conduct a meta-analysis or similar synthesis of the various epidemiologic-microbiologic health effects studies available in the world's literature. After establishing study characteristic and quality criteria, they identified a total of 27 (17 marine water and 10 freshwater) studies for a "systematic review"[9] including 24 cohort studies, 2 randomized trials, and 1 case-control study. From these studies, a subset was found to be amenable to the determination of relative risk of a health outcome, such as gastrointestinal, respiratory, skin, ear, or eye effects.

For gastrointestinal illness, several indicators showed significant associations with the levels of the following indicators in recreational water: fecal streptococci (enterococci), fecal coliforms, E. coli, total coliforms (marine water only), enteroviruses (marine water only), and coliphages (freshwater only) in water. When regression analysis was used to examine the log relative risk of illness against indicator level in water, positive associations were found for fecal streptococci (enterococci). The authors concluded that E. coli and enterococci were the "best" indicators of gastrointestinal illness in marine water, while there was no best (consistent) indicator of gastrointestinal illness in freshwater. A log (base 10) unit increase in enterococci was associated with a 1.34 (range 1.00-1.75) increase in relative risk in marine waters and a log (base 10) unit increase in E. coli was associated with a 2.12 (range 0.93-4.85) increase in relative risk in freshwater. It was also noted that enteroviruses and bacteriophages may be promising indicators to predict risk of gastrointestinal illness, but there are too few studies to

---

[9]A systematic review involves a predefined rigorous review of existing studies, may or may not include meta-analyses, and can be exemplified by the Cochrane Review (http://www.cochrane.org). Rothman and Greenland (1998) define meta-analysis as a "statistical analysis of a collection of studies, especially an analysis in which studies are the primary unit of analysis. Meta-analysis methods thus focus on contrasting and combining results from different studies, in the hopes of identifying consistent patterns and sources of disagreement among those results."

establish their utility at this time. Overall, the study results supported the use of enterococci in marine waters at EPA levels.

None of the commonly used microbial indicators consistently predicted risks of respiratory illness, but relative risks of skin disorders tended to increase with several indicators, including fecal streptococci (enterococci), fecal coliforms, and *E. coli.* The results of these studies provide encouraging evidence that predictive associations exist or can be found between various swimming-associated health effects and various microbial indicators or pathogens in recreational bathing waters. However, Wade and colleagues identified several major research gaps that need to be addressed regarding the use of indicators for recreational waters, including the following:

1. studies of immunocompromised populations;
2. studies in other sensitive/vulnerable subpopulations such as children and the elderly;
3. determination of etiologies by analysis of clinical specimens;
4. additional rigorously conducted epidemiologic studies such as observational studies that have standardized definitions of exposure and health outcomes and standardized methods, as well as randomized trials to establish etiology;
5. additional studies using enteric viruses and bacteriophages as water quality indicators;
6. use of combinations of water quality indicators to assess overall health risks; and
7. analysis of the effects of study location and climate on results.

The results of the systematic review of recreational water epidemiologic studies by Wade and colleagues (2003) provided several informative observations and led to some important conclusions from the authors, which are supported by the committee: (1) the analysis documented that a more thorough meta-analysis of many of the international studies on recreational water quality and health effects is both possible and able to provide useful data to further interpretation and related decision making; (2) it pointed out both study design and data weaknesses and gaps that can be filled in future epidemiologic-microbiological studies of recreational water quality and health; (3) it indicated that bacterial indicators such as enterococci, *E. coli,* and fecal streptococci could provide reliable estimates of water quality that are predictive of human health risks under some, but not all, water quality conditions (e.g., statistically, enterococci followed by *E. coli* were the best indicators in marine waters, but there were insufficient data to make similar conclusions about freshwaters); and (4) it also provided evidence that other microbial fecal indicators, such as coliphages and certain pathogens were predictive of human health risks, despite the fact that few studies included these water quality tests. Therefore, these other microbial indicators deserve further consideration in future studies.

EPA's BEACH Act Studies

As part of EPA's Action Plan for Beaches and Recreational Waters and legislative assistance from the Beaches Environmental Assessment and Coastal Health (BEACH) Act of 2000 (see also Chapter 1), EPA is conducting annual national survey(s) on state and local microbial water quality monitoring efforts, and will begin collecting health effects data (EPA, 1999). CDC is collaborating with EPA on the health effects studies, which use a prospective cohort design. Several Great Lakes and marine sites will be evaluated, including one that served first as the pilot site. Health outcomes to be studied include both enteric and nonenteric (e.g., respiratory illnesses, dermatitis, eye and ear infections) disease. The pilot study was conducted at one freshwater recreational beach (Indiana Dunes National Lakeshore) during the summer of 2002 to evaluate public response rates, to test the questionnaire, and to establish the study's operational protocols. Two beaches were studied during the summer of 2003, one in Indiana and one in Ohio. Enrollment criteria included all persons on the beach regardless of gender and age. Biological specimens were not collected. However, EPA still hopes to collect these specimens, including stool, serum, and saliva in some subset of the study population.

Sites will include both freshwater and marine beaches but not tropical or subtropical recreational waters. Indicators that might be used include enterococci and *E. coli* but not total and fecal coliforms. Other potential microbial and chemical indicators are still being considered but will focus solely on human sources of fecal contamination. Although nonpoint sources of contamination (e.g., fecal contamination from nonhuman sources, runoff, rainfall) will not be addressed due to lack of funding, they should be included in future epidemiologic studies of recreational water exposure (see also Chapter 4). In addition, habitual users of recreational water, such as professional surfers, will not be studied despite the fact that knowledge is lacking on the epidemiology of chronic or recurrent illness in these populations (e.g., eye and ear infections). EPA's goal is to have a water quality test that will provide results within two hours so that a determination can be made to close the beach if deemed necessary prior to the time visitors are expected to begin arriving (Alfred Dufour, EPA, personal communication, 2002).

QUANTITATIVE MICROBIAL RISK ASSESSMENT

Historically, as noted throughout this report, acceptable microbial levels for evidence of pathogen risk in drinking water, contact recreational waters, and shellfish harvesting waters have been set using indicator organisms, most often the coliform (either total or fecal) group. The recognition of many of the pathogens responsible for waterborne disease from microbiological and epidemiologic investigations, the advent of better methods for direct measurement of pathogens in water (Gerba and Rose, 1990; Gregory, 1994; Leong, 1983; Ongerth, 1989; Rose,

Overview   Detail of analysis phase

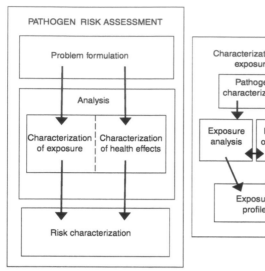
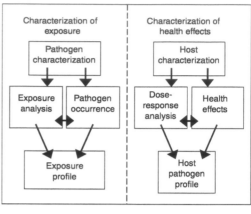

FIGURE 2-3 Schematic of ILSI microbial risk assessment protocol. SOURCE: Adapted from ILSI, 1996.

1990; Rose et al., 1991a,b; Sobsey, 1999, 2001), and the development of risk assessment paradigms for developing risk management systems and setting environmental standards (NRC, 1983, 1989; Silbergeld, 1993) provide a basis for the application of quantitative microbial risk assessment to the development of risk criteria for establishing microbial standards of acceptable water quality. These analytic advances provide a rational basis for either validating, revising, refuting, supplementing, or replacing traditional microbial indicator measurements.

The quantitative microbiological risk assessment approach follows the framework proposed for (chemical) risk assessment in the seminal 1983 NRC report *Risk Assessment in the Federal Government: Managing the Process*, which includes the following basic steps: hazard assessment, exposure assessment, dose-response analysis, risk characterization, and risk management. Alternative but similar protocols have been published—for example, by the International Life Sciences Institute (ILSI, 1996, 2000)—that are specifically designed to apply to waterborne pathogens. A schematic of the ILSI protocol is shown in Figure 2-3. Notably, this protocol more clearly emphasizes the interrelationships between the technical and policy-making components surrounding the risk assessment process, particularly at the problem formulation stage.

Several substantive differences exist, however, between the assessment of risk from microorganisms and the assessment of risk from chemicals, including the following:

- Exposure to microorganisms from water generally involves the ingestion of low numbers (up to tens or hundreds) of microorganisms. Exposure to chemical agents even at very low doses involves quite larger numbers (thousands or much greater numbers) of molecules.[10] At low numbers (as with microorganisms), there may be large differences between individuals with respect to the actual number of organisms ingested induced by pure statistical sampling variability, while with large numbers (as with chemical agents), this source of variability is quite small with respect to other sources. Thus, the assessment of exposure, and dose-response with microorganisms must consider this intrinsic sampling and exposure variability, while chemical risk assessment can ignore this phenomenon.

- For microorganisms, there is strong biological information to indicate that as few as one microorganism has the potential to cause harm (Haas et al., 1999b). That is, there is a non-zero probability that one organism can initiate infection. For chemical agents, it may be (depending on the agent and the mode of action) that far more units (molecules) are necessary to provoke an effect. In the case of microbial agents, it is generally the case[11] that an ingested microorganism has the potential to multiply within the body and thereby produce sufficient microorganisms in vivo to result in illness. This does not mean that ingesting a single organism in and of itself will always produce illness since an organism may be killed by defense processes (e.g., the acidity of the gastrointestinal tract, the immune system) prior to reproducing in sufficient amounts to have an adverse effect. However, one organism potentially (if it and a sufficient number of its progeny survive) has the biological potential to produce an effect.

- Individuals' microbial exposure may have subsequent impact on the broader population (including individuals that do not ingest pathogenic microorganisms from water). Once infected (even if not symptomatic), an individual may infect others and cause others to become ill through person-to-person contact and other transmission routes unrelated to water. This is called secondary spread, and the degree of such spread depends on the organism (its infectivity, excretion pattern, and intensity and duration of contagion) and the behavioral aspects of infected individuals.[12] Often the extent and magnitude of such effects are difficult

---

[10]Consider the exposure to 1 ng (10^{-9} g) of a chemical with a molecular weight of 100. This is the exposure to 10^{-11} moles of substance. Since there are 6.02×10^{23} molecules in a mole (Avogadro's number), this amounts to an exposure to 6×10^{12} molecules (6 trillion molecules).

[11]The exception being microorganisms that produce toxins as they grow in the environment, such as algae, and where the ingested *toxins*, rather than the ingested cells themselves, result in an adverse effect. However, as noted in Chapter 1, blue-green algae and their toxins are specifically excluded from the study charge.

[12]For example, infected adults are believed to have better hygienic practices, such as handwashing, compared to children, and therefore infected adults may produce fewer secondary cases than children. In addition, the number of susceptible persons with whom an infected person may come in contact is an important factor.

to assess, however they may represent an important element of public health risk estimates. Another potential population factor is that prior exposure to a particular microorganism (via water or other routes) may induce partial or complete immunity in an individual. That is, the individual may become less susceptible (partial immunity) or completely resistant (complete immunity) to subsequent exposures. This immunity may be of permanent or temporary duration. In the case of *Cryptosporidium*, rechallenges of individuals with oocysts a year after prior exposure has been found to confer partial immunity (i.e., a shift in the dose-response curve) to an additional exposure (Chappell et al., 1999). Whether such immunity persists for a longer duration and whether a similar effect is operative with other pathogens are not well established. However, this type of information would help make a quantitative microbial risk assessment more accurate, and such data can be used in more comprehensive, dynamic, risk assessments (Eisenberg et al., 1996, 1998).

It should be noted that quantitative microbial risk assessment, like risk assessment in general, has many inputs that are uncertain. These include uncertainty about the best dose-response model for the pathogen or indicator organism of interest and its behavior in the low dose region, assumptions about water consumption and other water-related exposures, and uncertainty about occurrence and concentration of pathogens or indicators in water. In addition, there may be variable host susceptibility and immunity to infection, etc. (some of which may be clarified with increasing knowledge of genetic determinants of host susceptibility to certain microbial pathogens). However, such limitations to current knowledge should not prohibit QMRA from being conducted, but rather (as with all risk assessments), it must be recognized that such analyses need to be updated as the state of knowledge evolves.

Case Studies

Since the advent of QMRA in the 1980s, there have been a number of articles published showing the application of this method for recreational waters (e.g., Haas, 1983a; 1986) and drinking water; two case studies for drinking water exposure are described below.

Risk from Ingestion of Giardia in Drinking Water

Using data from human volunteer studies, Regli et al. (1991) developed a dose-response relationship for infection from ingestion of *Giardia lamblia* that was compared to attack rates observed in waterborne outbreaks (Rose et al., 1991b) to assess the likelihood that an infected person would become ill. Researchers used a target risk of 1 infection in 10,000 persons per year—which was regarded as acceptable by EPA in the Surface Water Treatment Rule (SWTR; see also

Chapter 1)—and a daily average water consumption of 2 liters per person per day to estimate that an acceptable finished water concentration would be 6.75×10^{-6} per liter (i.e., one organism in 148,000 liters). Since verification of such a low level of microbial occurrence constitutes a technological impossibility, a graphical relationship between the microbial quality of source water and the number of logs of reduction required to reduce the microbial level to the acceptable value was developed.

In the initial SWTR a tiered treatment requirement was proposed that would incorporate this approach. However, in the final promulgated regulation a single fixed value of log reduction (3 $\log_{10}$, 99.9 percent) was required, based on an estimated upper value of source water microbial levels across the United States (EPA, 1989). In addition, because this approach did not adequately address other contaminants such as viruses and *Cryptosporidium*, the ICR (see Table 1-1), which was promulgated later (EPA, 1996), focused on source water monitoring for pathogens.

New York City: Cryptosporidium

The current New York City water system uses chlorination alone and has been exempted from filtration under a Memorandum of Agreement signed on January 21, 1997, between New York City, New York State, EPA, and other regional and environmental organizations. An intensive watershed protection and monitoring program has been mandated to ensure the water quality of its Catskill and Delaware ambient surface water supplies. A previous NRC committee performed a study on the effectiveness of this program in ensuring water quality in the future (NRC, 2000; see also Appendix B). As part of this study, monitoring data for *Cryptosporidium parvum* were used to conduct a risk assessment for consumers of water from the Catskill and Delaware systems. A dose-response relationship developed from human feeding studies was employed (Haas et al., 1996). Based on consideration of the variability and uncertainty of the inputs, it was concluded that the estimated risk to consumers from *Cryptosporidium parvum* infection was in excess of 1/10,000 per year, and thus—if this level was to be regarded as "acceptable"—additional reduction of oocyst levels would be necessary.

Acceptance by International Organizations

In the field of water quality, WHO has recently developed an overall framework for guideline and standard setting in all of its water-related activities—including drinking water, recreational water, and exposure to effluents and sludges from the agricultural use of such materials—using microbial risk assessment as a foundation (Fewtrell and Bartram, 2001). Specifically, the framework suggests that while risk assessment is a central element of water quality guideline and standard development, particular guidelines may be framed in terms of indicators (rather than pathogens) and the entire process is in need of continual refinement based on environmental monitoring data and public health surveillance (see Figure 2-4; Bartram et al., 2001).

Outside the water field, microbial risk assessment has increasingly been adopted both in the United States and internationally as a paradigm for developing standards for food safety. It is outside the scope of this report to review the field of food risk assessment; however, several recent studies describe and review such developments (e.g., Buchanan and Whiting, 1996; Hoornstra and Notermans, 2001; Jaykus, 1996; Ranta and Maijala, 2002).

Data Requirements

One of the key needs for QMRA is dose-response information. In the initial applications of this technique, reliance was placed on human dose-response information. For example, studies were done in the 1950s on human response to

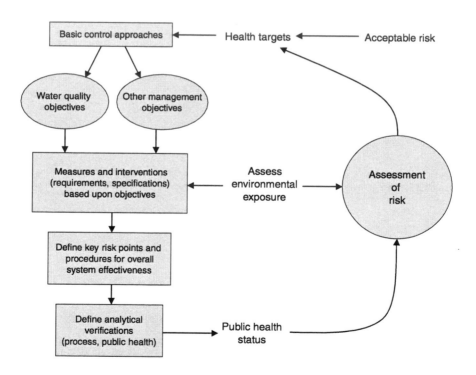

FIGURE 2-4 Conceptual framework for development of water-related microbial standards. SOURCE: Adapted from Bartram et al., 2001.

ingestion of *Salmonella* and other enteric bacterial pathogens (June et al., 1953; McCullough and Eisele, 1951a,b,c,d); somewhat later, human trials on response to viral ingestion were performed (Minor et al., 1981; Ward et al., 1986). Human volunteer trials of exposure to *Cryptosporidium* have been reported and are still being conducted (Chappell et al., 1996; Dupont et al., 1995; Okhuysen et al., 1998, 1999; Teunis et al., 1997, 2002a,b). However, it is increasingly less likely, owing to ethical and logistical concerns, that human volunteer data could be used to develop such information in the future. Therefore, alternative sources of such information must be developed.

In particular, the use of alternative sources of data will be important to understand the role and magnitude of human and microorganism variability in influencing the risk. From human feeding trials with *Cryptosporidium*, these factors appear highly significant (Teunis et al., 2002a,b); however, the use of different methods to obtain such information will be required with other microorganisms.

One alternative approach may be the increased use of animal models. These have been found useful in understanding the dose-response relationship of *Listeria monocytogenes* and *Escherichia coli* O157:H7 (Haas et al., 1999a, 2000). In general, the use of animal models is considered appropriate if the mechanisms and pathways for the processes of infection and disease are likely to be the same in experimental animals and humans and there are some waterborne pathogens for which this is the case. For example, Havellar et al. (2001) developed and evaluated a rat experimental model to study dose-response relationships of the enteropathogenic bacterium *Salmonella enterica* serovar Enteritidis. The authors concluded that the rat model is a sensitive and reproducible tool for studying the effects of oral exposure to *Salmonella* Enteritidis over a wide dose range and allows controlled quantification of different factors related to the host, pathogen, and food matrix in initial stages of infection by these bacterial pathogens. That study demonstrated that animal model systems of human infection and disease have advantages over human studies, such as the ability to examine the events leading to infection and disease at the cell, tissues, and organ level, including pathophysiological mechanisms and pathways. It also demonstrated the ability to score for mortality as an end point. The validation of animal models requires, in addition to a competent animal species, data from human outbreak studies in which attack rates and exposure are reliably estimated. For risk assessment, it is therefore particularly important in the use of epidemiologic outbreak data that greater effort be devoted to dose reconstruction. It should be noted that demonstrating that a particular animal is a competent species is a complex task, and therefore for newly emerging (or recognized) pathogens such development may require a significant research effort.

A second approach would be to use information obtained directly from the epidemiologic study of an outbreak to develop a dose-response relationship. For example, a drinking water outbreak of *Giardia* showed a graded response between attack rate and self-reported glasses of water consumed (Istre et al., 1984).

This was subsequently analyzed using an exponential dose-response relationship (between attack rate and glasses of water) (Rose et al., 1991b). However, the unavailability of a measurement of concentration of the pathogens at the time of exposure prevented the full development of a dose-response relationship.

A third approach to obtaining alternative sources of data is to examine in a risk assessment context historical dose-response data for enteric pathogens that have been given to humans. For example, numerous studies were done in which human volunteers, both adults and children, were challenged with different doses of hepatitis A virus and scored for infection and illness (Ward et al., 1958). These data have yet to be examined using QMRA techniques. Although the doses administered to the volunteers were not measured directly, they can be estimated on the basis of the expected concentrations of pathogens in the sample (e.g., an acute phase serum), based on more recent analytical information of measurements of hepatitis A virus in sera, and a distribution of the concentration can be put into dose-response models.

Relationship of QMRA to Microbial Indicators

The application of indicators or direct pathogen monitoring provides data that can be used within the QMRA framework to set criteria for establishing water quality standards and define the potential public health risk. Each approach, described below, has its own merits, difficulties, and uncertainties.

Direct Pathogen Monitoring

For any particular pathogen, QMRA may be used to develop a risk-related criterion. However, such criteria have to be considered carefully based on the methodology used to measure the particular pathogens in the water environment. Measurements can be based on infectivity or culturability, viability, physical presence (as detected by microscopy), or detection of microbial components (nucleic acids, proteins, or specific antigens; see Chapter 5 and Appendix C for further information). Estimations and interpretations of risk must consider how well these different measurement techniques detect infectious microorganisms that pose human health risks. However, as described earlier, the concentration of a pathogen in water at the point of exposure that would be allowed as an "acceptable risk" is likely to be much lower than can be practically and reliably detected, as with finished drinking water (Regli et al., 1991; Rose et al., 1991a,b). Thus, alternative approaches toward implementing risk-derived guidelines may be necessary. One approach would be to include key pathogen monitoring requirements in National Pollutant Discharge Elimination System (NPDES; see Table 1-2) permits where the discharge may be affecting key designated bodies of water. Historically, pathogen requirements have rarely been included in wastewater discharge permits (NRC, 2000). The formal computation of a risk assessment based on pathogen monitoring data also would enable those interested in designing novel sensor and analysis approaches to understand the required level of sensitivity of such systems if they are to be used in controlling risk from exposure to water-borne pathogens.

Indicators and Dose-Response Relationships

As discussed previously, the basic philosophy behind the EPA recreational water quality criteria is to develop human dose-response relationships between a measured indicator and a measured health effect (presumably due to pathogens in the water) of interest via direct epidemiologic investigation (Cabelli, 1983; Dufour, 1984; Seyfried et al., 1985a,b). From these investigations, the direct health risk from exposure to microbially contaminated recreational water can be determined as a function of a dose metric in terms of indicator concentrations in the water. This approach effectively condenses the dose-response and exposure assessment steps of risk assessment into a single functional relationship, by assuming that the exposure (i.e., amount of water ingested) is the same in epidemiologic investigations and in situations where risk is desired to be controlled.

Indicator to Pathogen Ratio

As discussed throughout this chapter, there is considerable evidence that the risk of becoming infected and ill from ingesting waterborne pathogens increases as the numbers or dose of pathogens increases (i.e., a dose-response relationship). Furthermore, it is generally but not always the case that the greater the number of indicator organisms in water and other media, the greater the number of pathogens (see also Chapter 1). In some cases, these indicator-to-pathogen relationships are sufficiently robust that they have been published in peer-reviewed journals, such as the relationship between enteroviruses and F+ coliphage indicators in ambient freshwaters in the Netherlands (Havelaar et al., 2003). While these relationships between indicator organisms and pathogens can change due to variable pathogen or indicator concentration in fecal sources and ambient waters, they do exist at any point in time. One approach to microbial water quality assessment, conceptualized many years ago by Fuhs (1975), would be to develop a thorough analytical monitoring program and a systematic quantitative relationship between pathogen concentrations and indicator concentrations. A more recent and rigorous explication on this concept has shown that indicator to pathogen ratios can then be used to modify the usual dose-response relationships such that the risk of exposure can be determined based on indicator concentrations (Lopez-Pila and Szewzyk, 2000). This approach has application to point sources of pollution discharged to water bodies and to the development of pathogen-to-indicator ratios in sewage and stormwater, combined sewer overflows, and so on. Modeling the impact at the site of exposure (e.g., the beach) would require detailed knowledge such as flow and microbial quality about the sources of the pathogens as well as the incorporation of various decay processes (perhaps as influenced by temperature, turbidity, and sunlight) that affect the indicators and the pathogens during transport. Assessing other factors including dilution and accumulation in sediments may also be necessary.

Future Directions for QMRA

To date, most applications of QMRA have focused on the prediction of primary infections or illnesses resulting from exposure to a contaminated medium (water, food, etc.). However, it is clear that for at least some illnesses, significant impact results from secondary transmission (Kappus et al., 1982; Mac Kenzie et al., 1994; Morens et al., 1979). In recent years, mathematical models have been increasingly applied to understanding of disease transmission process—including the processes of secondary transmission and immunity (Eisenberg et al., 1996, 1998). However, such approaches require a variety of data that are not readily available, including intensity and duration of contagion, duration and strength of immunity, and so on. The increased use of such dynamic mathematical frameworks in a sensitivity analysis to help determine the potentially most influential parameters for which there are data gaps, and to use such knowledge in focused epidemiologic investigations to fill these gaps, appears to have merit. The models must be used to inform data gathering, as well as be informed by data. To fully understand long-term and endemic risks associated with certain waterborne illnesses, it will also be necessary to develop models that account for pathogen dynamics in nonhuman reservoirs and survival in water bodies. Feedback between modelers and experimentalists will also be needed to develop data necessary for better quantitative understanding of microbial risk.

SUMMARY: CONCLUSIONS AND RECOMMENDATIONS

Health effects assessments for waterborne pathogens can be based on a number of approaches. Each approach has strengths and weaknesses, and all have been or are being used to document and quantify the health risks of microbes in water.

Epidemiologic methods are a well-established and essential tool for determining linkage between the presence of identified waterborne pathogens and their indicators and human disease. However, the significant cost and methodological difficulty of designing, conducting, and interpreting such studies have limited their use.

The comprehensiveness of investigations of waterborne disease outbreak in the United States varies by the type of outbreak and by state, and results are compiled in CDC's surveillance system. However, this system has low sensitivity and does not consistently provide information that links indicator and pathogen data with adverse health outcomes. This gap occurs because most outbreak investigations include primarily the epidemiologic component, which concentrates on linking illness to water and might include determination of the agent in clinical specimens, but tends to neglect the environmental component, which would include the determination of water quality through measurement of indicator and pathogen occurrence in water. This gap occurs more frequently with outbreaks associated with drinking water than with those associated with recreational water. In addition, 40-50 percent of identified outbreaks are of unknown etiology.

Under the SDWA Amendments of 1996, recently completed (though largely unpublished at the time this report was finalized) epidemiologic studies of drinking water and endemic disease have focused on establishing associations between water consumption and gastrointestinal illness. Thus far, they have not established a good correlation between indicators of waterborne pathogens, the pathogens themselves, and adverse human health effects, although some earlier studies have shown an association between tap water and endemic gastrointestinal illness with attributable fractions ranging between 14-40 percent. To have adequate statistical power to address the epidemiologic association of health outcomes with specific indicators and specific waterborne pathogens, the study sample needs to be large, leading to significant costs. In addition, methodologic complexities as well as difficulty in interpretation of results have limited the use of some of the studies.

In contrast, epidemiologic studies involving recreational bathing waters have shown predictive associations between several swimming-associated health effects and various microbial indicators or pathogens. A systematic review and meta-analysis of recreational waterborne studies (both freshwater and marine) confirmed that indicators can provide reliable estimates of water quality that are predictive of human health risks under some, but not all, water quality conditions, and the committee supports several conclusions provided in that study as related to this report.

Under the BEACH Act, the recently initiated EPA study of midwestern and eastern freshwater beaches is commendable, but limited in scope to the study of point-source contamination and acute disease; it does not yet include western regions or ocean beaches. Knowledge is lacking about the epidemiology of chronic or recurrent illness (i.e., gastrointestinal, respiratory, dermatologic illnesses) associated with habitual users of recreational waters subject to point and nonpoint source microbial contamination, and knowledge of the epidemiology of disease outbreaks associated with use of tropical and subtropical recreational waters and ocean beaches is fragmentary.

Quantitative microbiological risk assessment follows the traditional framework proposed for chemical risk assessment with several substantive differences. QMRA is a useful tool for identifying data gaps, especially models that include infectious disease parameters such as immunity. However, some of the key needs for QMRA are dose-response and exposure information (e.g., intensity and duration of contagion), which are often lacking. In some cases, impacts from such population level phenomena may dramatically alter projected estimates of human risk.

Building on its conclusions, the committee makes several recommendations regarding future directions for epidemiologic and microbiological research as related to health effects assessment of waterborne pathogens and their indicators. The committee first recommends that EPA and CDC take a greater leadership role in such efforts, and fund and work with stakeholders and academic researchers in the following areas:

• CDC should actively work with state and local health departments to encourage testing for pathogens (especially viruses and parasites) in clinical specimens during waterborne outbreak investigations.

• Standardized laboratory methods for clinical specimens as well as water samples which are both sensitive and specific must be developed for many viruses.

• CDC and EPA should actively work with state and local health departments to encourage collection and testing of environmental data (i.e., water quality data for source, finished, and distribution system waters that include indicators and pathogens) during waterborne outbreak investigations.

• Standardized protocols and definitions are needed for outbreak investigations and epidemiologic studies, especially to help ensure a comprehensive investigation or study that includes the collection of clinical, laboratory, and environmental data (including co-occurrence of pathogens and indicators).

• Epidemiologic studies should be conducted to (1) assess the effectiveness and validity of newly developed indicators or indicator approaches for determining poor microbial water quality and (2) assess the effectiveness of the indicators or indicator approaches at preventing and reducing human disease.

• Fewer but more comprehensive epidemiologic studies should be conducted rather than multiple small-scale studies that do not adequately address multiple risk factors and health outcomes when working within a fixed or constrained budget. More specifically, the link between pathogens and their potential indicators, and among pathogens, indicators, and adverse health outcomes, would be strengthened by including in comprehensive and adequately funded studies, epidemiologic measurements of health outcomes, measurements of pathogens in clinical specimens, as well as measurements of pathogens and their potential indicators in relevant water samples.

• Additional epidemiologic studies are needed to look at the association between water consumption and gastrointestinal illness in groundwater systems, and to correlate water quality data (pathogens and indicators) with health outcomes. Furthermore, these studies should include the collection of epidemiologic, clinical, laboratory, and environmental data whenever feasible.

• Health outcomes studied in association with drinking water exposure should not be limited to gastrointestinal illness (e.g., should consider respiratory and dermatological illnesses).

• The national estimate of waterborne disease should be expanded. Specifically, data have to be incorporated from sources other than randomized intervention trials and community trials (e.g., outbreaks, systematic reviews and meta-analyses, *Cryptosporidium* serologic data from NHANES, data from models derived from risk assessment).

• Additional epidemiologic studies should be conducted to determine the occurrence of chronic/recurrent disease attributable to waterborne pathogens in habitual users of recreational waters (e.g., surfers) from point and nonpoint sources of contamination.

• Studies of recreational waters should be carried out on a broader range of geographical and ecological sites, including tropical and subtropical waters and ocean beaches.

• Indicators being studied as part of the BEACH Act should not be limited to those than can yield results in two hours, as has been suggested.

• Since epidemiologic investigations are mandated as part of the BEACH Act, consistent scientific approaches should be used to monitor for various types of indicators as well as pathogens to establish dose-response relationships.

• Alternative sources to human volunteer data should be pursued to provide dose-response and exposure information for QMRA.

• Risk assessment with sensitivity analyses should be used to identify data gaps and help drive epidemiologic studies.

REFERENCES

Aragon, T.J., S. Novotny, W. Enanoria, D. Vugia, A. Khalakdina, and M. Katz. 2003. Endemic cryptosporidiosis and exposure to municipal tap water in persons with acquired immunodeficiency syndrome (AIDS): A case-control study. BioMed Central Public Health 2(3): 1-28.

AWWA (American Water Works Association). 1999. Two-City Cryptosporidium Study. American Water Works Association Report.

Bandaranayake, D.R., S.J. Turner, G.B. McBride, G.D. Lewis, and D.G. Till. 1995. Health Effects of Bathing at Selected New Zealand Marine Beaches. New Zealand.

Bartram, J., and G. Rees. 1999. Monitoring Bathing Waters: A Practical Guide to the Design and Implementation of Assessments and Monitoring Programmes. London: E. and F.N. Spon.

Bartram, J., L. Fewtrell, and T.A. Stenstrom. 2001. Harmonised assessment of risk and risk management for water-related infectious disease: an overview. Pp. 1-16 in Water Quality. Guidelines, Standards and Health: Assessment of Risk and Risk Management for Water-Releated Infectious Disease, L. Fewtrell and J. Bartram, eds. London: World Health Organization and IWA Publishing.

Barwick, R.S., D.A. Levy, G.F. Craun, M.J. Beach, and R.L. Calderon. 2000. Surveillance for waterborne-disease outbreaks - United States, 1997-1998. MMWR 49 (No. SS-4): 1-35.

Buchanan, R.L., and R.C. Whiting. 1996. Risk assessment and predictive microbiology. Journal of Food Protection Supplement: 31-36.

Cabelli, V.J., A.P. Dufour, L.J. McCabe, and M.A. Levin. 1982. Swimming-associated gastroenteritis and water quality. American Journal of Epidemiology 115(4): 606-616.

Cabelli, V.J. 1983. Health Effects Criteria for Marine Recreational Waters. Cincinnati, Ohio: EPA-600-1-80-031.

Cameron, J. 1943. Infective hepatitis. Quarterly Journal of Medicine 12: 139-155.

CDC (Centers for Disease Control and Prevention). 2001. Updated guidelines for evaluating public health systems - recommendations from the guidelines working group. MMWR 50 (No. RR-13): 1-35.

Chappell, C.L., P.C. Okhuysen, C.R. Sterling, and H.L. Dupont. 1996. *Cryptosporidium parvum*: Intensity of infection and oocyst excretion patterns in healthy volunteers. Journal of Infectious Diseases 173: 232-236.

Chappell, C.L., P.C. Okhuysen, C.R. Sterling, C. Wang, W. Jakubowski, and H.L. Dupont. 1999. Infectivity of *Cryptosporidium parvum* in healthy adults with pre-existing anti-*C. parvum* serum immunoglobulin G. American Journal of Tropical Medicine and Hygiene 60(1): 157-164.

Cheung, W.H.S., K.C.K. Chang, and R.P.S. Hung. 1989. Health effects of beach water pollution in Hong Kong. Epidemiology and Infection 105: 139-162.

Colford, J.M., Jr. 2002. Presentation to the Committee on Indicators for Waterborne Pathogens. September 4. Washington, D.C.

Colford, J.M., Jr., J.R. Rees, T.J. Wade, A. Khalakdina, J.F. Hilton, I.J. Ergas, S. Burns, A. Benker, C. Ma, C. Bowen, D.C. Mills, D.J. Vugia, D.D. Juranek, and D.A. Levy. 2002. Participant blinding and gastrointestinal illness in a randomized, controlled trial of an in-home drinking water intervention. Emerging Infectious Disease 8(1): 29-36.

Colford, J.M., Jr., T. Wade, S. Sandhu, C. Wright, S. Burns, A. Benker, S. Lee, A. Brookhart, M. van der Laan, and D.A. Levy. 2003. A randomized, blinded, crossover trial of an in-home drinking water intervention to reduce gastrointestinal illness. 15th Annual Meeting of the International Society of Environmental Epidemiology, Perth, Australia, September 26.

Corbett, S.J., G.L. Rubin, G.K. Curry, and D.G. Kleinbaum. 1993. The health effects of swimming at Sydney beaches. American Journal of Public Health 83: 1701-1706.

Cox, C.R. 1951. Acceptable standards for natural waters used for bathing. Proceedings of the American Society of Civil Engineering 77: 74.

Craun, G.F., ed. 1986. Waterborne Diseases in the United States. Boca Raton, Florida: CRC Press, Inc.

Dufour, A.P. 1984. Health Effects Criteria for Fresh Recreational Waters. Cincinnati, Ohio: EPA-600-1-84-004.

Dupont, H., C. Chappell, C. Sterling, P. Okhuysen, J. Rose, and W. Jakubowski. 1995. Infectivity of *Cryptosporidium parvum* in healthy volunteers. New England Journal of Medicine 332(13): 855.

Eisenberg, J.N., E.Y. Seto, A. Olivieri, and R.C. Spear. 1996. Quantifying water pathogen risk in an epidemiological framework. Risk Analysis 16(4): 549-563.

Eisenberg, J.N., E.Y. Seto, J.M. Colford, Jr., A. Olivieri, and R.C. Spear. 1998. An analysis of the Milwaukee cryptosporidiosis outbreak based on a dynamic model of the infection process. Epidemiology 9(3): 255-263.

Eisenberg, J.N.S., T.J. Wade, C.M. Vu, A. Hubbard, C.C. Wright, D. Levy, P. Jensen, and J.M. Colford, Jr. 2002a. Risk factors in HIV-associated diarrhoeal disease: The role of drinking water, medication and immune status. Epidemiology and Infection 128: 73-81.

Eisenberg, J.N.S., T.J. Wade, A. Hubbard, D.I. Abrams, R.J. Leiser, C.M. Vu, S. Saha, C.C. Wright, D.A. Levy, P. Jensen, and J.M. Colford, Jr. 2002b. Associations between water-treatment methods and diarrhoea in HIV-positive individuals. Epidemiology and Infection 129: 315-323.

EPA (U.S. Environmental Protection Agency). 1986. Bacteriological ambient water quality criteria availability. Federal Register 51(45): 8012.

EPA. 1989. Drinking Water; National Primary Drinking Water Regulations; Filtration, Disinfection; Turbidity; *Giardia lamblia*, Viruses, *Legionella*, and Heterotrophic Bacteria; Final Rule. Federal Register 54: 27486-27541.

EPA. 1996. Information Collection Rule. Federal Register 61(94): 24353-24388.

EPA. 1998. Interim Enhanced Surface Water Treatment Rule. Federal Register 63(241): 69477-69521.

EPA. 1999. EPA Action Plan for Beaches and Recreational Waters. Washington, D.C.: Office of Research and Development and Office of Water. EPA-600-R-98-079

Fattal, B., E. Peleg-Olvesky, T. Agurshy, and H.I. Shuval. 1987. The association between sea water pollution as measured by bacterial indicators and morbidity of bathers at Mediterranean beaches in Israel. Chemosphere 16: 565-570.

Ferley, J.P., D. Zimrou, F. Balducci, B. Baleux, P. Fera, G. Larbaigt, E. Jacq, B. Mossonnier, A. Blineau, and J. Boudot. 1989. Epidemiological significance of microbiological pollution criteria for river recreational waters. International Journal of Epidemiology 18(1): 198-205.

Fewtrell, L., A.F. Godfree, F. Jones, D. Kay, R.L. Salmon, and M.D. Wyer. 1992. Health effects of white-water canoeing. Lancet 339: 1587-1589.

Fewtrell, L., and J. Bartram, eds. 2001. Water Quality - Guidelines, Standards and Health: Assessment of Risk and Risk Management for Water-Releated Infectious Disease. London: World Health Organization and IWA Publishing.

Fleisher, J.M. 1990. Conducting recreational water quality surveys: some problems and suggested remedies. Marine Pollution Bulletin 21: 562-567.

Fleisher, J.M., F. Jones, D. Kay, R. Stanwell-Smith, M. Wyer, and R. Morano. 1993. Water and nonwater related risk factors for gastroenteritis among bathers exposed to sewage contaminated marine waters. International Journal of Epidemiology 22: 11.

Fleisher, J.M., D. Kay, R.L. Salmon, F. Jones, M.D. Wyer, and A.F. Godfree. 1996. Marine waters contaminated with domestic sewage: non-enteric illness associated with bather exposure in the United Kingdom. American Journal of Public Health 86: 1228-1234.

Frost, F.J., T. Muller, G.F. Craun, R.L. Calderon, and P.A. Roefer. 2001. Paired city *Cryptosporidium* serosurvey in the southwest USA. Epidemiology and Infection 126(2): 301-307.

Frost, F.J., T. Muller, G.F. Craun, W.B. Blockwood, and R.L. Calderon. 2002. Annals of Epidemiology 12(4): 222-227.

Fuhs, G.W. 1975. A probabilistic model of bathing beach safety. Science of the Total Environment 4: 165-175.

Gerba, C.P., and J.B. Rose. 1990. Viruses in source and drinking water. Pp. 380-396 in Drinking Water Microbiology, G.A. McFeters, ed. New York: Springer-Verlag.

Goldstein, S.T., D.D. Juranek, O. Ravenholt, A.W. Hightower, D.G. Martin, J.L. Mesnik, S.D. Griffiths, A.J. Bryant, R.R. Reich, and B.L. Herwaldt. 1996. Cryptosporidiosis: An outbreak associated with drinking water despite state-of-the-art water treatment. Annals of Internal Medicine 124(5): 459-468.

Gordis, L. 2000. Epidemiology. Pennsylvania: W.B. Saunders Co.

Gregory, J. 1994. *Cryptosporidium* in water - treatment and monitoring methods. Filtration and Separation 31(3): 283-289.

Haas, C.N. 1983a. Effect of effluent disinfection on risks of viral disease transmission via recreational exposure. Journal of the Water Pollution Control Federation 55: 1111-1116.

Haas, C.N. 1983b. Estimation of risk due to low doses of microorganisms: A comparison of alternative methodologies. American Journal of Epidemiology 118(4): 573-582.

Haas, C.N. 1986. Wastewater disinfection and infectious disease risks. CRC Critical Reviews in Environmental Control 17(1): 1-20.

Haas, C.N., C. Crockett, J.B. Rose, C. Gerba, and A. Fazil. 1996. Infectivity of *Cryptosporidium parvum* oocysts. Journal of the American Water Works Association 88(9): 131-136.

Haas, C.N., A. Thayyar-Madabusi, J.B. Rose, and C.P. Gerba. 1999a. Development and validation of dose-response relationship for *Listeria monocytogenes*. Quantitative Microbiology 1(1): 89-102.

Haas, C.N., J.B. Rose, and C.P. Gerba. 1999b. Quantitative Microbial Risk Assessment. New York: John Wiley & Sons, Inc.

Haas, C.N., A. Thayyar-Madabusi, J.B. Rose, and C.P. Gerba. 2000. Development of a dose-response relationship for *Escherichia coli* O157:H7. International Journal of Food Microbiology 56(2-3): 153-159.

Haile, R.W., J. Alamillo, K. Barrett, R. Cressey, J. Dermond, C. Ervin, A. Glasser, N. Harawa, P. Harmon, J. Harper, C. McGee, R.C. Millikan, M. Nides, and J.S. White. 1996. An Epidemiological Study of Possible Adverse Health Effects of Swimming in Santa Monica Bay. Final report of the Santa Monica Bay Restoration Project. Santa Monica, California.

Havelaar, A.H., J. Garssen, K. Takumi, M.A. Koedam, J.B. Dufrenne, F.M. van Leusden, L. de La Fonteyne, J.T. Bousema, and J.G. Vos. 2001. A rat model for dose-response relationships of *Salmonella enteritidis* infection. Journal of Applied Microbiology 91(3): 442-452.

Havelaar, A.H., M. van Olphen, and Y.C. Drost. 2003. F-specific RNA bacteriophages are adequate model organisms for enteric viruses in fresh water. Applied and Environmental Microbiology 59(9): 2956-2962.

Hellard, M.E., M.I. Sinclair, A.B. Forbes, and C.K. Fairley. 2001. A randomized, blinded, controlled trial investigating the gastrointestinal health effects of drinking water quality. Environmental Health Perspectives 109(8): 773-778.

Herikstad, H., S. Yang, T.J. Van Gilder, D. Vugia, J. Hadler, P. Blake, V. Deneen, B. Shiferaw, F.J. Angulo, and the FoodNet Working Group. 2002. A population-based estimate of the burden of diarrhoeal illness in the United States: FoodNet, 1996-1997. Epidemiology and Infection 129: 9-17.

Herwaldt, B.L., G.F. Craun, S.L. Stokes, and D.D. Juranek. 1991. Waterborne-disease outbreaks, 1989-1990. MMWR 40 (No. SS-3): 1-21.

Hodges, R.G., L.P. McCorkle, G.F. Badger, C. Curtis, J.H. Dingle, and W.S. Jordan, Jr. 1956. A study of illness in a group of Cleveland families. American Journal of Hygiene 64: 349-356.

Hoornstra, E. and S. Notermans. 2001. Quantitative microbiological risk assessment. International Journal of Food Microbiology 66(1-2): 21-29.

ILSI (International Life Sciences Institute) Risk Science Institute Pathogen Risk Assessment Working Group. 1996. A conceptual framework to assess the risks of human disease following exposure to pathogens. Risk Analysis 16(6): 841-848.

ILSI Risk Science Institute Water- and Foodborne Pathogen Risk Assessment Working Group. 2000. Revised Framework for Microbial Risk Assessment. Washington, D.C.: International Life Sciences Institute.

Istre, G.R., T.S. Dunlop, G.B. Gaspard and R.S. Hopkins. 1984. Waterborne giardiasis at a mountain resort: Evidence for acquired immunity. American Journal of Public Health 74(6): 602-604.

Jaykus, L.A. 1996. The application of quantitative risk assessment to microbial food safety risks. Critical Reviews in Microbiology 22(4): 279-293.

June, R.C., W.W. Ferguson, and M.T. Worfel. 1953. Experiments in feeding adult volunteers with *Escherichia coli* 55 B$_5$: A coliform organism associated with infant diarrhea. American Journal of Hygiene 57: 222-236.

Kappus, K.D., J.S. Marks, R.C. Holman, J.K. Bryant, C. Baker, G.W. Gary, and H.B. Greenberg. 1982. An outbreak of Norwalk gastroenteritis associated with swimming in a pool and secondary person to person transmission. American Journal of Epidemiology 116(5): 834-839.

Katz, M., and S. Plotkin. 1967. Minimal infective dose of attenuated poliovirus for man. American Journal of Public Health 57(10): 1837-1840.

Kay, D., J.M. Fleisher, R.L. Salmon, F. Jones, M.D. Wyer, A. Godfree, Z. Zelanauch-Jaquotte, and R. Shore. 1994. Predicting likelihood of gastroenteritis from sea bathing: Results from randomised exposure. Lancet 344: 905-909.

Kehr, R.W., and C.T. Butterfield. 1943. Notes on the relation between coliforms and enteric pathogens. Public Health Reports 58: 589.

Koprowski, H. 1956. Immunization against poliomyelitis with living attenuated virus. Journal of the American Medical Association 160: 954-966.

Kramer, M.H., B.L. Herwaldt, G.F. Craun, R.L. Calderon, and D.D. Juranek. 1996. Surveillance for waterborne-disease outbreaks - United States, 1993-1994. MMWR 45 (No. SS-1): 1-33.

Kueh, C.S.W., T.-Y. Tam, and D.C.J. Bassett. 1995. Epidemiological study of swimming-associated illnesses relating to bathing beach water quality. Water Science and Technology 31(5/6): 1-4.

Lainer, G. 1940. Zur frage der infecktiosität des ikterus. Wiener Klinische Wochenschrift 53: 601-604.

Langmuir, A.D. 1963. The surveillance of communicable diseases of national importance. New England Journal of Medicine 268: 182-192.

Last, J.M., ed. 2001. A Dictionary of Epidemiology, 4th Edition. New York: Oxford University Press.

Lavori, P.W., and J.L. Kelsey. 2002. Introduction and overview. Epidemiologic Reviews 24(1): 1-3.

Lee, S.H., D.A. Levy, G.F. Craun, M.J. Beach, and R.L. Calderon. 2002. Surveillance for waterborne-disease outbreaks - United States, 1999-2000. MMWR 51 (No. SS-8): 1-47.

Leong, L.Y.C. 1983. Removal and inactivation of viruses by treatment processes for potable water and wastewater—a review. Water Science and Technology 15: 91-114.

Levy, D.A., M.S. Bens, G.F. Craun, R.L. Calderon, and B.L. Herwaldt. 1998. Surveillance for waterborne-disease outbreaks - United States, 1995-1996. MMWR 47 (No. SS-5): 1-33.

Lightfoot, N.E. 1989. A prospective study of swimming related illness at six freshwater beaches in Southern Ontario, Ph.D. dissertation. University of Toronto.

Lopez-Pila, J.M., and R. Szewzyk. 2000. Estimating the infection risk in recreational waters from the faecal indicator concentration and from the ratio between pathogens and indicators. Water Research 34(17): 4195-4200.

MacCallum, F.O., and W.H. Bradley. 1944. Transmission of infective hepatitis to human volunteers: Effect on rheumatoid arthritis. Lancet 2: 228.

Mac Kenzie, W.R., W.L. Schell, B.A. Blair, D.G. Addiss, D.E. Peterson, N.J. Hozie, J.J. Kazmierczak, and J.P. Davis. 1994. Massive outbreak of waterborne *Cryptosporidium* infection in Milwaukee, Wisconsin: Recurrence of illness and risk of secondary transmission. Clinical Infectious Diseases 21: 57-62.

Matthews, J.N.S. 2000. An Introduction to Randomized Controlled Clinical Trials. London: Arnold Publishers.

McCullough, N.B., and C.W. Eisele. 1951a. Experimental human salmonellosis: I. pathogenicity of strains of *Salmonella meleagridis* and *Salmonella anatum* obtained from spray dried whole egg. Journal of Infectious Diseases 88: 278-289.

McCullough, N.B., and C.W. Eisele. 1951b. Experimental human salmonellosis: II. immunity studies following experimental illness with *Salmonella meleagridis* and *Salmonella anatum*. Journal of Infectious Diseases 66: 595-608.

McCullough, N.B., and C.W. Eisele. 1951c. Experimental human salmonellosis: III. pathogenicity of strains of *Salmonella newport*, *Salmonella derby* and *Salmonella bareilly* obtained from spray dried whole egg. Journal of Infectious Diseases 89: 209-213.

McCullough, N.B., and C.W. Eisele. 1951d. Experimental human salmonellosis: IV. pathogenicity of strains of *Salmonella pullorum* obtained from spray dried whole egg. Journal of Infectious Diseases 89: 259-265.

Mead, P.S., L. Slutsker, V. Dietz, L.F. McCaig, J.S. Bresee, C. Shapiro, P.M. Griffin, and R.V. Tauxe. 1999. Food-related illness and death in the United States. Emerging Infectious Disease 5(5): 607-625.

Medical Research Council and Council for Scientific and Industrial Research. 1995. Pathogenic Microorganisms/Epidemiological-Microbiological Study. Final Report 1991-1995. South Africa.

Meinert, C.L. 1986. Clinical Trials. New York: Oxford University Press.

Minor, T.E., C.I. Allen, A.A. Tsiatis, D.B. Nelson, and D.J. D'Alessio. 1981. Human infective dose determination for oral poliovirus type I vaccine in infants. Journal of Clinical Microbiology 13(2): 388-389.

Monto, A.S., and J.S. Koopman. 1980. The Tecumseh study. American Journal of Epidemiology 112(3): 323-333.

Moore, A.C., B.L. Herwaldt, G.F. Craun, R.L. Calderon, A.K. Highsmith, and D.D. Juranek. 1993. Surveillance for waterborne-disease outbreaks - United States, 1991-1992. MMWR 42 (No. SS-5): 1-22.

Morens, D.M., R.M. Zweighaft, T.M. Vernon, G.W. Gary, J.J. Eslien, B.T. Wood, R.C. Holman, and R. Dolin. 1979. A waterborne outbreak of gastroenteritis with secondary person to person spread. Lancet 1: 964-966.

Mujeriego, R., J.M. Bravo, and M.T. Feliu. 1982. Recreation in coastal waters: Public health implications. Vier Journee Etud. Pollutions, Cannes, Centre Internationale d'Exploration Scientifique de la Mer: 585-594.

Naumova, E.N., A.I. Egorov, R.D. Morris, and J.K. Griffiths. 2003. The elderly and waterborne Cryptosporidium infection: Gastroenteritis hospitalizations before and during the 1993 Milwaukee outbreak. Emerging Infectious Diseases 9(4): 418-425.

NRC (National Research Council). 1972. Drinking Water and Health, Volume 1. Washington, D.C.: National Academy Press.

NRC. 1983. Risk Assessment in the Federal Government: Managing the Process. Washington, D.C.: National Academy Press.

NRC. 1989. Drinking Water and Health, Volume 9. Selected Issues in Risk Assessment. Washington, D.C.: National Academy Press.

NRC. 2000. Watershed Management for Potable Water Supply: Assessing the New York City Strategy. Washington D.C.: National Academy Press.

NTAC (National Technical Advisory Committee). 1968. National Technical Advisory Committee (NTAC) Report of the Committee on Water Quality Criteria. Washington, D.C.

Nwachuku, N., G.F. Craun, and R.L. Calderon. 2002. How effective is the TCR in assessing outbreak vulnerability? Journal of the American Water Works Association 94(9): 88-96.

Okhuysen, P.C., C.L. Chappell, C.R. Sterling, W. Jakubowski, and H.L. DuPont. 1998. Susceptibility and serologic response of healthy adults to reinfection with Cryptosporidium parvum. Infection and Immunity 66(2): 441-443.

Okhuysen, P.C., C.L. Chappell, J.H. Crabb, C.R. Sterling, and H.L. DuPont. 1999. Virulence of three distinct Cryptosporidium parvum isolates for healthy adults. Journal of Infectious Disease 180(4): 1275-1281.

Ongerth, J.E. 1989. Giardia cyst concentrations in river water. Journal of the American Water Works Association 81(9): 81-86.

Payment, P., L. Richardson, J. Siemiatycki, R. Dewar, M. Edwardes, and E. Franco. 1991. A randomized trial to evaluate the risk of gastrointestinal disease due to consumption of drinking water meeting current microbiological standards. American Journal of Public Health 81: 703-708.

Payment, P., J. Siemiatycki, L. Richardson, G. Renaud, E. Franco, and M. Prevost. 1997. A prospective epidemiological study of gastrointestinal health effects due to the consumption of drinking water. International Journal of Environmental Health Resources 7: 5-31.

Payment, P., and M.S. Riley. 2002. Resolving the Global Burden of Gastrointestinal Illness: A Call to Action. Washington, D.C.: American Academy of Microbiology.

Pike, E.B. 1991. Health Effects of Sea Bathing (ET 9511) - Phase II. Studies at Ramsgate and Moreton, 1990 and 1991. Mendenham, U.K.: DOE 2736-M(P).

Pike, E.B. 1994. Health Effects of Sea Bathing (WMI 9021) - Phase III. Final report to the Department of the Environment. Mendenham, U.K.: DOE 3412/2.

Prescott, S.C., C.A. Winslow, and M. McCrady. 1945. Water Bacteriology, 6th Edition. New York: Wiley.

Prüss, A. 1998. Review of epidemiological studies on health effects from exposure to recreational water. International Journal of Epidemiology 27(1): 1-9.

Ranta, J., and R. Maijala. 2002. A probabilistic transmission model of *Salmonella* in the primary broiler production chain. Risk Analysis 22(1): 47-57.

Regli, S., J.B. Rose, C.N. Haas, and C.P. Gerba. 1991. Modeling risk from *Giardia* and viruses in drinking water. Journal of the American Water Works Association 83(11): 76-84.

Rentdorff, R.C. 1954a. The experimental transmission of human intestinal protozoan parasites. I. *Entamoeba coli* cysts given in capsules. American Journal of Hygiene 59: 196-208.

Rendtorff, R.C. 1954b. The experimental transmission of human intestinal protozoan parasites. II. *Giardia lamblia* cysts given in capsules. American Journal of Hygiene 59: 209-220.

Robinson, W.S. 1950. Ecological correlations and the behavior of individuals. American Sociological Review 15: 351-357.

Rose, J.B. 1990. Pathogens in water: Overview of methods, application limitations and data interpretation. Pp. 223-234 in Methods for the Investigation and Prevention of Waterborne Disease Outbreaks. G.F. Craun, ed. Cincinnati, Ohio: Health Effects Research Laboratory, EPA.

Rose, J.B., C.P. Gerba, and W. Jakubowski. 1991a. Survey of potable water supplies for *Cryptosporidium* and *Giardia*. Environmental Science and Technology 25: 1393-1400.

Rose, J.B., C.N. Haas, and S. Regli. 1991b. Risk assessment and the control of waterborne giardiasis. American Journal of Public Health 81: 709-713.

Rothman, K.J., and S. Greenland. 1998. Modern Epidemiology, 2nd Edition. Philadelphia, Pennsylvania: Lippincott-Raven Publishers.

Rothman, K.J. 2002. Epidemiology: An Introduction. New York: Oxford University Press.

Sabin, A.B. 1955. Behavior of chimpanzee - a virulent poliomyelitis viruses in experimentally infected human volunteers. American Journal of the Medical Sciences 230: 1-8.

Schlesselman, J.L. 1982. Case-Control Studies. New York: Oxford University Press.

Schwartz, J., R. Levin, and K. Hodge. 1997. Drinking water turbidity and pediatric hospital use for gastrointestinal illness in Philadelphia. Epidemiology 8(6): 615-620.

Schwartz, J., and R. Levin. 1999. Drinking water turbidity and health. Epidemiology 10(1): 86-90.

Schwartz, J., R. Levin, and R. Goldstein. 2000. Drinking water turbidity and gastrointestinal illness in the elderly of Philadelphia. Journal of Epidemiology and Community Health 54(1): 45-51.

Scott, W.J. 1951. Sanitary study of shore bathing waters. Bulletin on Hygiene 33: 353.

Seyfried, P.L., R.S. Tobin, N.E. Brown, and P.F. Ness. 1985a. A prospective study of swimming-related illness: I. swimming-associated health risk. American Journal of Public Health 75: 1068-1070.

Seyfried, P.L., R.S. Tobin, N.E. Brown, and P.F. Ness. 1985b. A prospective study of swimming-related illness: II. morbidity and the microbiological quality of water. American Journal of Public Health 75: 1071-1075.

Shaughnessy, H.J., R.C. Olsson, K. Bass, F. Friewer, and S.O. Levinson. 1946. Experimental human bacillary dysentery. Journal of the American Medical Association 132: 362-368.

Silbergeld, E.K. 1993. Risk assessment - the perspective and experience of United States environmentalists. Environmental Health Perspectives 101(2): 100-104.

Snow, J. 1854. On cholera. In The Sourcebook of Medical History, L. Clendening, ed. 1942. New York: Paul B. Hoeber, Inc.

Sobsey, M.D. 1999. Methods to identify and detect microbial contaminants in drinking water. Pp. 173-205 in Identifying Future Drinking Water Contaminants. Washington, D.C.: National Academy Press.

Sobsey, M.D. 2001. Microbial detection: implications for exposure, health effects, and control. Pp. 89-113 in Microbial Pathogens and Disinfection By-Products in Drinking Water: Health Effects and Management of Risks, G.F. Craun, F.S. Hauchman, and D.E. Robinson, eds. Washington, DC: ILSI Press.

Stavros, G. and I.H. Langford. 2002. Coastal Bathing Water Quality and Human Health Risks: A Review of Legislation, Policy and Epidemiology, with an Assessment of Current UK Water Quality, Proposed Standards, and Disease Burden in England and Wales. CSERGE Working Paper ECM 02-06. Centre for Social and Economic Research on the Global Environment, University of East Anglia, U.K.

Stevenson, A.H. 1953. Studies of bathing water quality. American Journal of Public Health 43(5): 529-538.

Teunis, P.F.M., G.J. Medema, L. Kruidenier, and A.H. Havelaar. 1997. Assessment of the risk of infection by *Cryptosporidium* in drinking water from a surface water source. Water Research 31(6): 1333-1346.

Teunis, P.F.M., C.L. Chappell, and P.C. Okhuysen. 2002a. *Cryptosporidium* dose response studies: Variation between hosts. Risk Analysis 22(3): 475-485.

Teunis, P.F.M., C.L. Chappell, and P.C. Okhuysen. 2002b. *Cryptosporidium* dose response studies: Variation between isolates. Risk Analysis 22(1): 175-183.

UNEP/WHO (United Nations Environment Programme/World Health Organization). 1991. Epidemiological studies related to environmental quality criteria for bathing waters, shellfish-growing waters and edible marine organisms (Activity D). Final Report on Project on Relationship Between Microbial Quality of Coastal Seawater and Rotavirus-Induced Gastroenteritis Among Bathers (1986-1988). MAP Technical Reports Series No. 46. Athens, Greece.

UNEP/WHO. 1991. Epidemiological studies related to environmental quality criteria for bathing waters, shellfish-growing waters and edible marine organisms (Activity D). Final Report on Epidemiological Study on Bathers from Selected Beaches in Malaga, Spain (1988-1989). MAP Technical Reports Series No. 53. Athens, Greece.

van Dijk, P.A.H., R.F Lacey, and E.B. Pike. 1996. Health Effects of Sea Bathing—Further Analysis of Data from UK Beach Surveys. Final report to the Department of the Environment. DoE 4126/3. Mendenham, U.K.

Voegt, H. 1942. Zur aetiologie der hepatitis epidemica. Münchener Medizinische Wochenschrift. 89: 76-79.

Wade, T.J., N. Pai, J.N.S. Eisenberg, and J.M. Colford, Jr. 2003. Do U.S. Environmental Protection Agency water quality guidelines for recreational waters prevent gastrointestinal illness? A systematic review and meta-analysis. Environmental Health Perspectives 111(8): 1102-1109.

Walker, E.L., and W.S. Sellards. 1913. Experimental entamoebic dysentery. Philippine Journal of Science and Biological Tropical Medicine 8: 253-331.

Ward, R., S. Krugman, J. Giles, M. Jacobs, and O. Bodansky. 1958. Infectious hepatitis: Studies of its natural history and prevention. The New England Journal of Medicine 258(9): 402-416.

Ward, R.L., D.L. Bernstein, E.C. Young, J.R. Sherwood, D.R. Knowlton, and G.M. Schiff. 1986. Human rotavirus studies in volunteers: Determination of infectious dose and serological response to infection. Journal of Infectious Diseases 154(5): 871.

WHO (World Health Organization). 1998. Guidelines for Safe Recreational-Water Environments. WHO/SDE/WSH/00.6. Geneva.

WHO. 1999. Annapolis Protocol. Health Based Monitoring of Recreational Waters: The Feasibility of a New Approach. WHO/SDE/WSH/99.1. Geneva.

WHO. 2001. Bathing Water Quality and Human Health: Faecal Pollution. Outcome of an Expert Consultation, Farnham, UK, April 2001. Cosponsored by Department of the Environment, Transport and the Regions, United Kingdom. WHO/SDE/WSH/01.2. Geneva.

WHO. 2003. Guidelines for Safe Recreational Water Environments, Volume 1, Coastal and Fresh Waters. Geneva.

3

Ecology and Evolution of Waterborne Pathogens and Indicator Organisms

INTRODUCTION

Past efforts to develop and implement indicators of waterborne pathogens have often given little or no consideration to the role of evolution in the ecology and natural history of waterborne pathogens of public health concern. Evolution is a powerful force and can act quickly, even over ecological timeframes, to bring about change in pathogenic and indicator microorganisms. Furthermore, although numerous studies exist on the pathogenicity of various waterborne pathogens few have sought to describe their life history or ecology. The interactions between pathogens and their hosts involve complex and diverse processes at the genetic, biochemical, phenotypic, population, and community levels, while the distribution and abundance of microorganisms in nature and their microbial processes are affected by both biotic and abiotic factors that act at different scales. To develop new and more effective indicators of waterborne pathogens it is important to better understand how both evolution and ecology interact with the genomes and natural history of waterborne pathogens and their indicators, if different from themselves. Failure to consider these effects may result in spurious conclusions that do not truly reflect the abundance and distribution of waterborne pathogens.

Most of the waterborne pathogens discussed in this report (see also Appendix A) are not native to the types of waterbodies addressed herein. Notable exceptions include various species of *Vibrio* and *Legionella* bacteria and protozoan parasites such as the free-living amoebae *Naegleria* and *Acanthamoeba*. Many microorganisms that are pathogenic to humans and animals enter ambient waters after import from various point and diffuse sources. Upon entry, new selective

---

BOX 3-1
Summary of Important Ecological and Evolutionary Questions That May Affect the Understanding of Various Indicators for Waterborne Pathogens and Infectious Diseases

1. What is the distribution and abundance of waterborne pathogens? Are these environmental reservoirs of pathogens biotic or abiotic?
2. What are the fates of freshwater pathogens when imported into marine or brackish waters?
3. Is the residence time of a pathogen sufficient to allow genetic exchange or change to occur?
4. What biotic and abiotic factors influence the viability and survivability of waterborne pathogens? Are there environmental conditions that promote genetic exchange or the acquisition of genetic elements that confer selective advantage under clinical conditions?
5. What effect do sampling and environmental variations have on the efficacy of indicators?

---

forces begin to act on these introduced or exotic microorganisms, whether eukaryotes or prokaryotes.

This chapter describes basic principles of ecology and evolution for waterborne viruses, bacteria, and protozoa (and yeasts and molds to a lesser extent) of public health concern as an aid to better understand how selective forces may alter one's ability to assess the microbial quality of water. Indeed, indicators of microbial water quality can be the pathogenic organisms themselves, other microorganisms, or other physical or chemical aspects of the aquatic environment (see Chapter 4 for further information), and any biological indicator is subject to evolutionary and ecological changes. The final section is a summary of the chapter and its conclusions and recommendations.

Answers to several sets of related and fundamental questions (summarized in Box 3-1) are imperative to facilitate the understanding of indicators of waterborne pathogens and emerging infectious diseases. These questions include but are not limited to the following:

1. What is the natural distribution and abundance of waterborne pathogens? Are there environmental reservoirs of these microorganisms and, if so, what environmental conditions promote their maintenance or growth? Are these environmental reservoirs biotic or abiotic (i.e., from the living or nonliving)? Can waterborne pathogens colonize and proliferate in sediments or within aquatic systems? The concepts of growth and regrowth are most often applied to water distribution systems and wastewater discharges (and their receiving waters), respectively.

Determining whether and how survival and growth occur under natural conditions is important in understanding whether an indicator is indicating "new" contamination. The ecological concept of "source/sink" (Pulliam and Danielson, 1991) needs to be better understood for waterborne pathogens. Are there populations of pathogens or indicator organisms in the environment (sources) that continually feed other habitats where the pathogens or indicators can be found (often at high densities) but cannot grow (sinks)?

2. What is the fate of freshwater pathogens that are transported into brackish or marine habitats and vice versa? The transition from fresh- to saltwater or the reverse is physiologically demanding, and microbial assemblages change both phenotypically and phylogenetically along salinity gradients. Given that freshwater has been imported into U.S. coastal waters for hundreds of years, along with the propensity of microbes to survive in novel environments, some freshwater pathogens might have adapted to increased salinity and some seawater pathogens might have adapted to reduced salinity. If so, flushes of these now "naturally" occurring bacteria may not be indicative of new inputs from either storms or saltwater intrusion but rather indicative of in situ bacterial growth.

3. Is the residence time of waterborne pathogens and indicators within a body of water sufficient for evolutionary mechanisms to alter the genetic composition of the pathogens? If so, could the genetic changes confound the reliability of the indicators or indicator mechanisms? Before selection can alter the genetics of a microorganism, the selective force must be applied for sufficient time and under the right conditions. Imported pathogens or pathogen indicator species gain or lose genetic traits under natural conditions—traits that may be the basis for detecting various indicators (e.g., β-galactosidase activity).

4. What biotic and abiotic factors influence the viability and survivability of waterborne pathogens and their indicators? Are there environmental conditions that promote genetic exchange or the acquisition of genetic elements that confer selective advantage under clinical conditions? For example, the increases in antibiotic and multiple antibiotic resistances may be influenced by physical conditions in the environment. What is the frequency of genetic exchange among native bacteria and introduced or imported bacteria?

5. What are the effects of sampling regime and environmental variation on the efficacy of indicators (see also Chapters 4 and 5)? Population, community, or genetic changes in space or time increase variability. Measures of statistical central tendency (i.e., means, medians, modes) are important in many aspects of science and ecology. However, because exposures at high extremes pose the greatest human health risks—and because of the immense economic component associated with waterborne pathogens and especially outbreaks (see also Chapter 2), including recreational losses and clinical costs—knowledge of simple means, medians, or modes is insufficient for making informed decisions about human health risks. Environmental variability occurs both spatially and temporally, and to understand ecological phenomena such variance must be estimated.

Many human pathogens and candidate indicators of fecal contamination also infect other host animals. Thus, nonhuman hosts may be the natural reservoirs of human pathogens and indicators. These additional ecological niches of pathogens and indicators have major implications for the following:

• the potential detection, load estimation, and tracking of fecal contamination sources;
• the ability to distinguish among and track or trace microbes of the same genus and species but from different sources;
• the ability of pathogens from different sources to cause infection and illness; and
• the potential for genetic exchange and evolution in microorganisms by coinfection of different strains or genotypes in a host animal or human or in the environment.

Identification of specific sources of pathogens or indicators is impossible unless advanced analytical methods, such as those described in Chapters 4 and 5, are used to genetically or phenotypically characterize the microorganisms. Because the same species of microorganism from different animal hosts or environmental reservoirs can differ greatly in human infectivity and the ability to cause disease, determining risks to human health requires the use of advanced analytical methods that are often well beyond the methods currently used for their detection in environmental waters. Furthermore, the continuous movement of microorganisms through different hosts and abiotic environmental media exerts selective pressures that are opportunities for genetic change leading to the emergence of new strains with different traits and health risks. Current analytical methods used to detect and quantify pathogenic and indicator microbes in water are limited in their ability to distinguish among genetically and phenotypically different organisms and to determine their sources or their human health risks.

Effects of Environmental Change

Environmental change at all scales, from local to global, influences microbial populations and indicator organisms. Large-scale or global changes in weather or climate are predicted to have major effects on waterborne or vector-borne diseases (Patz and Reisen, 2001; Patz et al., 2000). Past and continued alteration of forested areas (e.g., deforestation) and natural waters (e.g., water diversions such as dams and drainages of lakes, river diversions), road construction, commercial and residential development, and other disturbances change the ecological conditions of waterways. These changes often favor introduced over indigenous or "native" organisms at all levels of biological organization and can also result in changes in microbial diversity, the introduction of new or increased levels of pathogens and indicator organisms, and increased opportunities for hu-

---

BOX 3-2
The Cholera Paradigm

Colwell (1996) described the appearance of a new serogroup of *Vibrio cholerae* 0139 in 1992 in India. Cholera has had at least seven pandemics since 1817. This disease often disappears for decades and then reemerges with a vengeance. From 1926 to 1960, cholera was expected never to reach pandemic proportions because of the improvement in water supplies worldwide. Yet nature prevails, and in 1961 a new pandemic began and continues to this day. The responsible biotype of *V. cholerae* was designated El Tor 01. This particular biotype does not cause as severe disease as the classical type. However, in 1992 a new serogroup 0139 emerged in India. Evidence suggests that the new serogroup originates from genetic recombination, horizontal gene transfer, and subsequent acquisition of unique DNA. Furthermore, this new serogroup had completely replaced the *V. cholerae* 01 in Calcutta by 1993.

Various environmental factors have been implicated in the evolution of a new serogroup. The combination of increased inputs of nutrients to eutrophic conditions and association of the organism with shellfish, fish, and zooplankton created environmental reservoirs that could persist for extended periods of time. Thus, reintroduction was not necessary. The association with zooplankton, especially copepods, is central to understanding the dispersal and distribution of cholera. *Vibrio cholerae* preferentially attach to the chitinous exoskeleton of the copepods and thereby have the potential to be transported with ocean currents.

---

man exposure to native pathogens of that environment via water and other routes. Therefore, increases in disease-causing microorganisms would be predicted (see Box 3-2).

For example, certain aquatic ecosystem restoration projects that require construction of wetlands by legislation may affect the growth and distribution of waterborne pathogens. Lake inflows are controlled, in part, by littoral zones or lake margins, and such areas can greatly impact the thermal mediation of small or forested watersheds. Andradottir and Nepf (2000) suggested that littoral wetlands can actually raise the temperature of inflow during the summer and create surface intrusions rather than plunging inflows. In other words, density differences between surface and underlying water would cause warm water to flow above the cooler layers. Consequently, nutrients, contaminants, and pathogens that were previously in the underlying water enter the surface layer, thereby increasing the risk of human exposure in recreational water settings. Furthermore, warmer, nutrient-rich waters may favor growth of pathogens.

Lebaron et al. (1999) have shown that varying nutrient conditions in seawater affect bacterial communities directly and indirectly by stimulating either bacteria or various protozoans that selectively feed on the bacterial assemblage. The stimulation of protozoan fauna may be acute given their interaction with various pathogens (discussed later). In relatively simple mesocosms, bacterial assemblages could be affected by nutrient additions that promote increased growth and productivity. In complex environments, numerous and varied microhabitats (such as organic foams which are described later) exist that may provide the appropriate conditions for changes in microbial assemblages through either direct or indirect selection.

Implications for Indicators

The concept of indicators implies that certain characteristics of an organism (e.g., genes or gene products) are constant under varying environmental conditions. This major assumption is questionable and subject to verification. Although various (primarily bacterial) indicators have been historically effective in detecting and quantifying fecal contamination, they are not always reliable predictors of microbial water quality due largely to our lack of understanding of the basic ecology of waterborne pathogens and indicators. For example, total coliform counts and enterococci have been used as indicators of human fecal contamination for decades (see Chapter 1). However, there are nonhuman and naturally occurring coliforms and enterococci, and their presence confounds the results of the total coliform and enterococci tests.

All coliforms and enterococci do not have the same ecology. If one or more species of coliforms and enterococci had different biotic and abiotic sources and greater or lesser survivability than the indicator species or pathogen of concern, then their presence or absence would not be a reliable indicator of the source or survivability of that pathogen. Similarly, the use of *E. coli* as an indicator of human fecal contamination in areas where there are high numbers of naturally occurring or introduced *E. coli* would greatly overestimate a potential microbial contamination problem. Not recognizing alternative sources of indicator organisms could ignore their potential to detect and correctly characterize actual waterborne microbial contamination problems. More specifically, wastewater treatment processes, physical and chemical stressors, and biological antagonists, such as naturally occurring predators, can selectively affect the presence and survival of one "indicator" species, which in turn affects the implied correlation between the indicator and the target pathogen. Furthermore, gene products such as β-galactosidase or β-glucuronidase may not be produced or may be overproduced under various environmental conditions, thereby affecting indicator technologies based on the detection and quantification of these products.

Microbial species can change genetically under natural conditions in ways that can alter their ability to be detected by phenotypic or genotypic methods.

Some of these changes can be profound, with genomes increasing or decreasing in actual DNA content and changing phenotypic properties. Bacteria in aquatic systems have been shown to take up plasmids at fairly high rates. Fry and Day (1990) demonstrated that maximum uptake occurs within 24 hours but that transconjugants could be detected within the first three hours of their experiments. Recently, high mutation rates have been observed in stationary phase *E. coli* from various natural habitats (Loewe et al., 2003) and stressed aging colonies have also been shown to have increased mutagenesis (Bjedov et al., 2003). Both of these responses could result in increased adaptive responses and emergence of pathogenicity (Loewe et al., 2003). Notably, all of these mechanisms were shown to occur within 24 hours. In natural systems the residence times of introduced bacteria can be much longer than 24 hours, thus providing an opportunity for genetic changes either through acquisition of plasmids or by allowing mutations to take place under the selective pressures of the new habitat.

Various natural history and environmental aspects of pathogens and indicator organisms also contribute to their ability to be detected and monitored. Many of these aspects are discussed below because they directly relate to the ongoing public health challenge of developing and using better indicators for waterborne pathogenic viruses, bacteria, certain parasitic protozoa, and to a lesser extent—yeasts and molds.

VIRUSES

Introduction to Viruses and Their Properties

Virus-host interactions are fundamental to the biology and ecology of viruses because they are obligate intracellular parasites. Viruses are inert outside host cells, despite their persistence in the environment and their ability to infect another host when the opportunity arises. In this section, the ecology and evolution of viruses are considered, particularly for waterborne viruses that are human and animal pathogens or bacterial viruses that are potential indicators of fecal contamination.

Virus Composition, Basic Properties, and Diversity

Viruses are among the smallest and simplest microbes and are obligate intracellular parasites of host cells. They range from about 0.02 to 0.1 μm in size and consist of a nucleic acid surrounded by a protein coat or capsid. The capsid not only is protective but also functions as the structure for host cell attachment leading to infection, because it has specific chemical structures that recognize receptor sites on the host cell. Some viruses, although usually not the ones transmitted by fecally contaminated water, also possess an outermost lipoprotein membrane called the envelope. The envelope is usually a virus-modified host cell membrane containing virus-specific glycoproteins that is acquired as the virus exits the cell. Some of these glycoproteins in enveloped viruses are the chemical structures for attachment to host cell receptors. Viruses contain relatively small amounts of nucleic acid, usually from a few to several tens of nucleotide kilobases—enough information to encode a few to several tens of proteins. Despite this relative paucity of genetic information, viruses are genetically diverse, sometimes highly genetically variable, and quite capable of adapting to the changing conditions of their host cells and the host environment.

Viral Replication, Virus-Host Interactions, and Viral Evolution

The replication and evolution of viruses and their interactions with their hosts are strongly related to host fitness as both viruses and hosts coevolve. The ability of a virus to infect a particular host cell is primarily a function of the availability of the appropriate chemical structures on the surface of the virus and the host cell that allow for attachment to and penetration of the cell. These receptor-dependent interactions determine the virus host range, tissue tropisms (i.e., ability to infect cells of a particular tissue, such as intestinal, liver, or neurological tissues) for human and animal hosts, and thus the ability to cause certain kinds of infections and diseases. Despite the importance of cell surface receptors in the susceptibility of different cells or tissues to viral infection, the outcomes of viral infection— especially disease—are often mediated by additional events and other molecular interactions during virus replication (Bergelson, 2003; Dimitrov, 2000; Jindrak and Grubhoffer, 1999; McFadden, 1996; Mims et al., 2001; Ohka and Nomoto, 2001; Tyler and Nathanson, 2001).

Several outcomes of viral infection of host cells are possible: (1) virus multiplication leading to many progeny viruses with resulting cell lysis and death; (2) virus multiplication leading to many progeny viruses but cell survival; and (3) development of a stable relationship (at least temporarily) with the host cell with little or no virus multiplication—either as a discrete intracellular genetic element or as an integrated part of the host cell's genetic material. In the last situation, the virus genetic information is propagated as part of the cell when it divides, and a relationship of co-existence between the cell and the viral genome may form (lysogeny). Under some circumstances, however, the virus genetic material can become capable of initiating replication activities of the viral genome, leading to the production of progeny viruses, lysis, and death of the cell (the lytic cycle). In some cases, the course of the alternative events in viral infection and virus-host interaction, lysogeny (or integration) or the lytic (or cytopathogenic) cycle, are influenced by a number of virus, host, and environmental factors, such as temperature, pH, UV irradiation (sunlight), nutrients, and antagonists (toxicants).

At the human or animal host level, factors influencing the activation of latent viruses to a more active cytopathogenic cycle of events in virus infection and disease can include immune status, hormone levels, chemical (nutritional) cofactors, age, gender, and pregnancy. Therefore, the potential for, or likelihood of, viral infection and the potential outcomes of viral infection are complex and not easily predicted. In fact, some of the most studied viruses (e.g., hepatitis) are still not well understood, making reliable predictions of viral infection and disease outcomes at either the cellular or the population level difficult, if not impossible. Despite the variability and uncertainty of predicting waterborne virus infection and disease outcomes, studies of virus properties, virus-host interactions, virus infection and disease outcomes, and viral ecology and epidemiology have all helped to elucidate the natural history of viruses and virus risks to their hosts.

Virus strains that produce infectious viruses more rapidly and at higher yield are more likely to be successful if fitness is positively correlated with population size of the susceptible host. For many viruses the manifestation of disease in the host is rare, and most infections are unapparent or subclinical. Examples of such viruses are the polioviruses and the rotaviruses. Typically, these viruses infect the youngest members of the population who have previously not been infected. Unfortunately, such infections produce severe disease or death in a small proportion of the humans they infect, and the majority of infections in infants and young children are either subclinical (polioviruses) or mild and self-limiting (rotaviruses). However, the consequences of poliovirus infection are considered sufficiently profound in the small proportion of infected persons who develop paralytic disease or die that vaccination is considered essential and a global eradication for polio is under way by the World Health Organization (Hull and Aylward, 2001). Repeated rotavirus infections are common in infants and young children though most infections are not life-threatening, especially in healthy children in developed countries. However, rotavirus diarrhea does cause severe disease requiring hospitalization in a low proportion of infected infants and children in the United States and other developed countries (<1 percent of rotavirus infections) and there is a very low but non-zero risk of death from rotavirus infections (Parashar et al., 2003).

Hosts that recover from virus infections are immune to future infections, either temporarily or perhaps indefinitely. In the case of rotaviruses, immunity is transient, only partially protective, and even less protective against antigenically different rotaviruses that have considerable antigenic diversity (Jiang et al., 2002). In the case of polioviruses, infection is likely to result in long-lasting immunity that is protective against paralytic disease and mortality, although enteric infections that are subclinical or mild still occur in persons with immunity (Ghendon and Robertson, 1994). If primary (initial) poliovirus infection of a susceptible host does not occur until later in life, as an older child or an adult, the consequences of infection are likely to be severe disease or even death. For polioviruses, infection of infants and children is common in developing countries where poor sanitation and hygiene result in exposure early in life. However, in developed countries with improved hygiene and sanitation, virus exposure often does not occur until later in life so that the likelihood of severe disease and death as a result of infection is much greater (Evans, 1989; Pallansch and Roos, 2001; White and Fenner, 1994). The above examples serve to highlight the importance of host status and environmental conditions in the ecology and natural history of viruses, and to demonstrate that the "virulence" or pathophysiology of a virus depends on the status of the host and its environment.

Another example of the role of the host and its environment in the outcome of virus infection is hepatitis E virus (HEV). In developing countries, the members of the population at highest risk of severe illness and death are pregnant women. The mortality rate in this group can be as high as 25 percent (Aggarwal and Naik, 1997; Balayan, 1997; Emerson and Purcell, 2003; Hyams, 2002; Krawczynski et al., 2001). Yet, for most of the population in developing countries, HEV infection apparently occurs relatively early in life, with little illness incurred. Children are often asymptomatic and the mortality rate is between 0.1 and 4 percent (Grabow et al., 1994). Seroprevalence of HEV in developing countries ranges from 5 to upwards of 20 percent (Kamel et al., 1995; Mohanavalli, 2003). In developed countries such as the United States, HEV infection is rare and results in very few cases of disease (most traced to probable virus exposures in developing countries); seroprevalence is less than 5 percent (Bernal et al., 1996; Redlinger et al., 1998). Therefore, as with many other viruses, the pathophysiology of HEV varies with the health status of the host and with environmental conditions.

Viral Genetic Variability and Genetic Change

Viruses have evolved a variety of mechanisms that influence their host interactions and their ability to persist over time and in space. Viruses mutate spontaneously and without direct exposure to physical and chemical mutagens during replication in host cells. Mutation rates vary among different virus groups from high rates of 10^{-3} to 10^{-4} per incorporated nucleotide in the single-stranded RNA viruses to rates as low as 10^{-8} to 10^{-11} per incorporated nucleotide in some of the double-stranded DNA viruses (Domingo et al., 1999).

Genetic changes in viruses that involve relatively minor substitutions, insertions, or deletions of nucleotides as point or frameshift mutations can occur. Such changes are often referred to as genetic drifts, and if they occur in an expressed gene these changes are referred to as antigenic drift. Genetic and antigenic drifts can occur in response to selective pressures from host populations, such as immunity and genetic changes in host cells and whole hosts such as animals and plants. In some cases, genetic drift leads to more benign relationships between viruses and their hosts. At the other extreme, it can result in viruses with properties that have severe consequences, such as the reversion of attenuated poliovirus vaccine strains to a neurovirulence and the ability to cause paralytic disease in human hosts.

Effects of Virus Mutation on Hosts: Poliovirus Virulence, Attenuation, and Reversion to Virulence

Polioviruses are single-stranded RNA viruses belonging to the Picornaviridae family and the Enterovirus genus, and they consist of three genetically distinct types (I, II, and III). These viruses infect the gastrointestinal tract initially and can then spread via the bloodstream and lymphatic system to the central nervous system, thereby causing paralysis in their human hosts. The virus-specific factors responsible for the neurovirulence of polioviruses are still not fully understood at the genetic, protein, or virion (whole virus particle) level. Neurovirulence is mediated by the ability of the virus to successfully infect neurons and cause high levels of virus production and subsequently death of these cells (Ohka and Nomoto, 2001; Pallansch and Roos, 2001; Racaniello, 2001). Paralytic disease depends on the ability of the virus to infect cells of the central nervous system efficiently. The risks of paralytic disease to humans posed by wild-type, neurovirulent polioviruses, led to the selection of avirulent or attenuated polioviruses as vaccine strains in the mid-twentieth century. These live oral poliovirus vaccine strains differ from wild-type viruses because they have several different point mutations that are associated with the ability to infect neural cells. However, despite thorough knowledge of the complete nucleotide sequence of polioviruses for two decades, the cloning and expression of the cell surface receptor of the virus, the development and use of a transgenic (genetically modified) mouse model for neurovirulence, and considerable effort to identify neurovirulence mechanisms in cell culture and animal systems, these mechanisms have not been fully elucidated. However, it is becoming clear that neurovirulence depends on host factors as much as virus-specific factors and that virus-host interactions leading to neurovirulence are probably modulated by the host (Ohka and Nomoto, 2001; Yoneyama et al., 2001).

The attenuated live oral vaccine strains of poliovirus are also subject to back-mutations that cause reversion to wild-type viruses and paralytic poliomyelitis in vaccine recipients. Because virus mutation rates are high, there is rapid reversion of vaccine polioviruses to genotypes with neurovirulent properties among the excreted viruses of vaccine recipients. Serial transmission of vaccine strains of polioviruses among susceptible human hosts results in the accumulation of mutations, which can eventually lead to selection and further serial transmission of neurovirulent vaccine strains. This highly unfortunate outcome occurs when there is inadequate vaccine coverage of susceptible hosts over time, as occurred recently in the Dominican Republic and Haiti on the island of Hispaniola, the Philippines, and several other locations globally (Anonymous, 2002, 2003; Friedrich, 2000; Landaverde et al., 2001). Based on the extent of genetic change (about one to three percent), these viruses had apparently been spreading from person to person over one to two years or more.

Virus Mutation and Evolution by Exchange or Acquisition of Genetic Material

In addition to spontaneous point and frameshift mutations, the genetic composition of viruses can be altered by a number of mechanisms that involve virus-virus-host cell interactions (Domingo et al., 1999; Hendrix et al., 2000; Kaaden et al., 2002). That is, genetic changes in viruses can occur when two or more viruses coinfect host cells and exchange genetic information during replication. The genetic changes can involve major changes or substitutions in whole genes, genomic regions, or genome segments by mechanisms such as recombination and reassortment. Such changes can result in genetic and antigenic shifts that often have profound consequences for the natural history of viruses and their hosts. Examples include (1) the emergence of new strains of pandemic human influenza viruses by the creation of reassortant viruses from avian and human viruses by co-infection of swine, and (2) and the development of new strains of rotaviruses, either in nature or by experimental methods to produce reassortant rotavirus vaccine strains of human and either bovine or monkey origin (Baigent and McCauley, 2003; Bishop, 1996; Jiang et al., 2002; Webby and Webster, 2003).

Although viruses are often viewed as discrete entities that infect and interact with host cells alone, they can engage in genetic exchange and reproduction, directly or indirectly, within their host cells. Through coinfection, there can be evolutionary, cooperative, and competitive interactions among viruses. Intracellular interactions between coinfecting viruses are shown to be important in disease progression (e.g., herpesviruses, HIV; Holmes, 2001; Papathanasopoulos et al., 2003), and entire families of viruses rely on coinfection to complete their life cycle successfully (e.g., geminiviruses that infect plants; Gutierrez, 1999; Hanley-Bowdoin et al., 2000).

Genetic Recombination

In genetic recombination, coinfection with two different viruses results in the formation of new viruses whose genomes contain portions of each infecting virus that were created by the "crossover" event. Genetic recombination of polioviruses with other enteroviruses apparently occurred in the reversion of some vaccine strains to neurovirulence on the island of Hispaniola (Kew et al., 2002). Some of these viruses not only had back-mutations at critical sites associated with neurovirulence, but also had recombinations with other enteroviruses that may have increased their transmission rates in human hosts.

Viral Genetic Change and Evolution by Reassortment

Virus reassortment occurs when two or more viruses with segmented genomes simultaneously coinfect a host cell. The genomic units produced during replication are packaged randomly into virions, resulting in the formation of new progeny with combinations of genomic segments from each infecting virus. One of the best-studied examples of this phenomenon is the influenza A virus (Hay et al., 2001; Scholtissek, 1995; Webster et al., 1993). Reassortant strains produced when human and avian strains apparently coinfect pigs can have new combinations of surface antigens from each parent virus. These new hybrid viruses periodically emerge as pandemic strains. Similar antigenic shifts created by reassortment also occur in the enteric viruses known as rotaviruses (Bishop, 1996).

Viruses in Human and Animal Wastes and in the Aquatic Environment

Enteric viruses found in human and animal feces, sewage, and fecally contaminated water include not only enteric pathogens but also viruses that infect bacteria residing in the intestinal tracts of humans and other warm-blooded mammals that are called enteric bacteriophages. Some fecally shed viruses are respiratory pathogens that have been swallowed with respiratory exudates, that actually infected the enteric tract, or both. The aquatic environment also contains many other viruses that infect a variety of aquatic and terrestrial life ranging from prokaryotes to protozoans to plants and animals. The viruses shed in feces and present in sewage belong to a diverse range of taxonomic groups that have different genetic, morphological, and functional properties. Of the human enteric viruses, some belong to taxonomic groups containing single-stranded RNA (enteroviruses, caliciviruses, hepatitis A and E viruses, astroviruses, and coronaviruses); double-stranded, segmented RNA (reoviruses and rotaviruses); bisegmented and double-stranded RNA (picobirnaviruses); single-stranded DNA (parvoviruses); or double-stranded DNA (adenoviruses). The bacteriophages found in feces, sewage, and ambient water, while not pathogenic, are genetically and morphologically diverse.

Animal Reservoirs as Sources of Human Enteric Viruses

As noted previously, many human viral pathogens, including some waterborne enteric pathogens, also infect other animals and therefore have animal reservoirs (Enriquez et al., 2001; Weiss, 2003). These animals can potentially be important sources of virus released into aquatic environments leading to human exposure. More often than not, a particular virus infects only one animal species, however, there are some notable exceptions. For example, of the enteric viruses, reovirus type 3 can infect humans as well as a wide range of other mammals, including mice (Cohen et al., 1988). Reovirus 3 is an example of a virus that infects but causes little morbidity or mortality in its human hosts. Other human enteric viruses that infect animals are rotaviruses, hepatitis E virus, and probably caliciviruses (Desselberger et al., 2001; Emerson and Purcell, 2003; Enriquez et al., 2001; Smith et al., 2002). Caliciviruses that infect cattle and swine are genetically similar to certain subgroups of human caliciviruses. Porcine hepatitis E viruses are very similar to human hepatitis E viruses. Human HEV strains and porcine HEV strains have infected pigs and primates, respectively, in experiments (Clayson et al., 1995; Emerson and Purcell, 2003).

In addition to being reservoirs of human enteric viruses, animals also harbor enteric bacteriophages that are potential indicator viruses of fecal contamination. Somatic and male-specific coliphages (bacteriophages of *Eshcerichia coli*), *Salmonella* phages, and *Bacteriodes fragilis* phages can be found in human and animal feces. The viruses apparently infect the intestinal bacterial flora of a variety of feral, domestic and agricultural animals. The use of bacteriophages as indicators of fecal contamination of water has been considered seriously (see Chapter 4 for further information), and there is evidence that shows their predictive value for enteric viruses and fecal contamination by correlations between presence and levels of enteric viruses and bacteriophages and associations of bacteriophages with increased risks of viral illness (Chung et al., 1998; Havelaar, 1993; Wade et al., 2003).

There also appear to be genetic differences in the host ranges of at least some coliphages, such as the RNA and DNA containing male-specific (F+) coliphages. These differences in host range are dependent in part on host cell factors related to coliphage adsorption to the F pili of the host as well as other host-related factors during later events in virus replication (Miranda et al., 1997; Schuppli et al., 2000; Tomoeda et al., 1972). Of the four major subgroups of the F+ RNA coliphages, two of them (Groups II and III) are found primarily in human feces and sewage, one (Group IV) is found primarily in animal feces, and the last (Group I) is found in both human and animal feces and sewage (Hsu et al., 1995). Therefore, the ecology or natural history of at least some enteric bacteriophages appears to be related to the animal host of their host bacteria. The apparent animal-host specificity of these bacteriophages may be related to the bacterial host ranges of the phages themselves or to the animal host ranges of their bacteria, although the ecological aspects of these relationships have not been adequately studied. Regardless of the mechanisms, the diversity of enteric bacteriophages and their bacteria, and their occurrence in human and animal hosts, pose challenges to the development and application of bacteriophages as indicators of enteric viruses and fecal contamination. This is because the extent to which coliphages are able to reliably and quantitatively indicate the amount of fecal or sewage contamination in water depends on the concentrations and types of coliphages in different sources of fecal contamination or sewage, the absolute and relative stability, persistence, and resistance of the coliphages to water treatment processes, and the extent to which their properties can change depending upon the strain of host bacterium and its human or animal host. These factors influencing coliphage occurrence and properties are still being elucidated.

Stability, Survival, Effects of Physical and Chemical Agents, and Transport of Viruses

Some of the important properties of enteric viruses and bacteriophages that influence their environmental behavior and natural history include their small size, stability over a wide temperature and pH range, resistance to various chemical agents such as oxidants and proteolytic enzymes, and propensity to aggregate and adsorb to particles and surfaces. These properties allow some enteric viruses in feces and sewage to survive conventional sewage treatment processes and persist in environmental waters and their associated sediments.

Conventional sewage treatment systems employing primary and secondary treatment reduce enteric viruses by about 90 to 99 percent in the treated effluent (Leong, 1983). Many of the viruses removed from the effluent remain infectious in the resulting sludge or biosolids, which must be treated further to reduce the viruses and other pathogens (see also NRC, 2002). Chemical and physical disinfection processes vary greatly in their ability to inactivate enteric viruses. Appreciable virus reduction in sewage is achieved only when well-treated effluent is disinfected with free chlorine, ozone, chlorine, or high doses of UV radiation and/or when viruses are physically removed or inactivated by certain advanced wastewater treatment processes, such as membrane filtration or chemical coagulation. Because municipal sewage is often disinfected only by combined chlorine (a relatively weak oxidant), discharged sewage effluents often still contain relatively high concentrations of viruses (Griffin et al., 2003). Furthermore, sewage treatment plants often must bypass untreated sewage during wet weather by design, and many urban sewage systems still discharge such combined sewer overflows directly to receiving waters.

Because on-site wastewater treatment systems, typically septic tanks and subsurface drainfields, often inadequately reduce viruses and the wastes of feral, domestic, and agricultural animals are either untreated or inadequately treated, they can deliver substantial numbers of enteric viruses and other pathogens to ground- or surface waters (Borchardt et al., 2003; Powell et al., 2003; Scandura and Sobsey, 1996). Enteric viruses have been found on occasion in both surface and groundwaters used as drinking water sources and for primary contact recreation (Bellar et al., 1997; Donaldson et al., 2002; Hot et al., 2003; Jiang et al., 2001; Lipp et al., 2002; van Heerden et al., 2003).

Summary

Despite their relatively small size, limited genetic information, and relatively simple composition and structure, viruses are biologically complex, diverse, and highly adaptable to different environments and hosts. As obligate intracellular parasites, viruses multiply only in specific hosts. However, their host ranges can be either limited or broad and can change over time and space. Some human viruses also have animal reservoirs, and therefore animals can be sources of human viruses. Viruses evolve over time and do this by coevolving with their hosts. Virus-host interactions are complex and diverse, and they can have different outcomes ranging from virus proliferation with the death of the host to integration of the viral genome into the host cell without virus proliferation beyond cell division. Different viruses have different rates of mutation, but all viruses display genetic variability over time. Mutations can be minor (e.g., point mutation) and lead to genetic drift, or they can be major (gene substitutions or replacements) and lead to a genetic shift. Gene substitutions or replacements can occur by genetic recombination or reassortment when two or more viruses infect the same host cell. Furthermore, animal viruses and human viruses can coinfect cells to create new viruses (by recombination or reassortment) that are infectious to humans and have some properties from each original virus. Both minor genetic changes causing genetic drift and major genetic changes causing genetic shift can have profound effects on the relationships of viruses to their hosts. Such mutations can alter their virulence, either causing virulent viruses to become nonvirulent or the reverse.

Many human viruses can infect the enteric or respiratory tract, or both, and are a concern from exposures to contaminated water and other environmental media. Although they are inert in the environment, viruses can be stable, persist for long periods of time in environmental media, and be resistant to various physical and chemical agents, including disinfectants. In addition, viruses are so small that they are readily transported in water and wastes and can migrate through soils and other porous media. The persistence and transport of human enteric viruses in water and other environmental media constitute a public health concern because the viruses can retain their infectivity and cause human infection if humans ingest or otherwise come in contact with them in environmental media.

BACTERIA

Introduction to Bacteria and Their Properties

Bacterial waterborne pathogens and indicators vary in size from 0.2-2 μm and fall into at least two major groupings: (1) native opportunistic pathogens such as *Aeromonas* spp. and *Mycobacterium* spp. and (2) introduced pathogenic bacteria that are not "normally" found in a particular water system (e.g., *Shigella*) or other bacteria often found only at relatively low concentrations in natural waters and other environmental media (e.g., *Legionella, Clostridium*). It is important to note that waterborne bacterial pathogens and indicator organisms are only one small component of any aquatic microbial community which may also include heterotrophs, autotrophs, chemotrophs, and saprophytes. Furthermore, certain Gram-positive waterborne bacteria under certain environmental conditions can form endospores. With no metabolic activity, these specialized cells are able to survive extended periods of time in the environment compared to vegetative bacterial cells.

However, some "introduced" waterborne pathogenic bacteria can often be isolated from nearly pristine systems, thereby suggesting some "natural" low density (Fliermans et al., 1981; Hazen and Fliermans, 1979). Natural densities of pathogens are difficult to ascertain since most systems receive imports of bacteria through surface runoff from precipitation events, atmospheric dryfall, vertebrate and arthropod transport, and human activities. In highly disturbed systems, such as agriculture or water treatment discharges, imports of pathogenic bacteria would be expected to be much higher. For example, Lalitha and Gopakumar (2000) in a study of freshwater and brackish sediments, shellfish, and native fish in India found that 21 percent of all sediment samples contained *Clostridium botulinum*, 22 percent of the shellfish harbored *C. botulinum*, and between 2 and 8 percent of indigenous fish had *C. botulinum* on their surfaces.

Although some pathogenic bacteria exclusively inhabit humans, most also have environmental biotic reservoirs (are zoonotic), and these reservoirs can be important in the transmission of pathogens to other hosts. For example, a bacterial genus that has a substantial biotic habitat is *Campylobacter*. Both *C. jejuni* and *C. coli* are human gastrointestinal pathogens that are the major cause of bacterial diarrheal illness in many developed countries, and such outbreaks can be waterborne or foodborne (Rheinheimer, 1992). Waterborne outbreaks have been associated with community water supplies or untreated spring water, in which *Campylobacter* cells are viable for months. Outbreaks have also resulted from foods such as raw milk and poultry, although improper food handling is thought to account for the majority of endemic *Campylobacter* disease in the United States. *Campylobacter* is carried in a wide range of mammalian hosts, such as rabbits, cows, sheep, pigs, and chickens, as well as wild birds such as crows, gulls, pigeons, and migratory waterfowl. *Campylobacter* can be transmitted from aquatic sources to animals by direct contact or via carriage by birds or flies, and then spread between animals.

Aerobic Gram-negative bacteria, frequently found in water sources, are a common cause of hospital infection, particularly in intensive care units. Multidrug-resistant *Pseudomonas*, *Enterobacter*, *Acinetobacter*, *Klebsiella*, and *Stenotrophomonas* are particularly problematic (Denton and Kerr, 1998; Hanberger et al., 1999). These microorganisms are widespread in aquatic environments and may be introduced into hospitals by patients, staff, or visitors and become established in microenvironments such as sinks, showers, and ice machines. Apart from sporadic infections and outbreaks occurring in recreational or hospital settings, the ultimate sources of these microorganisms are not well known. Antibiotic use for growth promotion in animal agriculture and for treating infections in humans and agricultural animals accounts for the greatest amount of commercial antibiotic production in the United States (Levy, 1997, 1998). Antibiotics and other pharmaceutically active compounds have been found in ground- and surface waters, especially near human and agricultural animal waste sources, leading to further concerns about the selection of antibiotic-resistant bacteria in the aquatic environment, which is discussed in the next section.

Similarly, potentially pathogenic bacteria such as *Aeromonas*, *Escherichia coli*, or *Salmonella* all have substantial environmental reservoirs. *Aeromonas* species are frequently found in aquatic environments, and certain pathogenic strains (possessing specific virulence properties) cause human disease. However, the distinctions between nonpathogenic environmental strains of *Aeromonas* found in water and the pathogenic clinical strains of *Aeromonas* isolated from humans have not been established adequately, although some pathogenic strains have been isolated from water (Haburchak, 1996; Hazen and Fliermans, 1979). The spread of *E. coli* and *Salmonella* among human populations is mediated via foods contaminated by animal products. Notably, a dramatic increase in multidrug-resistant *Salmonella typhimurium* (phage type DT104) has been observed in the United States and the United Kingdom (CDC, 1997). Like *Campylobacter*, *S. typhimurium* is a ubiquitous zoonotic bacterium in nature and is found in wild birds, rodents, foxes, badgers, poultry, cattle, pigs, and sheep.

Bacteria have at least three novel evolutionary mechanisms that can facilitate their rapid response to many environmental changes through alteration of their genetic composition: (1) conjugation, (2) transduction, and (3) transformation (see Box 3-3). Of these, plasmid-mediated conjugation is the most common, though several bacterial genera, including *Campylobacter*, are naturally competent for DNA uptake through transformation (Wommack and Colwell, 2000). One example of the entry of foreign DNA into *Campylobacter* is a gene encoding for resistance to the antibiotic kanamycin, which was first identified in an *E. coli* strain also resistant to the antibiotics ampicillin, tetracycline, chloramphenicol, streptomycin, and erythromycin. The DNA sequence is identical to that from *Enterococcus faecalis*, and indicated the transfer of this resistance determinant from Gram-positive enterococcal or streptococcal bacteria to the Gram-negative *C. coli*. The gene was also found in *C. jejuni*, indicating the subsequent dissemination of kanamycin resistance among *Campylobacter* species. In some cases, plasmids conferring kanamycin resistance also provided resistance to tetracycline and chloramphenicol. The issue of antibiotic resistance in bacteria is discussed in the following section.

Antibiotic Resistance

Our understanding of the mechanisms that promote the selection and transmission of bacterial genes under various environmental conditions is critical to addressing long-term public health problems. For instance, exposure to heavy metals at concentrations above background may influence the frequency, abundance, and types of antibiotic resistance genes available in the environment, and these genes could subsequently be transmitted to waterborne pathogens of public health importance (McArthur and Tuckfield, 2000).

Esiobu et al. (2002) have shown that *Pseudomonas*, *Enterococcus*-like bacteria, and *Enterobacter* and *Burkholderia* species are the dominant reservoirs of certain antibiotic resistance genes in soil and water environments. Patterns of resistance were correlated with the abundance and types of bacterial species found in the various habitats. Movement of genes between and within these taxa has been demonstrated (Davison, 1999). Similarly, movement of resistance genes has been demonstrated between various "native" taxa and introduced bacteria (e.g., opportunistic and frank pathogens such as *Aeromonas* and *Campylobacter*, respectively).

While selection for tolerance or resistance to antibiotics from exposure to antibiotics is considerable in clinical and animal agricultural environments, there is increasing evidence that resistant phenotypes are being selected for in natural environments (Seveno et al., 2002). Contributing to the evolution of such resistance are transposons, which allow the movement of genes within cellular genomes and onto plasmids and bacteriophages where they can be more easily spread to neighboring cells (Liebert et al., 1999). Thus, the overall problem of antibiotic resistance and its impact on waterborne pathogens and indicators is one of genetic ecology (Mazel and Davies, 1999). An understanding of genetic ecology would require studies on the transfer of various genes under natural conditions as well as under stressed or disturbed conditions.

Kadavy et al. (2000) found high levels of antibiotic resistance in obligate commensal bacteria associated with flies living in the asphalt seeps of the Le Brea tar pits in California. They suggested that exposure to elevated levels of naturally occurring solvents may have resulted in the indirect selection of antibiotic resistance and that these bacteria are an environmental reservoir of antibiotic resistance genes. Selection acting on one set of genes (e.g., metal tolerance) may indirectly increase levels of other unrelated but linked genes. Such linked genes would then be available for transfer to other bacteria including waterborne pathogens.

Biological Interactions

Environmental Reservoirs

Critical to understanding the ecology of waterborne pathogens and indicators organisms is knowledge of various niches and habitats that promote or safeguard these microorganisms while they reside in a waterbody. Recent studies have shown unique biological interactions between certain prokaryotic and eukaryotic pathogens and other proto- and metazoans (Barker et al., 1999; Steinert et al., 1998). Winiecka-Krusnell and Linder (1999) have shown that free-living amoebae—which are well adapted to harsh or changeable environments such as desic-

BOX 3-3
Novel Evolutionary Mechanisms for Bacteria

At least three different mechanisms have been observed for the spread of genetic material among environmental and clinically relevant bacteria: conjugation, transformation, and transduction.

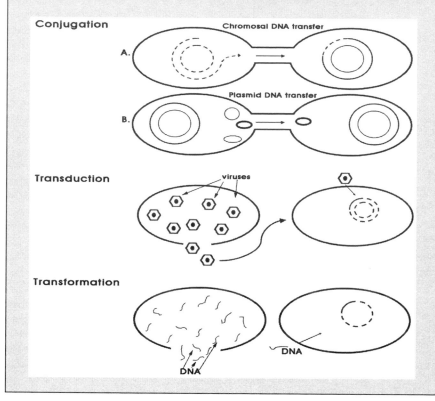

cation, elevated temperatures, and disinfectants—harbor bacteria intracellularly. Some bacteria can thus prevent intracellular destruction and can grow and survive within protozoa, finding both protection from adverse environmental conditions and protected modes of transportation. This interaction may also enhance their infectivity in mammals (Harb et al., 2000). For example, endosymbiotic or parasitic relationships between *Legionella* bacteria and their free-living algal and protozoan hosts allow not only for bacterial proliferation but also for protection from disinfection, thereby increasing their survival and ability to reach human hosts through drinking, recreational, and cooling tower waters. Therefore, proto-

1. *Conjugation.* Broad host-range conjugative plasmids can be transferred to a large number of Gram-positive and Gram-negative bacterial genera (Davison, 1999). Conjugative plasmids are typically large and can carry many different types of bacterial genes at once, including those for substrate metabolism, DNA repair, and resistance to heavy metals and/or antibiotics (R-plasmids). Conjugative plasmids therefore have a high capacity for disseminating plasmid-encoded traits throughout the environment. Some genes that confer antibiotic resistance are carried on plasmids.

2. *Transduction.* A second possible means for transmitting genetic material among bacteria is via bacteriophages (phages). Phages are extremely abundant in nature and in fact outnumber bacteria in aquatic systems (Wommack and Colwell, 2000). Lysogenic infection by phage is increasingly being implicated in the transfer of bacterial virulence factors and has been shown to encode such phenotypes as serum resistance, toxin expression, and host cell adherence or modification (Miao and Miller, 1999).

3. *Transformation.* Another possible mechanism for the spread of novel or new genes is natural transformation. Many bacteria are capable of uptake and incorporation of exogenous DNA and transfer of DNA among bacterial pathogens is well established. Transfer of antibiotic resistance by transformation of *Pseudomonas* and *Acinetobacter* with free DNA (in soil) has been demonstrated (Nielsen et al., 1997; Stewart and Sinigalliano, 1991). Transformation can also occur in aquatic environments, as with *Acinetobacter* grown in biofilms on sterile stones and dipped into a river, where free DNA concentrations would be expected to be relatively low (Williams et al., 1996). Natural transformation therefore appears to be a relatively efficient process under certain conditions and has the potential to allow the spread of genetic determinants such as metal and antibiotic resistance.

zoa play a role in the transition of bacteria from the environment to mammals including humans. In this regard, protozoa may be viewed as "biological gyms" where bacterial pathogens train for encounters with more evolved mammalian cells (Harb et al., 2000).

Bacterial "Trojan horses" thus become a mechanism for immediate survival and long-term reserve. Indeed, some anaerobic bacteria can survive and replicate under aerobic conditions in amoebae. In Tomov's study (Tomov et al., 1999), free-living *Mobiluncus curtisii* did not replicate and died in four to seven days whereas those grown with amoebae increased up to 1×10^6 colony forming units

(CFU) per mL over the same duration. If any single amoeba becomes infected with more than one strain or species of bacterium, the probability of gene exchange between bacteria increases considerably because of the increased probability of contact. Furthermore, this unique niche provides a mechanism for bacterial pathogen replication under normally adverse or inhibitory conditions.

Protozoa are not the only biological environmental reservoir for pathogenic organisms. Freshwater and marine mussels have been shown to harbor various bacterial pathogens. *Vibrio cholerae* non-O1, *Salmonella typhi*, *Escherichia coli*, and *Vibrio harvey* showed differential retention within a marine mussel under different environmental and culturing conditions (Marino et al., 1999). Such associations provide respite from selection imposed on free-living bacteria and increase the probability of gene exchange between strains, species or other taxa. Biological indicators of these pathogens that fail to identify these environmental reservoirs will be of little efficacy in tracking potential impacts or outbreaks.

Ecology of Plasmids

Bacteria in nature can and do acquire and lose genetic material through a variety of novel evolutionary mechanisms. Pathogenic bacteria introduced into aquatic systems could in theory, and do in practice, alter their genetic composition using these same mechanisms (see Box 3-3).

Although numerous papers and reviews have described the roles and exchange of plasmids, little attention has been given to their ecology (Sobecky, 1999). In fact, certain plasmids can be culled after environmental change wherein the benefit provided is no longer selectively advantageous. Plasmids confer varying levels of plasticity on cells and on entire microbial communities (Sobecky et al., 1997, 1998). Given the (re)emergence of new and old pathogens and related diseases (see Appendix A), it seems imperative to understand the acquisition and dissemination of numerous and diverse "natural" plasmids. Do bacteria "sample" available plasmids as an effective "hedge-bet" against future environmental change? What is the evolutionary cost for pathogens that take up environmentally derived plasmids? These and other questions have to be addressed so as to better monitor pathogens and bacterial indicators in the environment and enhance our ability to detect important strains or closely related, but nevertheless significantly different, bacteria.

Effect of Biodiversity on Pathogenic Microorganisms

Numerous studies have been undertaken to determine the effect of biodiversity on emergent properties of various systems. Biodiversity and evenness of bacterial species together may be an indicator of the overall condition of a particular system. For example, certain waterborne pathogens can be expected to be found directly below wastewater outfalls or feedlots. Outfall microbial biodiversity may be significantly reduced and the evenness may be skewed by pathogens contributed by the discharge. However, "indicator" microorganisms have been found to grow in uncontaminated systems under appropriate conditions (Gauthier and Archibald, 2001), and caution must be used in interpreting results for such indicators. Furthermore, as discussed previously, the biodiversity of higher organisms, especially protozoa, may facilitate the growth of pathogens. Most bacteria in transport in lotic aquatic ecosystems (e.g., stream riffles) are not active (Edwards et al., 1990) because the doubling times of the bacteria are slower than the flow rate of the water and selection would be incapable of causing changes in transported bacteria. Thus, transported bacteria may not be in a given location long enough for selection to act, unless the waterbody is static (i.e., lentic) or the introduced microorganisms are deposited in sediments. Additionally, attached bacteria and endosymbiotic bacteria would be subject to selection for extended time periods.

It is not always clear how long attached bacteria remain. Do strains or species that colonize persist or are they replaced by another species in the same way that terrestrial plant species replace one another in secondary succession? Wise et al. (1997) demonstrated that a particular strain of *Burkholderia cepacia* was maintained for at least 16 days in the biofilm of a blackwater, organically stained stream, but it is not known for how long that particular strain was present prior to sampling. However, between days 16 and 32 the dominant strain of *B. cepacia* was replaced by a genetically different strain. Although some evidence shows that biofilm development and maintenance follows a repeatable and predictable pattern, with certain groups of bacteria appearing and supplanting or replacing others, the details have not been adequately elucidated in ecological and evolutionary terms. Waterborne pathogens and indicators can integrate into biofilms under some conditions, and such incorporation could lead to protection, proliferation, and opportunities for gene exchange among different biofilm microbes. Furthermore, if pathogens can become integrated into biofilms and retained for sufficient time, they would be subject to selection.

Bacterial Persistence in the Environment and Detection by Culture Methods

The extent to which pathogenic and bacterial indicator organisms persist outside a vertebrate host is highly variable and depends on the type of pathogen and the environmental conditions (Mitchell, 1972). Evidence shows that there are numerous reservoirs in which these organisms can persist and even increase in number (see discussion above). The problem of new or reemerging diseases is due, in part, to evolution and selection of pathogens, which in turn is caused by changes in water quality. These changes include phenomena such as inputs of novel organic substances, disruption of natural linkages, removal of riparian zones, channelization, and removal of instream habitats (e.g., debris dams)—all of which combine to affect the biotic and abiotic interactions that have evolved for millennia. Bacteria and eukaryotes alike are then subject to new "harsh" environments. Interactions among species that normally do not occur have resulted in the panmixis of various genes and gene combinations (i.e., integrons and other transposable elements).

Other factors that must be considered regarding bacterial persistence in the environment, especially for bacteria from human or animal reservoirs, is the extent to which they are subjected to environmental stresses (such as extreme temperatures and pH levels, exposure to UV radiation in sunlight and toxic chemicals) that cause physiological stress and damage that is generally termed "injury." Injury can range in severity and the effects of such injury influence bacterial detection by culture and other methods, as well as bacterial infectivity for human or animal hosts.

Kurath and Morita (1983) called cells that could grow on media viable, but they recognized that most of the bacteria in their samples (>10 times the number of CFUs) had metabolic activity but did not grow on the culture plates. Bacteria that become injured by losing the ability to multiply (form colonies or grow in liquid media), but remain otherwise completely functional as individuals and metabolically active, have been termed "viable but non-culturable" (VBNC; Oliver, 1993). This condition may be due to nutrient deprivation or to the effects of a variety of environmental stresses (Roszak and Colwell, 1987). Many types of bacteria that are injured to varying extents and may be VBNC can be identified from samples using biochemical, immunological, and nucleic acid molecular techniques. Important unresolved questions about VBNC bacteria are what ecological role they play and whether or not they are infectious for human or animal hosts. In this regard, the mere presence of a bacterium, especially when detected by non-culture methods, does not necessarily imply ecological importance (Morita, 1997) or human health risk.

Several species of bacteria including frank human pathogens such as *Vibrio spp.*, *Escherichia coli*, *Campylobacter spp.*, *Salmonella spp.*, *Micrococcus*, and *Pseudomonas* have been found to be VBNC under a variety of conditions. A general concern is that many other waterborne bacterial pathogen and indicator species will be found that express this trait of non-culturablity and that this condition may confound the reliability of various microbial indicators that are based on culture techniques. However, as VBNC cells are metabolically active, indicators that measure some correlate or product of metabolism might be developed that are capable of monitoring these targets even when these cells cannot be cultured. Therefore, detection of bacteria by non-culture methods is both possible and a potentially useful measure of the presence and concentrations of these types of bacteria (see Chapter 5).

It is important to note that the environmental and public health significance of injured bacteria, especially those that are VBNC, remains controversial and uncertain (Bogosian and Bourneuf, 2001). As indicated in Chapter 5, there is considerable evidence that VBNC bacteria are not infectious for human or animals as well as some evidence that they are. Because of such conflicting evidence and the uncertainties of their public health significance, VBNC bacteria are not addressed or discussed in detail in this report. However, there are good reasons to address the relationships between injured bacteria and their detection by various biochemical, immunological, and nucleic acid methods, and these are covered in Chapter 5 and Appendix C.

Dispersal

Although bacteria and other microbes are widely dispersed in nature, not all bacteria are found everywhere. Whether transported and imported bacteria are capable of survival under new or novel environmental conditions is not known. In freshwater lotic ecosystems, many bacteria in transport are allochthonous, having originated from neighboring terrestrial systems and washed into the aquatic system. Many of these bacteria are not actively growing and presumably contribute little to any ecosystem process (Edwards et al., 1990). Because of the possibility of waterborne pathogens surviving and replicating in various environmental reservoirs however, an understanding of mechanisms of dispersal is important.

Bacteria and other microbes that successfully replicate within a system can take advantage of dispersal mechanisms to both move longitudinally within a waterbody and escape a waterbody. Bacteria can also use dispersal vectors such as formation of aerosols, invection, organic foams, arthropods, and vertebrates either actively or passively.

Abiotic Mechanisms of Dispersal

Long-distance dispersal of waterborne pathogens and bacterial indicators is dependent on the movement of bacteria within waterways and whether they can exit and survive outside the waterbody. Regarding the latter, bacteria can effectively escape the aquatic environment in several ways.

Aerosol Formation The formation of aerosols is a function of the geology of a watercourse. Any turbulence caused by rocks, boulders, and woody structures that make water splash or cause wave action results in the formation of aerosols. Depending on the size of the droplets, the aerosols are transported to varying degrees into the atmosphere. The types of bacterial species found in aerosols should be proportional to those normally found and those transported in the water. Thus, aerosol formation below a sewage treatment plant outfall would be expected to have higher proportions of enteric bacteria than aerosols created either upstream or far downstream of an outfall. Very little research has been conducted in the last two decades on aerosol formation and bacterial transport resulting from sewage treatment practices (e.g., EPA, 1980). However, Rosas et al. (1993) sampled the air over sewage treatment plants and at various distances from the plants in Mexico City and reported that the highest numbers of pathogenic microorganisms were closest to the plant. Furthermore, Rosas et al. (1997) isolated *E. coli* from settled dust and air samples in several indoor and outdoor residential environments in Mexico City. Notably, the heterogeneity of *E. coli* was represented by 89 serotypes, most isolated from settled-dust indoor samples, and 21 percent of these demonstrated antibiotic multiresistance.

Organic Foams Organic foams, foams formed from turbulence or wave action, can be found in both pristine and contaminated streams and beaches. These foams can contain up to three orders of magnitude higher concentrations of bacteria than the underlying water (Hamilton and Lenton, 1998). Bacteria aid in the formation of these foams, and selection may have favored this process as an aid in their dispersal. Air sampled immediately over naturally occurring foams had much higher densities of bacteria than air sampled over open water in two streams in South Carolina (J.V. McArthur, unpublished data), and the proportion of antibiotic-resistant bacteria was much higher in the foam than in the water. Since the numbers of bacteria were 1,000 times higher in the foam, many antibiotic-resistant bacteria were being released into the air as these foams broke apart.

Arthropods and Vertebrates

Both arthropods and vertebrates can assist in the transport and dispersal of bacteria in aquatic systems. The movement of juvenile or adult aquatic insects exiting the water through hatching may be one mechanism of moving waterborne bacteria out of the water and into the air. Insect activity may also increase the release of bacteria from biofilms (Leff et al., 1994), while fish have been shown to have many opportunistic pathogens associated with their surfaces (Pettibone et al., 1996; Son et al., 1997). For example, fish that feed in or disturb sediments have higher proportions of antibiotic-resistant *Aeromonas* bacteria on their surfaces than fish that feed primarily in the water column (J.V. McArthur, unpublished data).

Summary

Clearly, improved understanding of the ecology of waterborne bacterial pathogens is needed before more effective means of detecting them directly or through the use of indicator organisms can be implemented. Knowledge of environmental reservoirs, movement and dispersal, movement and uptake of various genes, species interactions, and other factors discovered through carefully planned investigations is needed before new and more effective indicators can be developed and implemented. Failure to consider the evolutionary ecology of these organisms may result in the development of methods that are effective under only a few environmental conditions or not at all. Selection has enabled bacteria to adapt to innumerable habitats and niches and it continues to modify bacterial genomes and genes, thus making the detection and identification of waterborne bacterial pathogens and indicators a moving target.

PROTOZOAN PARASITES

Ecology and Evolution of Parasites

To understand the requirements of indicators or indicator systems for waterborne pathogens, especially parasitic protozoa, it is important to first describe the ecology and evolutionary behavior of parasites.

Parasites and Population Ecology

Parasites, both protozoa and helminths (worms and flukes), have a complex population biology reflecting their diverse species and strains, their hosts, and the environment in which the parasites and the hosts reside. Parasite population ecology is described based on a nested hierarchy that identifies infrapopulations (all of the parasites of a single species in one host); suprapopulations (all of the parasites of a given species, in all stages of development, within all hosts of an ecosystem); and component populations (all of the infrapopulations of a species of parasite within all hosts of a given species in an ecosystem). The complexity of these associations is further complicated by the genetic diversity of the parasites, because many so-called "species" have genetic and phenotypic differences that are not reflected in the current taxonomy of a single genus and species.

Parasite populations are influenced by both density-dependent (i.e., regulated by the survival and reproduction of members of a population, including the immune response of the host and host mortality) and density-independent factors (regulated by external factors such as temperature, climate, and behavior). These density-independent factors are further complicated by both short- and long-term environmental changes that influence the presence and densities of the organisms over time and place.

Suprapopulation dynamics are influenced by both the density and the diversity of reservoir hosts. The impact of parasites on human hosts in a given geographical area will depend on the types and numbers of nonhuman hosts of the same parasite, such as feral, domestic, and agricultural animals. Host-parasite dynamics and host densities of nonhuman reservoir populations of the parasite influence the exposure risks and the flows of parasites through human populations. Parasites have a density-dependent impact on host populations and population dynamics by influencing per capita survival, reproduction, and fitness. The regulation of host populations by parasites has been described in quantitative terms using statistical models for the distributions of the parasites in their host populations and in the environment. A key consideration in these quantitative relationships is how and to what extent parasite distributions are aggregated (i.e., their nonregular and nonrandom distributions) in the host and in the environment.

Human populations and public health can be strongly influenced by parasite population dynamics, such as the presence and proliferation of schistosomes (flukes causing "swimmers itch") in surface waters used by humans for aquaculture, agriculture, recreation, and other purposes. There may be epizootic cycles of the parasite in other reservoir hosts that also influence human exposure risks for infection and illness. For example, the prevalence and aggregation of *Giardia lamblia* and *Cryptosporidium parvum* are probably influenced by the reproductive cycles of their host cattle, whereby calving season results in high infection rates and increased loads of cysts and oocysts into the environment. In some ecosystems and geographic locations, the extent of risks of human exposure to a parasite may depend on the population dynamics of the definitive host for the parasite. For example, in some locations a major risk for human exposure to *Toxoplasma gondii* may result from ingestion of the oocysts in the feces of felines, such as domestic and feral cats, as the definitive host or reservoir (see more below).

Non-density-dependent factors also are important in influencing parasite infection, pathogenicity, and virulence. These include age, immunity, nutritional status, sanitation, and behavior (such as eating habits and sexual activities). Age influences susceptibility to infection and severity of illness. Newborn animals are especially susceptible to infections with enteric protozoa such as *G. lamblia* and *C. parvum.* They become ill and often shed high concentrations of the parasites in feces. Immunity is important in protecting against parasite infection, at least temporarily. Human volunteer studies on *C. parvum* infectivity show that previously infected persons have a higher 50 percent infectious dose (shifted dose-response relationship), are protected from infection at lower exposure doses, and shed fewer oocysts when infected (Dann et al., 2000; Okhuysen et al., 1999, 2002; Teunis et al., 2002).

Geography influences host-parasite interactions at all scales ranging from global to very local. At the global scale, land mass fragmentation and movement and bodies of water can divide and separate parasites and their hosts. Such separations or barriers contribute to opportunities for changes in distribution and dispersal patterns and divergences in evolution. As scales decrease however, site-specific factors increasingly influence parasite distribution and host-parasite relationships. These include temperature, precipitation, soil type, vegetation, water quality, seasonal cycles, and availability of intermediate or alternative (non-human) hosts. Anthropogenic activities also can influence local loads of parasites. For example, as discussed elsewhere in this report, animal manures and sewage wastes can greatly influence local loads and concentrations of enteric parasites in water.

Evolutionary Aspects of Host-Parasite Relationships

Parasites and their hosts co-evolve under selective pressures that differ from those acting on free-living organisms. These pressures have shaped the ecology of parasites and their hosts over evolutionary time. A variety of interactions influence parasite pathogen effects on a host as well as host effects on the genetic and phenotypic properties of a parasite. The two organisms, parasite and host, live together—often one inside the other, sometimes cell inside cell, or even genome inside genome—and the duration of interactions can be prolonged. According to Combes (2001), two aspects of these interactions have played a major role in evolution. First, genetic information from the parasite can be expressed in the host phenotype, and vice versa. Because of the fundamental unity of the genetic code and the resemblance between signaling molecules in widely divergent organisms, a parasite can manipulate the physiology and behavior of its host to favor its own transmission or survival. Second, DNA can be exchanged between host and parasite, and such exchanges sometimes have occurred on a large scale, for example between eukaryotic cells and bacterial mitochondria. The invasion of genomes by transposable elements is a special case of gene exchange having important consequences for the variability of the host genome.

Phenotypic manipulation and exchange of genetic information can move host-parasite systems toward either symbiosis or greater adverse effects of the parasite on its host (Ewald, 1996). Furthermore, host-parasite associations can involve more than two partners. One host or its genome can use a second to manipulate the phenotype of a third genome. Every host-parasite system exerts pressure on its biotic environment, and thereby, parasites participate in the ecology and evolution of the biosphere.

Bush et al. (2001) state that "the essence of parasitism rests with the nature of host-parasite relationships." Parasitism is an ecological concept that requires consideration of the parasite, the host, and the environment. Fundamental to parasitism is host resistance and immune response. Complex interactions take place between the host and its many different cells, including those of the immune system. Host recognition of the invading parasite triggers a range of immune responses that influence infectivity and disease outcomes. Furthermore, these immune responses to parasite infection are influenced by the host's environment and other host-related factors, including genetics, age, gender, diet, physical environment, and behaviors. The host-parasite interaction can have outcomes ranging from successful elimination of the parasite with no adverse effects on the host to continued infection and invasion leading to immune responses that contribute to disease and death. Summarized below are a number of different host-parasite interactions and associations that highlight the importance of the host in the nature and outcomes of host-parasite relationships:

• *Evasion of host responses by mimicry and masking.* Parasites have evolved a number of mechanisms to overcome or counteract host immune responses and sustain infectivity. One of the most fundamental mechanisms by which parasites avoid host immune detection and responses is by mimicking key host cell macromolecules or by masking their foreign antigens with a coating of host macromolecules.

• *Intracellular localization.* A number of parasites enter host cells as a feature of their pathology and a mechanism of virulence. Examples of protozoan parasites that intracellularly invade the cells of their hosts are the amoebae *Entamoeba histolytica* and *Toxoplasma gondii*.

• *Transformation of surface antigens.* A number of parasites undergo antigenic variation as a mechanism of their pathology and as a virulence factor. Antigenic variation or change produces immunologically novel parasite strains or variants that enable the parasite to avoid or evade the host's immune response and also increases the abundance of parasites within an infected individual, thereby enhancing infectivity. Such antigenic variation affects the dynamics of parasite populations at both the between-host and the within-host levels. Thus, antigenic variation has a protective effect on the parasite at individual host, population, and community levels.

• *Direct suppression of host immune responses.* Many parasites cause direct suppression of the host immune system. The intimate relationship between parasite and host in immune suppression phenomena is indicated by the dominant role of host cytokines (low molecular weight proteins that function as mediators in immune systems), either through their physiopathological effects on the host or through a direct effect on the parasite.

• *Effects on apoptosis.* Programmed cell death (apoptosis) is a recently recognized mechanism of pathology and virulence dependent on host-parasite interactions by intracellular parasites. Apoptosis is an important regulator of the host's response during infection by a variety of intracellular protozoan parasites, and this phenomenon has recently been reviewed (Luder et al., 2001). Parasitic pathogens have evolved diverse strategies to induce or inhibit host cell apoptosis, thereby modulating the host's immune response, aiding dissemination within the host, or facilitating intracellular survival. The molecular and cell biological mechanisms of the pathogen-induced modulation of host-cell apoptosis and its effects on the parasite-host interaction and the pathogenesis of parasitic diseases are complex and only now being elucidated (e.g., for *Cryptosporidium parvum* and *Toxoplasma gondii*; Luder et al., 2001).

As described above, parasite-host interactions are manifest as both pathological effects and regulatory interactions involving host responses, including immune and other physiological responses, as well as genetic and adaptive responses. Such interactions must be considered not only at the level of individual human or other hosts, but also at the population and community levels and in the context of the environment and the biosphere. That is, host-parasite interactions must be considered in an ecological context and on an evolutionary basis for both the parasite and the host. Hosts and their parasites interact in ways that can be symbiotic and mutually beneficial at one extreme or deleterious and lethal at the other. In medical and public health parasitology, most of the attention is understandably focused on parasite-host interactions that are debilitating or lethal to the host and on understanding the molecular, biochemical, physiological, and immunological aspects of host-parasite relationships with the goal of designing and implementing prevention and control measures.

Mechanisms of Parasite Pathogenicity

The mechanisms by which parasites cause infection and disease are diverse and complex. The diversity and complexity of the ecology of human parasites and their ability to cause infection and disease constitute a sufficiently substantive area of science for entire books to have been devoted to the subject (e.g., Bogitsh and Cheng, 1998; Bush et al., 2001; Gilles, 2000; Scott and Smith, 1994). As summarized by those authorities, some of the main factors responsible for the pathogenicity and virulence of parasites are (1) direct mechanical effects, (2) biochemical effects, (3) and immunological effects. Not only are the known mechanisms of parasite virulence and pathogenicity diverse and poorly understood, but new mechanisms and factors continue to be discovered or become more fully recognized for their importance. Because of the importance of molecular, biochemical, and immunological factors, genomics and proteomics (the study of all proteins produced by an organism) are contributing greatly to elucidation of the mechanisms of pathogenesis, virulence, and host susceptibility of waterborne and other pathogens. However, the biochemical mechanisms and genetic basis of pathology and virulence are far from being known for the vast majority of parasites and are unlikely to be fully elucidated and quantified for many of them for quite some time.

Introduction to Protozoa and Their Properties

The protozoa are an ancient group of unicellular organisms (single-celled eukaryotes sized 3-30 µm) probably derived from unicellular algae, but most have subsequently lost their photosynthetic capabilities. Movement is accomplished through one of three modes: flagellae, ameboid locomotion, or cilia (Allen, 1987; Stossel, 1994). Although there are numerous free-living protozoa, some can be obligate parasites of humans as well as animals, are zoonotic (spread from animals to humans), and often spread through the fecal-to-oral route. As such, these are important organisms from a public health perspective and are associated with waterborne disease worldwide, including the United States (see also Chapters 1 and 2).

Parasitic protozoa have both a trophozoite (ameboid) and a sporozoite stage within the host (Anderson, 1988). Sporozoites, which are the only stage that can survive outside a host, are called either cysts or oocysts depending on the taxonomic level. Cysts are the sporozoa of parasitic protozoa that reproduce by simple, asexual cell division, whereas oocysts are sporozoa that have both sexual and reproductive stages (Fenchel, 1987). For enteric protozoa, cysts or oocysts are the only stages that can survive outside a host and are excreted in the feces of infected individuals. Water can be contaminated by these supplies of fecally-laden (oo)cysts. Another important parasitic group is the Microsporidia. The Microsporidia are obligate spore-forming parasites in which the only environmentally stable form is the spore (Roberts and Janovy, 1995).

Although several species of waterborne protozoa are of public health concern, this section focuses on the ecology, evolution, and basic biology of the following groups and genera: the free-living amoebae *Naegleria* and *Acanthamoeba*; the enteric protozoa *Giardia, Cryptosporidium, and Toxoplasma*; and a relatively newly recognized group in human infections, the Microsporidia. *Entamoeba histolytica* and related species are amoeboid enteric protozoans that remain a risk worldwide, but waterborne disease transmission of this parasite appears to be quite rare in the United States during the last 20 years and is not discussed extensively.

Life Cycles, Taxonomy, and Health Effects

While the protozoa (especially those discussed below) may differ in their specific life cycles, they all have in common the production of (oo)cysts or spores, which are the resistant stages found in the environment. The pathway forward to a new host depends on the movement of (oo)cysts and spores through the environment, with water playing a significant role. This movement is accomplished via excretion of large numbers of (oo)cysts and spores over extended periods of time, in some cases chronic infections, survival of resistant forms in the environment, resistance to water treatment, various biotic effects, and a low infectious dose (i.e., few organisms are necessary to initiate a new infection).

The free-living amoebae produce cysts (dormant forms that are characterized by environmentally resistant external coverings) that are the resting stage of these protozoa and are abundant in the environment. As such, amoebae are the main predators of bacteria in soil and in fresh- and marine water sediments (Rodriguezzaragoza, 1994). The reproductive trophozoite stage is released from the cyst (excystation), and the protozoa reproduce by simple, binary cell division. The free-living amoebae, which can be parasitic, include *Naegleria* and *Acanthamoeba*. In this case, humans are accidental hosts, via exposure to cysts through the eyes or nose. Free-living amoebae are also capable of harboring other pathogens (e.g., the bacteria *Legionella* and *Vibrio*; Harf, 1994). For example, Thom et al. (1992) demonstrated that various strains of *V. cholerae* survived better when *Naegleria* and *Acanthamoeba* were also present. Furthermore, strains of *V. cholerae* were isolated from cysts of *Naegleria*.

Naegleria fowleri is virtually ubiquitous in the aquatic environment and is considered an opportunistic pathogen of great public health concern. However, it is not a pathogen of the fecal-oral route, and the gastrointestinal system is not the primary target; rather, the most serious infections involve the central nervous system. There are two morphological forms of *Naegleria*, and successful diagnosis and appropriate therapy depend on precise laboratory identification and differentiation of pathogenic from nonpathogenic forms (Szenasi, 1998). *Acanthamoeba* are also normally found in soil or water and occur worldwide. They are not parasitic per se but can be pathogenic to humans. The most common mode of infection from *Naegleria* and *Acanthamoeba* for healthy individuals is swimming or diving into water inhabited by these amoebae and their accidental introduction into the nasal passages.

Naegleria spp. and *Acanthamoeba* spp. have similar life cycles, and both go through a trophozoite or vegetative stage where the amoebae proliferate asexually by fission. It is at this stage that they are pathogenic. If environmental conditions become adverse, both types of amoebae will encase themselves in a thick-walled cyst containing several ostioles or pores. The cysts are able to withstand desiccation and a host of other environmental stresses. When conditions are again favorable, excystation takes place with the new trophozoite emerging through an ostiole or pore (Martinez, 1985; Schuster, 2002).

There are several manifestations of infection from these free-living amoebae including rare but usually fatal cases of primary amoebic meningoencephalitis (PAM) caused by the genus *Naegleria* and most commonly the species *N. fowleri*, and granulomatous amebic encephalitis (GAE) most often caused by *Acanthamoeba* spp. In PAM, *Naegleria* trophozoites travel along the olfactory nerves and gain direct access to the central nervous system (CNS) where they quickly multiply and cause extensive damage by way of hemorrhagic necrosis, eventually destroying the olfactory bulb and the cerebral cortex. The victim generally dies within 3 to 10 days after the onset of symptoms and there is no effective treatment (Martinez, 1985; Schuster, 2002; Wiersma, 2002). In contrast to PAM, cases and deaths from GAE usually occur in immunocompromised individuals after weeks or months of CNS symptoms although both diseases are usually confirmed only after autopsy. In addition, the entry of *Acanthamoeba* can take place outside the nasal passages, and they can enter the bloodstream through a break in the skin or through the lungs before reaching the CNS. In addition to GAE, *Acanthamoeba* can also cause amoebic keratitis, a condition first noted in individuals with corneal trauma (Ma et al., 1981). It is more commonly associated with contact lens wearers who do not properly disinfect their lenses, thus allowing amoebae to proliferate on the lens. However, amoebic keratitis has not been shown to lead to CNS infection (Schuster, 2002).

Giardia are among the most primitive eukaryotic organisms in existence, being bilaterally symmetrical flagellated amoebae; a characteristic that is unique at their evolutionary level. They are obligate enteric parasites that undergo reproduction only after ingestion of cysts from fecally contaminated water, food, or hands. This parasite was first described by Antony von Leeuwenhoek from his own feces in 1681. *Giardia* have a broad range of environmental hosts including dogs, cats, sheep, and beavers. Between 1984 and 1990 there were approximately 25 outbreaks of this parasite that infected nearly 3,500 people (Marshall et al., 1997; Steiner et al., 1997). It is estimated that worldwide, 2.8 million people per year in both developed and undeveloped countries are infected with *Giardia* (Ali and Hill, 2003). Excystation occurs and trophozoites are released in the small intestine, where they attach to the microvillae and begin asexual reproduction. This is followed by the production of both immature sporozoites and more trophozoites in the large intestine, and finally maturation of immature sporozoites to cysts as they travel down the large intestine and exit into the environment in the feces. *Giardia lamblia* causes diarrhea and abdominal pain in infected persons around the world, in both industrialized and developing countries, and it is an important cause of morbidity in children and adults. However, the basic ecology of this parasite is not well understood.

Microsporidia are eukaryotic spore-forming obligate protozoan parasites that infect all animal groups (especially arthropods), including humans (Weiss, 2001). Microsporidia are usually transmitted by direct human-to-human contact (Mota et al., 2000); however, spores from Microsporidia are common in the environment including surface and drinking waters. These organisms have the smallest known eukaryotic genome and appear to reproduce both sexually and asexually; the latter by a budding process being most common results in a variety of types of spores. Although humans are not their natural hosts, Microsporidia are considered obligate parasites that require a host, and the most common infections in humans are found in the immunocompromised. There are more than 1,000 species of Microsporidia, of which 13 are presently known to infect humans (e.g., *Encephalitozoon cuniculi*, *E. intestinalis*, *E. hellem*, *Enterocytozoon bieneusi*). Two species of Microsporidia are associated with gastrointestinal disease in humans: *Enterocytozoon bieneusi* and *E. intestinalis* (Dowd et al., 1998a).

Cryptosporidium and *Toxoplasma* are obligate parasites and require a host to reproduce—the former as an intestinal parasite and the latter as a tissue parasite. *Cryptosporidium* completes its cycle in a single host in the intestinal tract (O'Donoghue, 1995). Oocysts are ingested and the parasite undergoes asexual and sexual reproduction in the intestinal tract. Oocysts are excreted in the feces as a result of reproduction and are extremely hardy. The organism has great genetic recombination abilities through these reproduction strategies and has co-evolved with its various mammalian hosts. Recent analyses of *Cryptosporidium* show significant host adaptation and the ability of genotypes to expand their host range (Xiao et al., 2002).

Cryptosporidiosis infections are ubiquitous. In industrialized and developing countries, 2.2 percent compared to 6.1 percent of hospital patients admitted for diarrhea pass oocysts, respectively (Guerrant, 1997). However, up to 30 and 50 percent of the U.S. population has antibodies to *C. parvum* (Frost et al., 2002; Isaac-Renton et al., 1999) compared to upwards of 90 percent of people in impoverished regions of developing countries (Guerrant, 1997; Zu et al., 1994).

Toxoplasma gondii may be one of the most common parasitic infections of man and other warm-blooded animals (Hill and Dubey, 2002). It has a worldwide distribution and can be found from Alaska to Australia. Hill and Dubey (2002) estimated that one-third of humanity has been exposed to this parasite. This parasite completes its entire life cycle in two hosts. Cats (feral and domestic) are the definitive host in they alone produce the oocysts. The parasite completes its life cycle by reproduction in the intestine of cats, and cysts are excreted in their feces. The parasite travels to the next host as an oocyst excreted in feces; however the oocyst (unlike *Cryptosporidium*) is not infectious upon excretion and requires a maturation phase in the environment that is on the order of days. Other mammals ingest the oocysts and become infected (Kitamoto and Tanabe, 1987). In rodents, the parasite undergoes the reproductive stages in muscle tissue, and upon ingestion by a cat the parasite is able to complete its life cycle. Humans exhibit a "dead-end" infection associated with partial life stages that results in fever. In humans, the primary health concern is for the fetus since the organism can cross the placental barrier in pregnant women and result in severe birth defects such as retardation (Lopez and Wilson, 2003).

Sources, Stability, and Survival

The environmental route of transmission for many protozoan parasites has made it necessary to develop new methods for their early and repeated detection (Smith, 1998). These parasites pose new and emerging threats because of their ability to survive in a variety of moist habitats including surface waters and because protozoan parasites were among the most frequently identified etiologic agents in waterborne disease outbreaks (see Figure 1-1; Marshall et al., 1997).

The parasitic protozoa are of particular interest in this chapter because they may have significant environmental reservoirs that harbor the (oo)cysts, including the free-living forms, in fresh- and marine waters. Furthermore, since ingestion is required for most diseases to develop, knowledge of potential vectors or sources is critical. All of the protozoa discussed in this chapter, except for *Naegleria* and *Acanthamoeba*, are zoonotic in nature; thus, a wide range of animals may serve as sources for the parasites.

Species of the genus *Naegleria* are found in stagnant bodies of freshwater such as lakes, slow-moving rivers, ditches, and non- or poorly chlorinated swimming pools. *Naegleria* are known to be thermophilic, and the incidence of infection follows a seasonal pattern occurring mostly in the summer months when water temperatures rise (Martinez, 1985). They have been isolated from environments with temperatures between 26.5 and 28°C and become more virulent between 30 and 37°C (John, 1993). *Acanthamoeba* spp. are ubiquitous and hardier than *Naegleria* and have been isolated from many different environments including ocean sediment and even dust. However, their growth is more prolific at lower temperatures (25 to 35°C) than *Naegleria* (Martinez, 1985).

Research on *Naegleria gruberi*, which differentiates from the amoebae to the flagellate form in less than 90 minutes, suggests that there are at least two "hold-points" where the cells hold for up to four hours awaiting additional stimulus (Fulton and Lai, 1998). Changes in temperature greater than 10°C will cause complete differentiation whereas smaller temperature changes result in the intermediate forms. If these hold-points are found for other related species such as *N. fowleri*, large-scale climate and weather change may affect the forms present in the environment and the potential infection rates.

The ecology of *Toxoplasma gondii* oocysts is diverse and varied. While the cat is the primary source for *Toxoplasma*, oocysts from terrestrial land animals (e.g., feral cats) have found their way into surface waters, resulting in major documented outbreaks in humans such as in the Greater Victoria area of British Columbia, Canada (Bowie et al., 1997; Isaac-Renton et al., 1998). Coastal freshwater runoff has been implicated in the infection of southern sea otters (Miller et al., 2002). These researchers found that sea otters sampled near areas of maximal freshwater runoff were three times more likely to be seropositive for *T. gondii* than otters sampled at low-flow areas. Serological evidence of *T. gondii* infection of deer mice living in the riparian zone of various watersheds suggests that the oocysts are being shed near the waters' edge (Aramini et al., 1999).

Besides sea otters, numerous other marine mammals have been infected with *T. gondii*. Although it is not known how all of these mammals become infected, there is evidence that oocysts can remain viable and infective after removal by eastern oysters (Lindsay et al., 2001). Invertebrates and vertebrates (including humans) may be infected by the handling and eating of oysters. The ability of *Toxoplasma* to survive under extremely broad environmental conditions has prompted research into the ranges of tolerance of different variables including temperature. The infectivity of *T. gondii* oocysts showed no loss at 10, 15, 20, and 25°C for 200 days (Dubey, 1998), and oocysts remained infective up to 54 months at 4°C and for 106 days at −5 and −10°C. Some oocysts remained infective at temperatures approaching 55°C. With tolerance ranges that broad, it is easy to see why these protozoa have a global distribution.

Very little is known about the sources of Microsporidia. In France during the summer of 1995, a waterborne outbreak of Microsporidia occurred with approximately 200 cases of disease (Sparfel et al., 1997). The causative species identified was *Enterocytozoon bieneusi*. Although fecal contamination of the drinking water was never detected, contamination from a nearby lake was suspected. In the United States there are minimal data on the occurrence of human strains of

Microsporidia in surface waters. Dowd et al. (1998b, 1999) described a polymerase chain reaction (PCR) method for detection and identification of Microsporidia (amplifying the small subunit ribosomal DNA of Microsporidia). They found isolates in sewage, surface waters, and groundwaters. The strain that was most often detected was *Enterocytozoon bieneusi*, which is a cause of diarrhea excreted from infected individuals into wastewater. Microsporidia spores have been shown to be stable in the environment and remain infective for days to weeks outside their hosts (Shadduck, 1989; Shadduck and Polley, 1978; Waller, 1979). Because of their small size (1 to 5 μm), they may be difficult to remove using conventional water filtration techniques, and there is a concern that these microorganisms may have an increased resistance to chlorine disinfection similar to *Cryptosporidium*. Initial studies using cell culture suggest that the spores may be more susceptible to disinfection (Wolk et al., 2000).

In contrast to Microsporidia, much is known about sources and survivability of *Cryptosporidium* and *Giardia* in the environment, especially the aquatic environment (see more below), where their (oo)cysts can survive for weeks or months (Robertson et al., 1992; Rose and Slifko, 1999). Both *Cryptosporidium* and *Giardia* are well adapted to environmental extremes, being able to survive temperatures ranging from 4 to 37°C and environments ranging from homeothermic animal bodies to thermally and chemically variable freshwater (Rose and Slifko, 1999). Species of *Giardia* exhibit a high degree of host specificity and as such have been named based on their normal host (e.g., *G. lamblia* inhabits humans, *G. muris* inhabits rodents, and *G. ardeae* inhabits birds). Recent phylogenetic analysis has shown that *G. lamblia* infects humans, and many other mammals can potentially become infected with the human type (Marshall et al., 1997). Thus, there is good evidence that *G. lamblia* has environmental reservoirs such as beavers and possibly muskrats.

Cryptosporidium parvum, also a zoonotic species, can infect an unusually wide range of mammals, including humans (O'Donoghue, 1995). The phylogenic relationships between the whole range of species and genotypes have shed light on the zoonotic transmission potential of *Cryptosporidium* (Egyed et al., 2003). Typing of isolates from various geographic regions and host origins has relied on direct DNA sequencing or selected genetic loci analysis using PCR. These studies have found mixtures of genotypes and species in feces of infected animals or humans (Morgan et al., 1999).

Two genotypes of *C. parvum* are of special interest: Genotype 1 only infects humans, whereas Genotype 2 occurs in a wide range of animals, including humans although it is not known if it results in different pathology (Peng et al., 1997; Rose et al., 2002). Indeed, the ability of *C. parvum* Type 2 to infect not only man, but domesticated animals such as cows, goats, sheep, pigs, horses, dogs, and cats and even wild animals such as mule deer may provide a tremendous evolutionary advantage. DNA-DNA hybridizations reveal little or no mixing between genotypes, suggesting cryptic species. A cryptic species shows little, if any, phenotypic divergence but differs genetically and as seen in this example shows no recombination occurring between the two genotypes.

The ability of *Cryptosporidium* to survive outside the host suggests that evolution favors strains that can generate and input many oocysts into the environment without seriously harming the host. Oocysts were found in three species of rodents in Poland (Bajer et al., 2002). Interestingly, there were significant differences between the *Cryptosporidium* species identified in rodents over time, though fewer older animals carried infection and there were marked seasonal differences. Rickard et al. (1999) tracked both *Cryptosporidium* and *Giardia* species in populations of white-tailed deer in the southern United States. These researchers found higher infection rates associated with *Cryptosporidium* than *Giardia*. As for rodents, the probability of protozoan infection decreased with increasing age of the deer.

A few studies have tried to determine primary sources of *Cryptosporidium* and *Giardia* cysts and oocycts in watersheds. Studies comparing agricultural and wildlife sources have shown that the lowest prevalence of *Giardia* and *Cryptosporidium* was found in wildlife (Heitman et al., 2002), while the highest concentrations were found in cattle feces. Given the potential runoff from agricultural sources into watersheds and waterways, this observation is significant. In a 17-month survey (Bodley-Tickell et al., 2002) of *Cryptosporidium* oocysts in surface waters draining a livestock operation, the parasites were found to be present year-round, with maximum concentrations and highest frequency of occurrence during autumn and winter. *Cryptosporidium* was also found in an isolated pond (no livestock), indicating that wild animals alone could transport or import oocysts to surface waters. Waterfowl can also disseminate infectious *C. parvum* and *Giardia* in the environment (Graczyk et al., 1998). Although this finding is not surprising, the widespread levels of contamination may be, since seven out of nine sites showed occurrence of *C. parvum* and all nine sites were positive for *Giardia* in the feces of migrating Canada geese. Given the size of the migrating waterfowl populations in the United States, this observation is of concern.

Cryptosporidium oocysts transported from river waters into estuarine and marine systems have been shown to be taken up and sequestered by filter-feeding invertebrates. Lowery et al. (2001) showed that the marine filter-feeding mussel (*Mytilus edulis*) collected from the shores of Belfast Lough in Northern Ireland had Genotype 1 oocysts of *C. parvum*. Thus, these filter-feeding invertebrates serve as environmental reservoirs of this protozoan. While the natural rate of release from these *Cryptosporidium* reservoirs is not known (because *Mytilus edulis* is consumed by humans), this aspect of the ecology of *C. parvum* may have direct public health ramifications.

Terrestrial insects have been implicated as both control agents and vectors of dissemination of *Cryptosporidium* oocysts (Dumoulin et al., 2000; Follet-Dumoulin et al., 2001; Mathison and Ditrich, 1999). Mathison and Ditrich (1999) demonstrated that many oocysts can pass safely through the mouth parts and gastrointestinal tracts of dung beetles; however, the majority are destroyed. Predation by coprophagous (dung eating) insects can be important in both reducing oocysts and spreading the oocysts that are not destroyed throughout the environment. In contrast, free-living ciliated protozoa, such as *Paramecium caudatum*, are capable of consuming up to 170 oocysts per hour (Stott et al., 2001). Thus, predation, under certain conditions, may affect the density of oocysts.

Although *Cryptosporidium* infects mammals, it is unable to infect other vertebrate groups such as fish, amphibians, and reptiles (Graczyk et al., 1996). This inability greatly limits the spread and dissemination of oocysts in surface waters. However, depending on the characteristics of the watershed, runoff can carry oocysts shed by infected mammals in major pulses. Once in the waterway, oocysts can be maintained until consumed either by humans or by other mammals.

Cryptosporidium and *Giardia* in Water

Most studies of waterborne protozoa of public health concern have focused on the occurrence of the enteric protozoa *Cryptosporidium* and *Giardia*. Although these two enteric protozoan parasites are not related taxonomically, they are closely related from an epidemiological, regulatory, and public health point of view. In addition, methods for their simultaneous detection in water have led to much information on their occurrence, more so than for any of the other protozoan parasites. In surveys from the early 1990s, it was reported that for 66 surface water treatment plants in 14 states and 1 Canadian province, 81 percent of the raw water samples had *Giardia* cysts and 87 percent tested positive for *Cryptosporidium* (LeChevallier et al., 1991a,b). Further reviews of the literature have found that most waters contain some level of cysts and oocysts ranging from a high in treated wastewaters (10^4 per 100 liters) to a low in pristine waters (0.1 per 100 liters) (Rose, 1997; Rose et al., 2001a; Slifko et al., 2000). Similar levels have been found throughout the United States and in Europe (Ong et al., 1996; Smith and Rose, 1998). As noted previously, domestic animals, cattle in particular, and sewage discharges have been identified as some of the primary sources of *Cryptosporidium* and *Giardia* in water.

The cysts and oocysts of *Giardia* and *Cryptosporidium* may also be transported from irrigation waters to row crops, some of which may be eaten raw. Thurston-Enriquez et al. (2002), using both molecular and immunofluorescent techniques, were able to detect oocysts of *Cryptosporidium* in 36 percent and cysts of *Giardia* in 60 percent of all irrigation water samples. They also found that 28 percent of samples had Microsporidia. Average concentrations for *Giardia* and *Cryptosporidium* varied from 559 cysts and 229 oocysts per 100 liters in samples from Central America to 25 cysts and <19 oocysts per 100 liters in the United States, respectively. These researchers demonstrated that agricultural irrigation waters may be a significant vector for the transmission of waterborne protozoa and corresponding diseases.

Weather can influence both the transport and the dissemination of these protozoa (Fayer, 2000; Fayer et al., 2000; Rose et al., 2001b). A study conducted in the United States demonstrated similar results with both wildlife and dairy farms contributing to *Cryptosporidium* oocysts in the watershed and implicated cold seasons as high-risk periods for oocyst contamination of surface waters (Jellison et al., 2002). High concentrations of oocysts during winter and autumn may be indicative of reduced predation. Alternatively, cold temperatures may actually preserve oocysts and thus provide sources during warmer weather. Skerrett and Holland (2000), in a temporal survey at five sites near Dublin, Ireland, showed that the maximum number of oocysts was found after a period of heavy rainfall associated with increased runoff. Knowledge of how changes in rainfall, snowmelt, and runoff affect the transport of protozoan parasites is critical.

Another important aspect of *Cryptosporidium* and *Giardia* (oo)cyst ecology that may affect their distribution and survival is their sedimentation velocity. These velocities are much too low to cause significant sedimentation in surface waters or reservoirs (Medema et al., 1998). However, both cysts and oocysts attach readily to organic particles, which greatly increase their sedimentation velocities. Attachment to particles will affect not only the deposition onto sediments but also the potential for consumption. Depending on the size of the particles, ciliated protozoa may not be able to graze the (oo)cysts. Although infection of fish, amphibians, and reptiles has not been shown, bottom-feeding fish may transport the (oo)cysts to new locations.

Summary

Parasite interactions with their hosts and the environment are diverse, complex, and continually evolving. These interactions involve a considerable amount of genetic exchange and selection, as well as host-dependent phenotypic expression. For many parasites, the host immune response is a major factor influencing pathogen-host interaction and its effects on the health of the host. In addition, parasites and their hosts coevolve and show a variety of population-based ecological interactions influencing population density, geographic location, and other factors related to the parasite and host environment.

Factors such as temperature, source of shedding (i.e., from hosts), rainfall, and predation can affect the survivability of these parasites in the aquatic environment. However, other aspects of the ecology of cysts and oocysts are not well understood. For example, why do low temperatures seem to promote survival, and why are the highest concentrations often found in autumn and winter months? Is there more than a proximate answer to this question? Does this observation confer any selective advantage on the parasites and increase their chance of infecting new hosts? In other words, are certain genetic variants better able to adapt to and survive the changing environment than others? If so, these strains would be expected to increase over time. Are there density-independent factors that affect the survival and chance of future infection? Factors such as genetic background, metabolic activity, or (oo)cyst development are possible examples. Alternatively, are density dependent factors such as predation more important?

While this discussion has focused on the environmental forms (i.e., cysts and [oo]cysts) of protozoans with little mention of the infectious forms, an improved understanding of the ecology of the infectious forms may be crucial in understanding the distribution and dissemination of the environmental forms. For example, Thompson and Lymbery (1996) discussed the genetic variability in parasites in the context of ecological interactions with the host. More information is needed on parasite variability and species recognition as related to parasite or host occurrence. There is a need for accurate parasite characterization to determine fundamental questions about zoonotic relationships and parasite transmissions. Ecological interactions between inter- and intraspecific levels within the host and the impact of these interactions on the evolutionary and clinical outcomes of parasitic infections also have to be studied.

YEASTS AND MOLDS

Introduction and Background

Yeasts and molds are collectively called fungi as a group and possess defined nuclear membranes that contain the chromosomes of the cells. There are more than 100,000 species of known fungi, although only a few are known to be human pathogens. Fungi are more than 10 to 100 times larger than bacteria. Molds are multicellular, complex organisms that produce sexual and asexual spores. They appear as cottony and fuzzy growth in food and other materials due to the growth of hyphae and mycelium, and many are spoilage organisms. Some molds are beneficial to humans because of their production of important antibiotics and fermentation of foods, while others may cause a variety of human diseases ranging from "athlete's foot" to aspergillosis.

Yeasts are single-celled fungi that are usually oval in shape and divide asexually by budding or sexually through production of spores. They are important in food fermentation, food spoilage, and several human diseases—especially *Candida albicans*, the common cause of various "yeast infections." Yeasts and molds are ubiquitous in the environment, including air, soil, food, and water.

Ecology and Evolution of Fungi and Their Role as Human Pathogens

Humans are continuously exposed to fungi from various environmental sources and often are colonized with fungi. On rare occasions, some fungi cause human infection and illness and most of these illnesses occur in immunocompromised hosts. Fungal infections or mycoses are classified according to the degree of tissue involvement and mode of entry into the host. These categories are (1) superficial (local in skin, hair, and nails); (2) subcutaneous (infection of the dermis, subcutaneous tissue, or adjacent structures); and (3) systemic (deep infections of internal organs). For a fungus to cause serious disease it usually has to actively invade tissues, especially deeper tissues, and become disseminated throughout the body (systemic).

Fungal infections also can be categorized as frank (can infect healthy, immunocompetent hosts) or opportunistic (can infect only the immunocompromised) pathogens. Frank fungal infections in the United States are uncommon, being confined to conditions such as candidiasis (thrush) and dermatophyte skin infections such as athlete's foot. However, in immunocompromised hosts, a variety of normally mild or nonpathogenic fungi can cause systemic and potentially fatal infections. Thus, the immune state of the host plays an important role in the infectivity and health effects of fungi.

The pathogenic mechanisms of fungi tend to be highly complex. This is because they arise in large part from adaptations of preexisting characteristics of the organisms' nonparasitic life-styles (van Burik and Magee, 2001). Most of the human pathogenic fungi are dimorphic (i.e., able to reversibly transition between yeast and hyphal forms) (Gow et al., 2002; Rooney and Klein, 2002). This dimorphism is an important attribute of fungi because the morphogenic change from one form to the other is often associated with host invasion and disease. The nature of the association between morphogenic transition, pathogenicity, and virulence is a subject of considerable interest and debate among experts, and efforts are being made to understand both the fungal and the host factors that influence host-fungi interactions and outcomes. Currently, the evidence suggests that the significance of morphological changes by fungi varies considerably in different fungal diseases.

For many fungi, other factors besides morphological changes play an important role in pathogenicity, as do the host's immune competence, immune response, and other physiological and constitutive states. As noted above, it is well established that the nature and extent of impairment of host defenses influences the pathology, severity, and outcomes of fungal infections, such that the clinical manifestations of disease are contingent upon the hosts' immune system and other host characteristics. Unlike bacteria, where virulence genes are often organized together into definable "pathogenicity islands" that are co-regulated, most fungal toxins are spread across the genome, unlinked, and independently regulated. The interaction of the fungus with its host and the properties of the host are also critical factors in the expression of virulence and the production of disease. Some protein components of fungi and human cells are functionally interchangeable (Brown, 2001). For example, human proteins with fundamental roles in the cell cycle, stress responses, gene regulation, protein localization, metabolism, and energy generation can functionally replace the corresponding proteins in the budding yeast *Saccharomyces cerevisiae*.

Fungi and Waterborne Pathogens

Although the potential role of yeasts and fungi as waterborne pathogens has not been systematically assessed, they are regularly isolated from water samples. Of the fungal colonies isolated on agar plates from water, typically about half are yeast and the other half mold colonies. Molds from the families *Oomycetes* and *Chytridiomycetes* can virtually always be found in fresh and saline water in the environment. While these molds are not generally pathogenic to humans, they can infect animals and cause disease (Alexopoulos et al., 1996).

Other potential waterborne yeast and mold pathogens include *Candida albicans*, *Geotrichum candidum*, and *Aspergillus fumigatus*. A severe disease called "swamp cancer" can occur when farmers are exposed to *Pythium insidiosum* in swampy environments (Deneke and Rogers, 1996). Although yeasts and molds are not considered a major source of human waterborne pathogens at the present time, their ubiquitous presence in water should not be ignored.

Summary

Fungi, yeasts, and molds are widely distributed in the environment and most are not human pathogens. For the relatively few that are known to be pathogenic to humans, the interactions between the pathogenic fungi and their hosts are intimate and complex and depend greatly on fungal and host properties as well as environmental conditions. Most experts agree that the properties of the fungus and the host seem to contribute about equally to the outcomes of exposure, infection, and disease and that there are multiple factors at work on the part of both participants (fungus and host). The majority of fungi are opportunistic pathogens, which serves to emphasize the fundamentally important properties of the host in fungal pathogenesis and virulence. A comprehensive analysis that considers the multiple properties of the fungus and the host and the complex manner in which they interact in various environmental settings is necessary in order to understand the host-parasite interaction and gain insights into the factors responsible for exposure, infection, disease, and the resulting health effects in humans.

SUMMARY: CONCLUSIONS AND RECOMMENDATIONS

The ecology and evolution of waterborne pathogens have important implications for the emergence and reemergence of those pathogens of public health concern. The concept of using indicators for waterborne pathogens implies that certain characteristics of microorganisms such as genes and gene products remain constant under varying environmental conditions. As discussed in this chapter however, this assumption cannot always be relied on.

The effectiveness of indicator technologies that are based on the detection of some aspect of the biology or chemistry of a living organism (whether a pathogen or an indicator microorganism) may decrease over time because of evolutionary changes in the target organism. Natural and artificial selection may alter the structure, function, and production of biological molecules or cause other changes in the organism that affect the ability of the indicator to detect it. Therefore, it is important to understand the effects of the environment on these targets and organisms. In other words, will an indicator be effective in a wide range of geographic locations, habitats, and under different environmental conditions? How rapidly do waterborne infectious microorganisms change in the environment, and are such changes in indicator organisms reflective of changes in the infectious organism (i.e., do they have parallel evolutionary trajectories)?

Existing and candidate indicator organisms should have ecologies and responses to environmental variations similar to those of the pathogenic organisms that they are supposed to be indicating. Furthermore, environmental changes may lead to changes in selective pressures resulting in new strains of pathogens with different traits. These reservoirs can be important in their transmission to other hosts. Genetic materials can be gained or lost during evolution. This gain or loss affects not only the effectiveness of a particular indicator but also one's ability to detect pathogens directly. The presence and interaction of biotic and abiotic reservoirs that offer environmental refuge may affect the survivability and pathogenicity of the pathogen. Understanding the ecology and evolution of pathogens will provide insights into their pathways of transmission, modes of distribution, potential to reemerge in the future, or emergence in other environments.

The committee makes the following recommendations to improve the understanding of the ecology and evolution of waterborne pathogens and the development of new and effective indicators of microbial contamination:

• Natural background density of waterborne pathogens should be established to differentiate between native opportunistic pathogens and introduced pathogens.

• Efforts should be made to differentiate between indicators and pathogens that are native to the environment and those that are introduced from external sources, such as human and animal wastes.

• Because some waterborne pathogens or indicator organisms may survive and replicate in various environmental reservoirs independently of each other, an improved understanding of the ecology and natural history of microbial indicators and pathogens and the mechanisms of their persistence, proliferation, and dispersal should be sought.

• Advanced analytical methods should be used to help distinguish between introduced pathogenic and naturally occurring nonpathogenic strains of waterborne microorganisms and to characterize the emergence of new strains of pathogens as a result of genetic change.

• Bacteria, viruses, and protozoa have evolved mechanisms that facilitate their rapid response to environmental changes. These mechanisms may influence the infectivity and pathogenicity of the organism. Therefore, additional research is needed on microbial evolutionary ecology to address long-term public health issues.

• The ecology of waterborne pathogens should be assessed in relation to modern agricultural practices and other anthropogenic activities, such as urbanization. Animal wastes from agriculture and urban sewage, runoff, and stormwater are major contributors to both human pathogenic and nonpathogenic strains of microbes, and the wide use of antibiotics in animal agriculture and in human and veterinary therapy leads to selection for antibiotic-resistant phenotypes.

• Research in genetic ecology is needed to address issues of bacterial resistance to antibiotics, disinfectants, and other chemicals (such as heavy metals) and the regulation and transferability of these resistance traits either independently or together as sets of multiple resistance genes. The factors that select for increased resistance to these agents in natural populations of bacteria need to be elucidated as do the factors influencing the natural transfer of these resistance traits to waterborne pathogens, indicators, and other aquatic microorganisms.

• Research is needed to develop a better understanding of the ecology and natural history of both the environmental and infectious stages of pathogens and the parallel stages of indicator organisms to grasp how the organisms are distributed in nature; how they persist and accumulate in water, other environmental media, and in animal reservoirs; and how dissemination of the environmental form occurs, especially human exposures.

• Genetic and phenotypic characterization of pathogenic viral, bacterial, and protozoan parasites is needed to elucidate zoonotic relationships with their hosts and factors influencing waterborne transmission to humans.

• Given the ubiquity of yeasts and molds in water samples, research should be conducted to clarify their role in the transmission of waterborne diseases.

REFERENCES

Aggarwal, R., and S.R. Naik. 1997. Epidemiology of hepatitis E: Past, present and future. Tropical Gastroenterology 18: 49-56.

Alexopoulos, C.J., C.W. Mims, and M. Blackwell. 1996. Introductory Mycology. New York: John Wiley and Sons.

Ali, S.A., and D.R. Hill. 2003. *Giardia intestinalis.* Current Opinion in Infectious Diseases 16(5): 453-460.

Allen, R.D. 1987. The microtubulues as an intracellular engine. Scientific American 256: 42-49.

Anderson, O.R. 1988. Comparative Protozoology: Ecology, Physiology, Life History. New York: Springer-Verlag.

Andradottir, H.O., and H.M. Nepf. 2000. Thermal mediation by littoral wetlands and impact on lake intrusion depth. Water Resources Research 36: 725-735.

Anonymous. 2002. From the Centers for Disease Control and Prevention. Acute flaccid paralysis associated with circulating vaccine-derived poliovirus - Philippines, 2001. Journal of the American Medical Association 287(3): 311.

Anonymous. 2003. Laboratory surveillance for wild and vaccine-derived polioviruses, January 2002-June 2003. MMWR 52(38): 913-916.

Aramini, J.J., C. Stephen, J.P. Dubey, C. Engelstofdt, H. Schwantje, and C.S. Ribble. 1999. Potential contamination of drinking water with *Toxoplasma gondii* oocysts. Epidemiology and Infection 122: 305-315.

Baigent, S.J., and J.W. McCauley. 2003. Influenza type A in humans, mammals and birds: Determinants of virus virulence, host-range and interspecies transmission. Bioessays. 25(7): 657-671.

Bajer, A., M. Bednarska, A. Pawelczyk, J.M. Behnke, F.S. Gilbert, and E. Sinski. 2002. Prevalence and abundance of *Cryptosporidium parvum* and *Giardia* spp. in wild rural rodents from the Mazury Lake District region of Poland. Parasitology 125: 21-34.

Balayan, M.S. 1997. Epidemiology of hepatitis E virus infection. Journal of Viral Hepatitis 4: 155-165.

Barker, J., T.J. Humphrey, and M.W.R. Brown. 1999. Survival of *Escherichia coli* O157 in a soil protozoan: Implications for disease. FEMS Microbiology Letters 173: 291-295.

Bellar, M., A. Ellis, S.H. Lee, M.A. Drebot, S.A. Jenkerson, E. Funk, M.D. Sobsey, O.D. Simmons III, S.S. Monroe, T. Ando, J. Noel, M. Petric, J. Hockin, J.P. Middaugh, and J.S. Spika. 1997. Outbreak of viral gastroenteritis due to a contaminated well: International consequences. Journal of the American Medical Association 278(7): 563-568.

Bergelson, J.M. 2003. Virus interactions with mucosal surfaces: Alternative receptors, alternative pathways. Current Opinion in Microbiology 2003 6(4): 386-391.

Bernal, W., H.M. Smith, and R. Williams. 1996. A community prevalence study of antibodies to hepatitis A and E in inner-city London. Journal of Medical Virology 49(3): 230-234.

Bishop, R.F. 1996. Natural history of human rotavirus infection. Archives of Virology supplementum 12: 119-128.

Bjedov, I., O. Tenaillon, B. Gérard, V. Souza, E. Denamur, M. Radman, F. Taddei, and I. Matic. 2003. Stress-induced mutagenesis in bacteria. Science 300: 1404-1409.

Bodley-Tickell, A.T., S.E. Kitchen, and A.P. Sturdee. 2002. Occurrence of *Cryptosporidium* in agricultural surface waters during an annual farming cycle in lowland UK. Water Research 36: 1880-1886.

Bogitsh, B.J., and T.C. Cheng. 1998. Human Parasitology. New York: Academic Press.

Bogosian, G., and E.V. Bourneuf. 2001. A matter of bacteria life and death. EMBO Reports 2(9): 770-774.

Borchardt, M.A., P.D. Bertz, S.K. Spencer, and D.A. Battigelli. 2003. Incidence of enteric viruses in groundwater from household wells in Wisconsin. Applied and Environmental Microbiology 69(2): 1172-1180.

Bowie, W.R., A.S. King, D.H. Werker, J.L. Isaac-Renton, A. Bell, S.B. Eng, and S.A. Marion. 1997. Outbreak of toxoplasmosis associated with municipal drinking water. Lancet 350: 173-177.

Brown, A.J.P. 2001. Fungal pathogens—the devil is in the detail. Microbiology Today 29: 120-122.

Bush, A.O., J.C. Fernández, G.W. Esch, and J.R. Seed. 2001. Parasitism: The Diversity and Ecology of Animal Parasites. Cambridge: Cambridge University Press.

CDC (Centers for Disease Control and Prevention). 1997. Multidrug-resistant *Salmonella* serotype typhimurium—United States, 1996. MMWR 46(14): 308-310.

Chung, H., L.-A. Jaykus, G. Lovelace, and M.D. Sobsey. 1998. Bacteriophages and bacteria as indicators of enteric viruses in oysters and their harvest waters. Water Science and Technology 38(12): 37-44.

Clayson, E.T., B.L. Innis, K.S. Myint, S. Narupiti, D.W. Vaughn, S. Giri, P. Ranabhat, and M.P. Shrestha. 1995. Detection of hepatitis E virus infections among domestic swine in the Kathmandu Valley of Nepal. American Journal of Tropical Medicine and Hygiene 53(3): 228-232.

Cohen, J.A., W.V. Williams, and M.I. Greene. 1988. Molecular aspects of reovirus-host cell interaction. Microbiological Science 5(9): 265-270.

Colwell, R.R. 1996. Global climate change and infectious disease: The cholera paradigm. Science 274: 2025-2031.

Combes, C. 2001. Parasitism. The Ecology and Evolution of Intimate Interactions. Chicago, Illinois: University of Chicago Press.

Dann, S.M., P.C. Okhuysen, H.L. DuPont, and C.L. Chappell. 2000. Fecal IgA response to reinfection with *Cryptosporidium parvum* in healthy volunteers. American Journal of Tropical Medicine and Hygiene 62(3): 670.

Davison, J. 1999. Genetic exchange between bacteria in the environment. Plasmid 42(2): 73-91.

Deneke, E.S., and A.L. Rogers. 1996. Medical Mycology and Human Mycoses. Delmonte, California: Star Publication Company.

Denton, M., and K.G. Kerr. 1998. Microbiological and clinical aspects of infection associated with *Stenotrophomonas maltophilia*. Clinical Microbiology Reviews 11: 57-80.

Desselberger, U., M. Iturriza-Gomara, and J.J. Gray. 2001. Rotavirus epidemiology and surveillance. Novartis Foundation Symposium 238: 125-152.

Dimitrov, D.S. 2000. Cell biology of virus entry. Cell 101: 697-702.

Domingo, E., R.G. Webster, and J.J. Holland. 1999. Origin and Evolution of Viruses. New York: Academic Press.

Donaldson, K.A., D.W. Griffin, and J.H. Paul. 2002. Detection, quantitation and identification of enteroviruses from surface waters and sponge tissue from the Florida Keys using real-time RT-PCR. Water Research 36(10): 2505-2514.

Dowd, S., C. Gerba, F. Enriquez, and I. Pepper. 1998a. PCR amplification and species determination of Microsporidia in formalin-fixed feces after immunomagnetic separation. Applied and Environmental Microbiology 64: 333-336.

Dowd, S., C. Gerba, and I. Pepper. 1998b. Confirmation of the human-pathogenic Microsporidia *Enterocytozoon bieneusi, Encephalitozoon intestinalis*, and *Vittaforma corneae* in water. Applied and Environmental Microbiology 64: 3332-3335.

Dowd, S.E., C.P. Gerba, and I. Pepper. 1999. Evaluation of methodologies including immunofluorescent assay (IFA) and the polymerase chain reaction (PCR) for detection of human pathogenic Microsporidia in water. Journal of Microbiological Methods 35(1): 43-52.

Dubey, D.P. 1998. *Toxoplasma gondii* oocyst survival under defined temperatures. Journal of Parasitology 84: 862-865.

Dumoulin, A., K. Guyot, E. Lelievre, E. Dei-Cas, and J.C. Cailliez. 2000. *Cryptosporidium* and wildlife: a risk for humans? Parasite-Journal de la Societe Francaise de Parasitologie 7: 167-172.

Edwards, R.T., J.L. Meyer, and S.E.G. Findlay. 1990. The relative contribution of benthic and suspended bacteria to system biomass, production and metabolism in a low-gradient blackwater river. Journal of the North American Benthological Society 9: 216-228.

Egyed, Z., T. Sreter, Z. Szell, and I. Varga. 2003. Characterization of *Cryptosporidium* spp.—recent developments and future needs. Veterinary Parasitology 111(2): 103-114.

Emerson, S.U., and R.H. Purcell. 2003. Hepatitis E virus. Reviews in Medical Virology 13(3): 145-154.

Enriquez, C., N. Nwachuku, and C.P. Gerba. 2001. Direct exposure to animal enteric pathogens. Reviews in Environmental Health 16(2): 117-131.

EPA (U.S. Environmental Protection Agency). 1980. Wastewater Aerosols and Disease: Proceedings of a Symposium, H. Pahren and W. Jakubowski, eds. Cincinnati, Ohio: Health Effects Research Laboratory. EPA-600-9-80-028.

Esiobu, N., L. Armenta, and J. Ike. 2002. Antibiotic resistance in soil and water environments. International Journal of Environmental Health Research 12: 133-144.

Evans, A.S. 1989. Viral Infection of Humans: Epidemiology and Control, 3rd Edition. New York: Plenum Medical Book Company.

Ewald, P.W. 1996. Evolution of Infectious Disease. Oxford, U.K.: Oxford University Press.

Fayer, R. 2000. Global change and emerging infectious diseases. Journal of Parasitology 86: 1174-1181.

Fayer, R., U. Morgan, and S.J. Upton. 2000. Epidemiology of *Cryptosporidium*: Transmission, detection and identification. International Journal of Parasitology 30(12-13): 1305-1322.

Fenchel, T. 1987. Ecology of Protozoa: The Biology of Free-Living Phagotrophic Protists. Madison, Wisconsin: Science Tech Publishers.

Fliermans, C.B., W.B. Cherry, L.H. Orrison, S.J. Smith, D.L. Tison, and D.H. Pope. 1981. Ecological distribution of *Legionella pneumophila*. Applied and Environmental Microbiology 41: 9-16.

Follet-Dumoulin, A., K. Guyot, S. Duchatelle, B. Bourel, F. Guilberrt, E. Dei-Cas, D. Gosset, and J.C. Cailiez. 2001. Involvement of insects in the dissemination of *Cryptosporidium* in the environment. Journal of Eukaryotic Microbiology 48(Suppl. 1): 36S.

Friedrich, F. 2000. Molecular evolution of oral poliovirus vaccine strains during multiplication in humans and possible implications for global eradication of poliovirus. Acta Virology 44(2): 109-117.

Frost, F.J., T. Muller, G.F. Craun, W.B. Lockwood, and R.L. Calderon. 2002. Serological evidence of endemic waterborne cryptosporidium infections. Annals of Epidemiology 12(4): 222-227.

Fry, J.C., and M.J. Day. 1990. Plasmid transfer in the epilithon. Pp. 55-80 in Bacterial Genetics in Natural Environments, J.C. Fry and M.J. Day, eds. London: Chapman and Hall.

Fulton, C., and E.Y. Lai. 1998. Stable intermediates and holdpoints in the rapid differentiation of *Naegleria*. Experimental Cell Research 242: 429-438.

Gauthier, F., and F. Archibald. 2001. The ecology of "fecal indicator" bacteria commonly found in pulp and paper mill water systems. Water Research 35: 2207-2218.

Ghendon, Y., and S.E. Robertson. 1994. Interrupting the transmission of wild polioviruses with vaccines: Immunological considerations. Bulletin of the World Health Organization 72(6):973-983.

Gilles, H.M. 2000. Protozoal Diseases. London: Edward Arnold (Hodder Arnold).

Gow, N.A.R., A.J.P. Brown, and F.C. Odds. 2002. Fungal morphogenesis and host invasion. Current Opinion in Microbiology 5(4): 366-371.

Grabow, W.O.K., M.O. Favorov, N.S. Khudyakova, M.B. Taylor, and H.A. Fields. 1994. Hepatitis E seroprevalence in selected individuals in South Africa. Journal of Medical Virology 44:384-388.

Graczyk, T.K., R. Fayer, and M.R. Cranfield. 1996. *Cryptosporidium parvum* is not transmissible to fish, amphibians, or reptiles. Journal of Parasitology 82(5): 748-751.

Graczyk, T.K., R. Fayer, J.M. Trout, E.J. Lewis, C.A. Farley, I. Sulaiman, and A.A. Lal. 1998. *Giardia* sp. cysts and infectious *Cryptosporidium parvum* oocysts in the feces of migratory Canada geese (*Branta candensis*). Applied and Environmental Microbiology 64: 2736-2738.

Griffin, D.W., K.A. Donaldson, J.H. Paul, and J.B. Rose. 2003. Pathogenic human viruses in coastal waters. Clinical Microbiology Reviews 16(1): 129-143.

Guerrant, R.L. 1997. Cryptosporidiosis: An emerging, highly infectious threat. Emerging Infectious Diseases 3(1): 51-57.

Gutierrez, C. 1999. Geminivirus DNA replication. Cell and Molecular Life Sciences 56(3-4): 313-329.

Haburchak, D.R. 1996. *Aeromonas hydrophilia*: An underappreciated danger to fishermen. Infections in Medicine 13: 893-896.

Hamilton, W.D., and T.M. Lenton. 1998. Spora and gaia: How microbes fly with their clouds. Ethology, Ecology, and Evolution 10: 1-16.

Hanberger, H., J. Garcia-Rodriguez, M. Gobernado, H. Goossens, L.E. Nilsson, L. Struelens, and the French and Portuguese ICU Study Groups. 1999. Antibiotic susceptibility among aerobic Gram-negative bacilli in intensive care units in 5 European countries. Journal of the American Medical Association 281: 67-71.

Hanley-Bowdoin, L., S.B. Settlage, B.M. Orozco, S. Nagar, and D. Robertson. 2000. Geminiviruses: Models for plant DNA replication, transcription, and cell cycle regulation. Critical Reviews in Biochemistry and Molecular Biology 35(2): 105-140.

Harb, O.S., L.Y. Gao, and Y. Abu Kwaik. 2000. From protozoa to mammalian cells: A new paradigm in the life cycle of intracellular bacterial pathogens. Environmental Microbiology 2: 251-265.

Harf, C. 1994. Free-living ameba: interactions with environmental pathogenic bacteria. Endocytobiosis and Cell Research 10: 167-183.

Havelaar, A.H. 1993. Bacteriophages as models of human enteric viruses in the environment. American Society for Microbiology News 59: 614-619.

Hay, A.J., V. Gregory, A.R. Douglas, and Y.P. Lin. 2001. The evolution of human influenza viruses. Philosophical Transactions - Royal Society of London Series B Biological Sciences 356(1416): 1861-1870.

Hazen, T.C., and C.B. Fliermans. 1979. Distribution of *Aeromonas hydrophila* in natural and man-made thermal effluents. Applied and Environmental Microbiology 38: 166-168.

Heitman, T.L., L.M. Frederick, J.R. Viste, N.J. Guselle, U.M. Morgan, R.C.A. Thompson, and M.E. Olson. 2002. Prevalence of *Giardia* and *Cryptosporidium* and characterization of *Cryptosporidium* spp. isolated from wildlife, human, and agricultural sources in the North Saskatchewan River Basin in Alberta, Canada. Canadian Journal of Microbiology 48: 530-541.

Hendrix, R.W., J.G. Lawrence, G.F. Hatfull, and S. Casjens. 2000. The origins and ongoing evolution of viruses. Trends in Microbiology 8(11): 504-508.

Hill, D., and J.P. Dubey. 2002. *Toxoplasma gondii*: Transmission, diagnosis and prevention. Clinical Microbiology and Infection 8: 634-640.

Holmes, E.C. 2001. On the origin and evolution of the human immunodeficiency virus (HIV). Biological Reviews of the Cambridge Philosophy Society 76(2): 239-254.

Hot, D., O. Legeay, J. Jacques, C. Gantzer, Y. Caudrelier, K. Guyard, M. Lange, and L. Andreoletti. 2003. Detection of somatic phages, infectious enteroviruses and enterovirus genomes as indicators of human enteric viral pollution in surface water. Water Research 37(19): 4703-4710.

Hsu, F.C., Y.S. Shieh, J. van Duin, M.J. Beekwilder, and M.D. Sobsey. 1995. Genotyping male-specific RNA coliphages by hybridization with oligonucleotide probes. Applied and Environmental Microbiology 61(11): 3960-3966.

Hull, H.F., and R.B. Aylward. 2001. Progress towards global polio eradication. Vaccine 19(31): 4378-4384.

Hyams, K.C. 2002. New perspectives on hepatitis E. Current Gastroenterology Reports 4(4): 302-307.

Hyder, S.L., and M.M. Streitfeld. 1978. Transfer of erythromycin resistance from clinically isolated lysogenic strains of *Streptococcus pyogenes* via their endogenous phage. Journal of Infectious Diseases 138(3): 281-286.

Issac-Renton, J., W.R. Bowie, A. King, G.S. Irwin, C.S. Ong, C.P. Fung, M.O. Shokeir, and J.P. Dubey. 1998. Detection of *Toxoplasma gondii* oocysts in drinking water. Applied and Environmental Microbiology 64: 2278-2280.

Isaac-Renton, J., J. Blatherwick, W.R. Bowie, M. Fyfe, M. Khan, A. Li, A. King, M. McLean, L. Medd, W. Moorehead, C.S. Ong, and W. Robertson. 1999. Epidemic and endemic seroprevalence of antibodies to *Cryptosporidium* and *Giardia* in residents of three communities with different drinking water supplies. American Journal of Tropical Medicine and Hygiene 60(4): 578-583.

Jellison, K.L., H.F. Hemond, and D.B. Schauer. 2002. Sources and species of *Crypotsporidium* oocysts in the Wachusett Reservoir watershed. Applied and Environmental Microbiology 68: 569-575.

Jiang, B., J.R. Gentsch, and R.I. Glass. 2002. The role of serum antibodies in the protection against rotavirus disease: An overview. Clinical Infectious Diseases 34: 1351-1361.

Jiang, S., R. Noble, and W. Chu. 2001. Human adenoviruses and coliphages in urban runoff-impacted coastal waters of Southern California. Applied and Environmental Microbiology 67(1): 179-184.

Jindrak, L., and L. Grubhoffer. 1999. Animal virus receptors. Folia Microbiology (Praha) 44(5): 467-486.

John, D.T. 1993. Opportunistically pathogenic free-living ameba. Pp. 143-246 in Parasitic Protozoa, Volume 3, 2nd Edition, J.P. Kreier and J.R. Baker, eds. New York: Academic Press.

Kaaden, O.R., W. Eichhorn, and S. Essbauer. 2002. Recent developments in the epidemiology of virus diseases. Journal of Veterinary Medicine, Series B. 49(1): 3-6.

Kadavy, D.R., J.M. Hornby, T. Haverkost, and K.W. Nickerson. 2000. Natural antibiotic resistance of bacteria isolated from larvae of the oil fly, *Helaeomyia petrolei*. Applied and Environmental Microbiology 66: 4615-4619.

Kamel, M.A., H. Troonen, H.P. Kapprell, A. el-Ayady, and F.D. Miller. 1995. Seroepidemiology of hepatitis E virus in the Egyptian Nile Delta. Journal of Medical Virology 47(4): 399-403.

Kew, O., V. Morris-Glasgow, M. Landaverde, C. Burns, J. Shaw, Z. Garib, J. Andre, E. Blackman, C.J. Freeman, J. Jorba, R. Sutter, G. Tambini, L. Venczel, C. Pedreira, F. Laender, H. Shimizu, T. Yoneyama, T. Miyamura, H. van Der Avoort, M.S. Oberste, D. Kilpatrick, S. Cochi, M. Pallansch, and C. de Quadros. 2002. Outbreak of poliomyelitis in Hispaniola associated with circulating type 1 vaccine-derived poliovirus. Science 296(5566): 356-359.

Kitamoto, I., and K. Tanabe. 1987. Secretion by *Toxoplasma gondii* of an antigen that appears to become associated with the parasitophorous vacuole membrane upon invasion of the host cell. Journal of Cell Science 88: 231-239.

Krawczynski, K., S. Kamili, and R. Aggarwal. 2001. Global epidemiology and medical aspects of hepatitis E. Forum (Genova) 11(2): 166-179.

Kurath, G., and R.Y. Morita. 1983. Starvation-survival physiological studies of a marine *Pseudomonas* sp. Applied and Environmental Microbiology 45:1206-1211.

Lalitha, K.V., and K. Gopakumar. 2000. Distribution and ecology of *Clostridium botulinum* in fish and aquatic environments of a tropical region. Food Microbiology 17(5): 535-541.

Landaverde, M., L. Venczel, and C.A. de Quadros. 2001. Poliomyelitis outbreak caused by vaccine-derived virus in Haiti and the Dominican Republic. Pan American Journal of Public Health 9(4): 272-274.

Lebaron, P., P. Servais, M. Troussellier, C. Courties, J Vives-Rego, G. Muyzer, L. Bernard, T. Guindulain, H. Schafer, and E. Stackebrandt. 1999. Changes in bacterial community structure in seawater mesocosms differing in their nutrient status. Aquatic Microbial Ecology 19: 255-267.

LeChevallier, M.W., W.D. Norton, and R.G. Lee. 1991a. *Giardia* and *Cryptosporidium spp.* in filtered drinking water supplies. Applied and Environmental Microbiology 57(9):2617-2621.

LeChevallier, M.W., W.D. Norton, and R.G. Lee. 1991b. Occurrence of *Giardia* and *Cryptosporidium* spp. in surface-water supplies. Applied and Environmental Microbiology 57(9): 2610-2616.

Leff, L.G., J.V. McArthur, J.L. Meyer, and L.J. Shimkets. 1994. Effect of macroinvertebrates on detachment of bacteria from biofilms in stream microcosms. Journal of the North American Benthological Society 13: 74-79.

Leong, L.Y.C. 1983. Removal and inactivation of viruses by treatment processes for potable water and wastewater—a review. Water Science and Technology 15:91-114.

Levy, S.B. 1997. Antibiotic resistance: An ecological imbalance. Pp. 1-14 in Antibiotic Resistance: Origins, Evolution, Selection and Spread. Ciba Foundation Symposium 207. Chichester, U.K.: Wiley and Sons.

Levy, S.B. 1998. Multidrug resistance—a sign of the times. New England Journal of Medicine 338: 1376-1378.

Liebert, C.A., R.M. Hall, and A.O. Summers. 1999. Transposon Tn21, flagship of the floating genome. Microbiology and Molecular Biology Reviews 63(3): 507-522.

Lindsay, D.S., K.K. Phelps, S.A. Smith, G. Flick, S.S. Sumner, and J.P. Dubey. 2001. Removal of *Toxoplasma gondii* oocysts from sea water by eastern oysters (*Crassostrea virginica*). Journal of Eukaryotic Microbiology 48 (Supp. 1): 197S-198S.

Lipp, E.K., J.L. Jarrell, D.W. Griffin, J. Lukasik, J. Jacukiewicz, and J.B. Rose. 2002. Preliminary evidence for human fecal contamination in corals of the Florida Keys, USA. Marine Pollution Bulletin 44(7): 666-670.

Loewe, L., V. Textor, and S. Scherer. 2003. High deleterious genomic mutation rate in stationary phase of *Escherichia coli*. Science 302: 1558-1559.

Lopez, J.J., and M. Wilson. 2003. Congenital toxoplasmosis. American Family Physician 67(10): 2131-2138.

Lowery, C.J., P. Nugent, J.E. Moore, B.C. Milliar, X. Xiru, and J.S.G. Dooley. 2001. PCR-IMS detection and molecular typing of *Cryptosporidium parvum* recovered from a recreational river source and an associated mussel (*Mytilus edulis*) bed in Northern Ireland. Epidemiology and Infection 127: 545-553.

Luder, C.G.K., U. Gross, and M.F. Lopes. 2001. Intracellular protozoan parasites and apoptosis: diverse strategies to modulate parasite-host interactions. Trends in Parasitology 17(10): 480-486.

Ma, P., E. Willart, K.B. Juechter, and A.R. Stevens. 1981. A case of keratitis due to *Acanthameba* in New York, New York and features of 10 cases. Journal of Infectious Diseases 143: 662-667.

Marino, A., G. Crisafi, T.L. Maugeri, A. Nostro, and V. Alonzo. 1999. Uptake and retention of *Vibrio harvey* by mussels in seawater. Microbiologica 22: 129-138.

Marshall, M.M., D. Naumovitz, Y. Ortega, and C.R. Sterling. 1997. Waterborne protozoan pathogens. Clinical Microbiology Reviews 10: 67-85.

Martinez, A.J. 1985. Free-Living Amebas: Natural History, Prevention, Diagnosis, Pathology and Treatment of the Disease. Boca Raton, Florida: CRC Press.

Mathison, B.A., and O. Ditrich. 1999. The fate of *Cryptosporidium parvum* oocysts ingested by dung beetles and their possible role in the dissemination of cryptosporidiosis. Journal of Parasitology 85: 678-681.

Mazel, D., and J. Davies. 1999. Antibiotic resistance in microbes. Cellular and Molecular Life Sciences 56: 742-754.

McArthur, J.V., and R.C. Tuckfield. 2000. Spatial patterns in antibiotic resistance among stream bacteria: Effects of industrial pollution. Applied Environmental Microbiology 66: 3722-3726.

McFadden, G. 1996. Viroceptors, Virokines and Related Immune Modulators Encoded by DNA Viruses. New York: Springer-Verlag.

Medema, G.J., F.M. Schets, P.F.M. Teunis, and A.H. Havelaar. 1998. Sedimentation of free and attached *Cryptosporidium* oocysts and *Giardia* cysts in water. Applied and Environmental Microbiology 64: 4460-4466.

Miao, E.A., and S.I. Miller. 1999. Bacteriophages in the evolution of pathogen-host interactions. Proceedings of the National Academy of Sciences 96: 9452-9454.

Miller, M.A., I.A. Gardner, C. Kreuder, D.M. Paradies, K.R. Worcester, D.A. Jessup, E. Dodd, M.D. Harris, J.A. Ames, A.E. Packham, and P.A. Conrad. 2002. Coastal freshwater runoff is a risk factor for *Toxoplasma gondii* infection of southern sea otters (*Enhydra lutris nereis*). International Journal for Parasitology 32: 997-1006.

Mims, C.A., A. Nash, and J. Stephen. 2001. Mims' Pathogenesis of Infectious Disease, 5th Edition. New York: Academic Press.

Miranda, G., D. Schuppli, I. Barrera, C. Hausherr, J.M. Sogo, and H. Weber. 1997. Recognition of bacteriophage Qbeta plus strand RNA as a template by Qbeta replicase: Role of RNA interactions mediated by ribosomal proteins S1 and host factor. Journal of Molecular Biology 267(5): 1089-1103.

Mitchell, R. 1972. Ecological control of microbial imbalances. Pp. 273-288 in Water Pollution Microbiology, R. Mitchell, ed. New York: Wiley Interscience.

Mohanavalli, B., E. Dhevahi, T. Menon, S. Malathi, and S.P. Thyagarajan. 2003. Prevalence of antibodies to hepatitis A and hepatitis E virus in urban school children in Chennai. Indian Pediatrics 40(4): 328-331.

Morgan, U.M., L. Xiao, R. Fayer, A.A. Lal, and R.C. Thompson. 1999. Variation in *Cryptosporidium*: Towards a taxonomic revision of the genus. International Journal of Parasitology 29(11): 1733-1751.

Morita, R.Y. 1997. Bacteria in Oligotrophic Environments. New York: Chapman Hall.

Mota, P., C. Rauch, and S.C. Edberg. 2000. Microsporidia and Cyclospora: Epidemiology and assessment of risk from the environment. Critical Reviews in Microbiology 26: 69-90.

Nielsen, K.M., M.D. van Weerelt, T.N. Berg, A.M. Bones, A.N. Hagler, and J.D. van Elsas. 1997. Natural transformation and availability of transforming DNA to *Acinetobacter calcoaceticus* in soil microcosms. Applied Environmental Microbiology 63(5): 1945-1952.

NRC (National Research Council). 2002. Biosolids Applied to Land: Advancing Standards and Practices. Washington, D.C.: National Academy Press.

O'Donoghue, P.J. 1995. *Cryptosporidium* and cryptosporidiosis in man and animals. International Journal for Parasitology 25: 139-195.

Ohka, S., and A. Nomoto. 2001. The molecular basis of poliovirus neurovirulence. Developments in Biological Standardization (Basel) 105: 51-58.

Okhuysen, P.C., C.L. Chappell, and H.L. DuPont. 1999. Virulence of three distinct *Cryptosporidium parvum* isolates for healthy adults. Journal of Infectious Diseases 180(4): 1275.

Okhuysen, P.C., and C.L. Chappell. 2002. Cryptosporidium virulence determinants—are we there yet? International Journal for Parasitology 32(5): 517-525.

Oliver, J.D. 1993. Formation of viable but nonculturable cells. Pp. 239-276 in Starvation in Bacteria, S. Kjelleberg, ed. New York: Plenum Press.

Ong, C., W. Moorehead, A. Ross, and J. Isaac-Renton. 1996. Studies of *Giardia* spp. and *Cryptosporidium* spp. in two adjacent watersheds. Applied and Environmental Microbiology 62(8): 2798-2805.

Pallansch, M.A., and R.P. Roos. 2001. Enteroviruses: Polioviruses, coxsackieviruses, echoviruses, and newer enteroviruses. Pp. 723-775 in Fields Virology, B.N. Fields, P.M. Howley, D.E. Griffin, R.A. Lamb, M.A. Martin, B. Roizman, S.E. Straus, and D.M. Knipe, eds. Philadelphia, Pennsylvania: Lippincott Williams & Wilkins Publishers.

Papathanasopoulos, M.A., G.M. Hunt, and C.T. Tiemessen. 2003. Evolution and diversity of HIV-1 in Africa—a review. Virus Genes 26(2): 151-163.

Parashar, U.D., E.G. Hummelman, J.S. Bresee, M.A. Miller, and R.I. Glass. 2003. Global illness and deaths caused by rotavirus disease in children. Emerging Infectious Diseases 9(5): 565-571.

Patz, J.A., D. Engelberg, and J. Last. 2000. The effects of changing weather on public health. Annual Reviews of Public Health 21: 271-307.

Patz, J.A., and W.K. Reisen. 2001. Immunology, climate change and vector-borne diseases. Trends in Immunology 22: 171-172.

Peng, M.M., L. Xiao, A.R. Freeman, M.J. Arrowood, A.A. Escalante, A.C. Weltman, C.S. Ong, W.R. Mac Kenzie, A.A. Lal, and C.B. Beard. 1997. Genetic polymorphism among *Cryptosporidium parvum* isolates: Evidence of two distinct human transmission cycles. Emerging Infectious Disease 3(4): 567-573.

Pettibone, G.W., J.P. Mear, and B.M. Campbell. 1996. Incidence of antibiotic and metal resistance and plasmid carriage in Aeromonas isolated from brown bullhead (*Ictalurus nebulosus*). Letters in Applied Microbiology 23(4): 234.

Powell, K.L., R.G. Taylor, A.A. Cronin, M.H. Barrett, S. Pedley, J. Sellwood, S.A. Trowsdale, and D.N. Lerner. 2003. Microbial contamination of two urban sandstone aquifers in the UK. Water Research 37(2): 339-352.

Pulliam, H.R., and B.J. Danielson. 1991. Sources, sinks, and habitat selection: A landscape perspective on population-dynamics. American Naturalist 137(Supplement S): S50-S66.

Racaniello, V. 2001. Picornaviridae: The viruses and their replication. Pp. 685-722 in Fields Virology, B.N. Fields , P.M. Howley, D.E. Griffin, R.A. Lamb, M.A. Martin, B. Roizman, S.E. Straus, and D.M. Knipe, eds. Philadelphia, Pennsylvania: Lippincott Williams & Wilkins Publishers.

Redlinger, T., K. O'Rourke, L. Nickey, and G. Martinez. 1998. Elevated hepatitis A and E seroprevalence rates in a Texas/Mexico border community. Texas Medicine 94(5): 68-71.

Rheinheimer, G. 1992. Aquatic Microbiology. 4th Edition. West Sussex, U.K.: John Wiley and Sons, Ltd.

Rickard, L.G., C. Siefker, C.R. Boyle, and E.J. Gentz. 1999. The prevalence of *Cryptosporidium* and *Giardia* spp. in fecal samples from free-ranging white-tailed deer (*Odocoileus virginianus*) in southeastern United States. Journal of Veterinary Diagnostic Investigation 11: 65-72.

Roberts, L.S., and J.J. Janovy, Jr. 1995. Foundations of Parasitology, 5th Edition. Dubuque, Iowa: William C. Brown Publishers.

Robertson, L.J., A.T. Campbell, and H.V. Smith. 1992. Survival of *Cryptosporidium parvum* oocysts under various environmental pressures. Applied and Environmental Microbiology 58(11): 3494-3500.

Rodriguezzaragoza, S. 1994. Ecology of free-living amebas. Critical Reviews in Microbiology 20: 225-241.

Rooney, P.J., and B.S. Klein. 2002. Linking fungal morphogenesis with virulence. Cellular Microbiology 4(3): 127-138.

Rosas, I., C. Calderon, M. Ulloa, and J. Lacey. 1993. Abundance of airborne *Penicillium* CFU in relation to urbanization in Mexico City. Applied and Environmental Microbiology 59: 2648-2652.

Rosas, I., E. Salinas, A. Yela, E. Calva, C. Eslava, and A. Cravioto. 1997. *Escherichia coli* in settled-dust and air samples collected in residential environments in Mexico City. Applied and Environmental Microbiology 63: 4093-4095.

Rose, J.B. 1997. Environmental ecology of *Cryptosporidium* and public health implications. Annual Review of Public Health 18: 135-161.

Rose, J.B., and T.R. Slifko. 1999. *Giardia, Cryptosoridium*, and *Cyclospora* and their impact on foods: A review. Journal of Food Protection 62(9): 1059-1070.

Rose, J.B., D.E. Huffman, K. Riley, S.R. Farrah, J.O. Lukasik, and C.L. Harman. 2001a. Reduction of enteric microorganisms at the Upper Occoquan Sewage Authority Water Reclamation Plant. Water Environmental Resources 73(6): 711-720.

Rose, J.B., P.R. Epstein, E.K. Lipp, B.H. Sherman, S.M. Bernard, and J.A. Patz. 2001b. Climate variability and change in the United States: Potential impacts on water- and foodborne diseases caused by microbiologic agents. Environmental Health Perspectives 109(Supp. 2): 211-221.

Rose, J.B., D.E. Huffman, and A. Gennaccaro. 2002. Risk and control of waterborne cryptosporidiosis. FEMS Microbiology Reviews 26: 113-123.

Roszak, D.B., and R.R. Colwell. 1987. Metabolic activity of bacterial cells enumerated by direct viable count. Applied and Environmental Microbiology 53: 2889-2983.

Scandura, J.E., and M.D. Sobsey. 1996. Viral and bacterial contamination of groundwater from on-site sewage treatment systems. Water Science and Technology 35(11-12): 141-146.

Scholtissek, C. 1995. Molecular evolution of influenza viruses. Virus Genes 11(2-3): 209-215.

Schuppli, D., J. Georgijevic, and H. Weber. 2000. Synergism of mutations in bacteriophage Qbeta RNA affecting host factor dependence of Qbeta replicase. Journal of Molecular Biology 295(2): 149-154.

Schuster, F.L. 2002. Cultivation of pathogenic and opportunistic free-living amebas. Clinical Microbiology Reviews 13(3): 342-354.

Scott, M.E., and G. Smith. 1994. Parasitic and Infectious Diseases: Epidemiology and Ecology. New York: Academic Press.

Seveno, N.A., D. Kallifidas, K. Smalla, J.D. van Elsas, J.M. Collard, A.D. Karagouni, and E.M.H. Wellington. 2002. Occurrence and reservoirs of antibiotic resistance genes in the environment. Reviews in Medical Microbiology 13: 15-27.

Shadduck, J.A., and M.B. Polley. 1978. Some factors influencing the *in vitro* infectivity and replication of *Encephalitozoon cuniculi*. Journal of Protozoology 25: 491-496.

Shadduck, J.A. 1989. Human microsporidiosis and AIDS. Reviews in Infectious Diseases 11: 203-207.

Skerrett, H.E., and C.V. Holland. 2000. The occurrence of *Cryptosporidium* in environmental waters in the greater Dublin area. Water Research 34: 3755-3760.

Slifko, T.A., H.V. Smith, and J.B. Rose. 2000. Emerging parasite zoonoses associated with water and food. International Journal of Parasitology 30: 1379-1393.

Smith, A.W., D.O. Matson, D.A. Stein, D.E. Skilling, A.D. Kroeker, T. Berke, and P.L. Iversen. 2002. Antisense treatment of Caliciviridae: An emerging disease agent of animals and humans. Current Opinions in Molecular Therapy 4(2): 177-184.

Smith, H.V., and J.B. Rose. 1998. Waterborne cryptosporidiosis current status. Parasitology Today 14(1): 14-22.

Sobecky, P.A., T.J. Mincer, M.C. Chang, and D.R. Helinski. 1997. Plasmids isolated from marine sediment microbial communities contain replicon and incompatibility regions unrelated to those of known plasmid groups. Applied and Environmental Microbiology 63: 888-895.

Sobecky, P.A., T.J. Mincer, M.C. Chang, A. Toukdarian, and D.R. Helinski. 1998. Isolation of broad-host-range replicons from marine sediment bacteria. Applied and Environmental Microbiology 64: 2822-2830.

Sobecky, P.A. 1999. Plasmid ecology of marine sediment microbial communities. Hydrobiologia 401: 9-18.

Son, R., G. Rusul, A.M. Sahilah, A. Zainuri, A.R. Raha, and I. Salmah. 1997. Antibiotic resistance and plasmid profile of *Aeromonas hydrophila* isolates from cultured fish, telapia (*Telapia mossambica*). Letters in Applied Microbiology 24(6): 479-482.

Sparfel, J.M., C. Sarafati, O. Ligoury, B. Caroff, N. Dumoutier, B. Gueglio, E. Billaud, F. Raffi, J.M. Molin, M. Miegeville, and F. Derouin. 1997. Detection of Microsporidia and identification of *Enterocytozoon bieneusi* in surface water by filtration followed by specific PCR. Journal of Eukaryotic Microbiology 44: 78.

Steiner, T.S., N.M. Theilman, and R.L. Guerrant. 1997. Protozoal agents: What are the dangers for the public water supply? Annual Review of Medicine 48: 329-340.

Steinert, M., K. Birkness, E. White, B. Fields, and F. Quinn. 1998. *Mycobacterium avium* bacilli grow saprozoically in co-culture with *Acanthamaoeba polyphaga* and survive within cyst walls. Applied and Environmental Microbiology 64: 2256-2261.

Stewart, G.J., and C.D. Sinigalliano. 1991. Exchange of chromosomal markers by natural transformation between the soil isolate, *Pseudomonas stutzeri* JM300, and the marine isolate, *Pseudomonas stutzeri* strain ZoBell. Antonie Van Leeuwenhoek 59(1): 19-25.

Stossel, T.F. 1994. The machinery of cell crawling. Scientific American 271: 54-63.

Stott, R., E. May, E. Matsushita, and A. Warren. 2001. Protozoan predation as a mechanism for the removal of *Cryptosporidium* oocysts from wastewaters in constructed wetlands. Water Science and Technology 44: 191-198.

Szenasi, Z., T. Endo, K. Yagita, and B. Nagy. 1998. Isolation, identification and increasing importance of "free-living" amoebae causing human disease. Journal of Medical Microbiology 47: 5-16.

Teunis, P.F.M., C.L. Chappell, and P.C. Okhuysen. 2002. *Cryptosporidium* dose-response studies: Variation between hosts. Risk Analysis 22(3): 475-486.

Thom, S., D. Warhurst, and B.S. Drasar. 1992. Association of *Vibrio cholerae* with freshwater amebas. Journal of Medical Microbiology 36: 303-306.

Thompson, R.C.A., and A.J. Lymbery. 1996. Genetic variability in parasites and host-parasite interactions. Parasitology 112(Supp. S): S7-S22.

Thurston-Enriquez, J.A., P. Watt, S.E. Dowd, J. Enriquez, I.L Pepper, and C.P. Gerba. 2002. Detection of protozoa parasites and Microsporidia in irrigation waters used for crop production. Journal of Food Protection 65: 378-382.

Tomoeda, M., A. Shuta, and M. Inuzuka. 1972. Studies on sex pili: Mutants of the sex factor F in *Escherichia coli* defective in bacteriophage-adsorbing function of F pili. Journal of Bacteriology 112(3): 1358-1363.

Tomov, A.T., E.D. Tsvetkova, I.A. Tomova, L.I. Michailova, and V.K. Kassovski. 1999. Persistence and multiplication of obligate anaerobe bacteria in amebae under aerobic conditions. Anaerobe 5: 19-23.

Tyler, K.L., and N. Nathanson. 2001. Pathogenesis of viral infections. Pp. 199-243 in Fields Virology, B.N. Fields, P.M. Howley, D.E. Griffin, R.A. Lamb, M.A. Martin, B. Roizman, S.E. Straus, and D.M. Knipe, eds. Philadelphia, Pennsylvania: Lippincott Williams & Wilkins Publishers.

van Burik, J.A.H., and P.T. Magee. 2001. Aspects of fungal pathogenesis in humans. Annual Review of Microbiology 55: 743-772.

Van Heerden, J., M.M. Ehlers, W.B. Van Zyl, and W.O. Grabow. 2003. Incidence of adenoviruses in raw and treated water. Water Research 37(15): 3704-3708.

Wade, T.J., N. Pai, J.N. Eisenberg, J.M. Colford, Jr. 2003. Do U.S. Environmental Protection Agency water quality guidelines for recreational waters prevent gastrointestinal illness? A systematic review and meta-analysis. Environmental Health Perspectives 111(8): 1102-1109.

Waller, T. 1979. Sensitivity of *Encephalitozoon cuniculi* to various temperatures, disinfectants and drugs. Lab Animals 13: 227-230.

Webby, R.J., and R.G. Webster. 2003. Are we ready for pandemic influenza? Science 302(5650): 1519-1522.

Webster, R.G., S.M. Wright, M.R. Castrucci, W.J. Bean, and Y. Kawaoka. 1993. Influenza - a model of an emerging virus disease. Intervirology 35(1-4): 16-25.

Weiss, L.M. 2001. Microsporidia: Emerging pathogenic protists. Acta Tropica 78: 89-102.

Weiss, R.A. 2003. Cross-species infections. Current Topics in Microbiology and Immunology 278: 47-71.

White, D.O., and F.J. Fenner. 1994. Medical Virology, 4th Edition. New York: Academic Press.

Wiersma, S. 2002. Primary Amebic Meningoencephalitis. Tallahassee, Florida: Florida Department of Health.

Williams, H.G., M.J. Day, J.C. Fry, and G.J. Stewart. 1996. Natural transformation in river epilithon. Applied Environmental Microbiology 62: 2994-2998.

Winiecka-Krusnell, J., and E. Linder. 1999. Free-living amoebae protecting *Legionella* in water: The tip of an iceberg? Scandinavian Journal of Infectious Diseases 31: 383-385.

Wise, M.G., J.V. McArthur, C. Wheat, and L.J. Shimkets. 1997. Temporal variation in genetic diversity and structure of a lotic population of *Burkholderia* (*Pseudomonas*) *cepacia*. Applied and Environmental Microbiology 63: 1505-1514.

Wolk, D.M., C.H. Johnson, E.W. Rice, M.M. Marshall, K.F. Grahn, C.B. Plummer, and C.R. Sterling. 2000. A spore counting method and cell culture model for chlorine disinfection studies of *Encephalitozoon* syn. *Septata intestinalis*. Applied and Environmental Microbiology 66: 1266-1273.

Wommack, K.E., and R.R. Colwell. 2000. Virioplankton: Viruses in aquatic ecosystems. Microbiology and Molecular Biology Review 64: 69-114.

Xiao, L., I.M. Sulaiman, U.M. Ryan, L. Zhou, E.R. Atwill, M.L. Tischler, X. Zhang, R. Fayer, and A.A. Lal. 2002. Host adaptation and host-parasite co-evolution in *Cryptosporidium*: Implications for taxonomy and public health. International Journal of Parasitology 32: 1773-1785.

Yoneyama, T., H. Yoshida, H. Shimizu, K. Yoshii, N. Nagata, O. Kew, and T. Miyamura. 2001. Neurovirulence of Sabin 1-derived poliovirus isolated from an immunodeficient patient with prolonged viral excretion. Developments in Biological Standardization (Basel) 105: 93-98.

Zu, S.X., J.F. Li, L.J. Barrett, R. Fayer, S.Y. Shu, J.F. McAuliffe, J.K. Roche, and R.L. Guerrant. 1994. Seroepidemiologic study of *Cryptosporidium* infection in children from rural communities of Anhui, China and Fortaleza, Brazil. American Journal of Tropical Medicine and Hygiene 51(1): 1-10.

4

Attributes and Application of Indicators

INTRODUCTION

Microbial water quality indicators are used in a variety of ways within public health risk assessment frameworks, including assessment of potential hazard, exposure assessment, contaminant source identification, and evaluating effectiveness of risk reduction actions. The most desirable indicator attributes, and therefore the most appropriate indicators, naturally depend on their manner of use. This chapter describes desirable attributes of an indicator, typical applications of indicators, indicator attributes that are appropriate for such applications, and provides an assessment of whether current indicators and indicator approaches are meeting the needs of each application. The chapter ends with a summary of its conclusions and recommendations.

INDICATOR ATTRIBUTES

For almost 40 years, Bonde's (1966) attributes of an ideal indicator have served as an effective model of how a fecal contamination index for public health risk and treatment efficiency should function (Box 4-1). Three of Bonde's attributes (1, 2, and 4) address the relationship between indictor organisms and pathogens of concern, while the remaining five describe desirable properties associated with quantifying the indicator. However, Bonde's attributes of an ideal indicator must be refined to continue their relevance to public health protection because the development and increasing availability of new measurement methods necessitates the separation of criteria for evaluating indicators and detection

---

BOX 4-1
Bonde's (1966) Criteria for an Ideal Indicator

An ideal indicator should

1. Be present whenever the pathogens are present;
2. Be present only when the presence of pathogens is an imminent danger (i.e., they must not proliferate to any greater extent in the aqueous environment);
3. Occur in much greater numbers than the pathogens;
4. Be more resistant to disinfectants and to the aqueous environment than the pathogens;
5. Grow readily on simple media;
6. Yield characteristic and simple reactions enabling as far as possible an unambiguous identification of the group;
7. Be randomly distributed in the sample to be examined, or it should be possible to obtain a uniform distribution by simple homogenization procedures; and
8. Grow widely independent of other organisms present, when inculcated in artificial media (i.e., indicator bacteria should not be seriously inhibited in their growth by the presence of other bacteria).

---

methods. Historic definitions of microbial indicators, such as coliforms, have been tied to the methods used to measure them. Newly available methods (particularly molecular methods; see Chapter 5 and Appendix C) allow more specificity in the taxonomic grouping of microorganisms that are measured. More importantly, a variety of new methods are becoming increasingly available, providing several options for measuring each indicator group. Thus, separate criteria allow one to choose the indicator with the most desirable biological attributes for a given application and then match this with a measurement method that best meets the need of the application. Box 4-2 lists desirable biological attributes of indicators and Box 4-3 lists desirable attributes of methods.

Biological Attributes

The most important biological attribute is a strong quantitative relationship between indicator concentration and the degree of public health risk. This relationship has been demonstrated primarily through epidemiologic studies for recreational exposures (Cabelli et al., 1979; Cheung et al., 1990; Seyfried et al., 1985a,b; Zmirou et al., 1987). An alternative means of demonstrating the relationship to health risk is through correlation between prospective indicator concentration and pathogen levels (Gerba et al., 1979; Labelle et al., 1980; Lipp et

BOX 4-2
Desirable Biological Attributes of Indicators

- Correlated to health risk
- Similar (or greater) survival to pathogens
  - Ultraviolet exposure
  - Temperature
  - Salinity
  - Predation by indigenous flora
  - Desiccation
  - Freezing
  - Biologic survival mechanisms
    - Sporulation
    - Cyst and other latency mechanisms
    - Arrested metabolism (viable but non-culturable)
    - Shock proteins and other biochemical survival strategies
  - Response to disinfectants
- Similar (or greater) transport to pathogens
  - Filtration
  - Sedimentation or settling
  - Adsorption to particles
- Present in greater numbers than pathogens
- Specific to a fecal source or identifiable as to source of origin

BOX 4-3
Desirable Attributes of Methods

- Specificity to desired target organism
  - Independent of matrix effects
- Broad applicability
- Precision
- Adequate sensitivity
- Rapidity of results
- Quantifiable
- Measures viability or infectivity
- Logistical feasibility
  - Training and personnel requirements
  - Utility in field
  - Cost
  - Volume requirements

al., 2001a; Robertson, 1984; Seyfried et al., 1984). The latter approach is used less frequently because assays for pathogens are specific to individual agents or classes of agents (e.g., enteroviruses) and correlation with a single pathogen, or subset of pathogens, does not establish a relationship with all illness-causing agents or their risks to human health (their health effects).

The next two desirable biological attributes are similarity in survival and transport characteristics of the indicator to those of the pathogen(s) of interest. If there is differential transport or survival, the relationship between pathogen and indicator concentrations will change at varying distances from the source and over different times in the environment, making it difficult to select a critical indicator concentration on which to make public health decisions (Griffin et al., 2001). For example, differences in viral and bacterial transport through soils and aquifers have been found to affect assessment of water quality impacts from septic systems (Harden et al., 2003). If there is differential survival, it is generally preferable that the indicator be more resilient than the pathogens so as to be protective of public health. However, exceptionally long survival of potential indicators, such as spore-forming *Clostridium perfringens*, may render them too overprotective or nondiscriminatory because they may be present at concentrations mistakenly considered to be indicative of a health risk long after the pathogens have declined to levels not considered a risk.

The next desirable attribute is that the indicator be present at densities that are detectable with an easily sampled volume. It is always possible to measure lower concentrations of indicators through use of high-volume collection strategies, but it is typically preferable for indicators to be present at high enough density to be detected easily in sample volumes that are convenient to collect and transport to a laboratory for analysis. Pathogens are excreted by infected individuals in numbers per gram of feces are comparable to that of coliforms (Gerba, 2001). However, domestic wastewater contains a mixture of excreta from a variety of people, many of whom are not infected with a pathogen but excrete coliforms and other microbial indicators. Thus, the indicators are present in wastewater at densities several thousand times higher than that of most pathogens, including enteric viruses and protozoa (Feachem et al., 1983; Rose et al., 2001).

The final desirable biological attribute is source-specificity. Indicators that are specific to animal digestive systems are preferable to those that occur naturally in the ambient environment, because the dichotomy of sources may lead to different risk potential depending on the nature of the source. A similar, though lesser, concern exists when the indicator occurs in the gut flora of numerous animal species, because of the difference in pathogen types and concentrations excreted among species. Some indicator microorganisms, while not source specific, have genotypic or phenotypic properties that allow distinction as to whether the fecal source is human or animal (Simpson et al., 2002). Other indicators even allow for identification of particular animal species contributing to the fecal contamination, which can be used to indicate the degree or type of risk. For example, the proximity of cattle to a water source could indicate a concern regarding *Cryptosporidium* and *Escherichia coli* O157:H7 because these pathogens are common in cattle (LeChevallier et al., 1999a,b).

Attributes of Methods

The attributes of a method that should be considered are not independent of one another, and these relationships are described in the following text. One of the most important method attributes is specificity, or ability to measure the target indicator organism in an unbiased manner. Specificity may be directed at microorganism groups (e.g., coliforms, cultivatable enteroviruses), genera (e.g., *Giardia*), species (e.g., *Cryptosporidium parvum*), or subtypes (e.g., *E. coli* O157:H7). Specificity can also be described on a biochemical, antigenic, or genetic basis.

In most cases, the specificity concern is for false positives, in which a confounding organism reacts similarly in the test and yields incorrectly high results. Among newer methods, Pisciotta et al. (2002) suggest that coliform measurements can be confounded with *Vibrio cholerae* counts in subtropical environments when using chromogenic substrate techniques. However, there are cases in which false negatives are of concern, such as when high levels of heterotrophic plate count microorganisms may, in some instances, interfere with the detection of coliforms (Allen, 1977; Edberg and Smith, 1989).

Lack of specificity can also be introduced from matrix interferences. Many waters that are tested for microbiological quality are saline, or turbid, or have a high organic content, all of which have the potential to interfere with some indicator measurement methods (Geldreich, 1978). For example, tannic and humic acids from decaying plant material can interfere with some molecular methods. Filtration methods are particularly susceptible to high suspended solid load, which can cause clogging or clumping. Low levels of residual chlorine can produce sublethal injury to coliforms, interfering with their enumeration on highly differential media (Camper and McFeters, 1979; McFeters et al., 1986), although this will be of greater concern in treated water monitoring systems. It is also desirable for a method to have broad applicability to a number of geographic locations (tropical waters versus temperate waters), various types of watersheds (e.g., point source and nonpoint source inputs), and different water matrices.

Preferred methods will also measure indicator concentrations precisely, which is particularly important when decisions must be made on a limited number of samples. Method precision includes not only repeatability with a laboratory, but variability across laboratories. Generally, greater precision is better, but in particular the precision must meet the needs for the decision-making process. Multiple tube fermentation, which has been one of the most frequently used indicator methods, is based on a statistical approach to estimating concentrations and has a coefficient of variation equal to more than half the mean (Noble et al.,

2003a), yet interlaboratory variability has been found to be acceptable for most applications.

Sensitivity is the lower limit of detection of an indicator in a certain sample volume and has implications for precision. The needed sensitivity may be risk based, technology based or management based. Methods that amplify or concentrate the target are typically more sensitive (e.g., culture, polymerase chain reaction [PCR], filtration). Methods may be quite amenable to changes in the sensitivity (e.g., membrane filtration and fecal coliform cultivation) but at some point they become technologically limited (e.g., via clogging of the filter and masking of the bacteria). Sensitivity is also affected by the sample volume, particularly if the target indicator concentration is low relative to the volume analyzed and detection is reduced to a "Poissonian sampling" process (see Chapter 5 and Figure 5-5 for further information). Although sensitivity concerns can be overcome by processing larger sample volumes, this can affect logistical feasibility in some applications.

It may not be necessary in all cases to be quantifiable. In some applications, presence/absence information may suffice, particularly since counting can be tedious, adds expense, and typically increases the time of the assay. However, quantification increases precision and is necessary in most applications associated with assessing public health risk.

The speed of the method is an important characteristic, particularly when warning systems (discussed later) are involved and human exposure continues to occur during the laboratory analysis period. Methods vary widely in their speed; with faster molecular methods soon becoming available to replace traditional culture-based methods (see Chapter 5 and Appendix C for further discussion). Culture-based methods often take several days to complete, whereas molecular methods take hours or less. However, hybrid approaches employing brief culture periods (to ensure the culturability or infectivity of the microbe) coupled with rapid molecular detection have the potential to rapidly detect and quantify culturable microbes in environmental samples. This has been particularly useful in decreasing the time for virus detection in cell culture (Reynolds et al., 1996).

Many indicator methods that are able to produce results rapidly do so by measuring molecular properties that do not address viability or infectivity. Thus, high indicator counts may be recorded in areas where chemical or physical agents have been effective at inactivating pathogens. Viability or infectivity is an important issue because the epidemiologic studies on which current standards are based have all been conducted with culture-based methods, and it is not clear how well those epidemiologic relationships will hold if nonviable indicators are included in the counts.

Logistical feasibility will often govern the indicator method of choice. Cost concerns can be important when large numbers of samples are needed for screening purposes, but they may be less important when the consequences to be addressed have major impacts on human health risk, such as the risk of an outbreak or a high burden of disease related to the exposure. Costs include not just labor and materials, but also capital and training costs. Many of the new measurement technologies require large initial investments because the equipment and personnel necessary to implement them are not already in place. Moreover, simpler methods with proven field utility and small volume requirements may be preferred when applications are most appropriately implemented on-site using typically less well-trained personnel, such as lifeguards.

Finally, although not considered a method attribute per se, all methods are amenable to some form of ad hoc or "official" standardization (see Chapter 5 for a full discussion of the importance of and approaches for standardizing and validating microbiological methods) over time and with increasing implementation.

INDICATOR APPLICATIONS

Measurement-based Warning Systems

One of the most frequent applications of indicators is in public health warning systems. Warning systems include measurement of indicators to assess whether there is a likelihood that pathogenic microorganisms are present at unacceptable risk levels. Warning systems may be related to ingestion of treated drinking water, recreational water contact, or shellfish consumption. Risk levels are codified through enforceable standards, which may be based on a single sample maximum level, an average or median concentration for a specified period of time, or a maximum frequency of samples over a threshold. When a standard is exceeded, actions are taken to reduce exposure, such as increased treatment levels for drinking water, shellfish bed closures, or warnings to avoid recreational water contact. Because drinking water warning systems focus on treatment effectiveness, which is largely outside the scope of this study, this section focuses on the recreational contact warning system. Box 4-4 provides some comparisons and contrasts between recreational and drinking water warning systems.

For recreational bathing waters, the U.S. Environmental Protection Agency (EPA) recommends the use of enterococci in marine water and *E. coli* in freshwater, based on epidemiologic evidence (see Chapter 2; EPA, 1986). Many states follow EPA's recommendation for freshwater, although there are considerable differences among standards for marine water (see Table 1-4), with several states still using fecal coliforms and more having no standards at all. California uses a multiple-indicator approach including enterococci, fecal coliforms, and total coliforms (see Box 4-5). Hawaii augments enterococci with the use of *Clostridium perfringens*, primarily because of the problem of regrowth associated with coliform bacteria in tropical environments (Fujioka, 2001). Although EPA (1986) also recommends action limits for each of these indicators, there remain considerable differences in standards among states (see Table 4-1), leading to differential levels of public health protection. The goal of the Beaches Environmental

---

BOX 4-4
Treated Drinking Water and
Recreational Water Monitoring Systems

Drinking water warning systems typically focus on treatment adequacy and integrity of the distribution system, rather than on source water quality. They differ from recreational water contact systems in three primary ways:

1. There is zero tolerance for fecal coliforms or *Escherichia coli* in treated drinking water, the presence of which is considered compelling evidence of unacceptable health risk requiring immediate action. However, background levels of microorganisms from natural sources have to be accounted for in monitoring ambient water systems.
2. Sampling frequency is higher and is typically linked to the size of the population served. For example, water supplies serving 50,000 people typically test 2 samples a day, whereas water supplies serving as many as 2.5 million people typically test 420 samples a month, or about 14 samples per day. In contrast, weekly to monthly sampling is typical for ambient recreational waters.
3. Drinking water systems make greater use of rapid real-time physical and chemical surrogates than recreational water systems, such as turbidity and chlorine residual and maintenance of a positive distribution system pressure. This is because they focus on treatment effectiveness as barriers against contamination rather than on natural variability in input sources.

Assessment and Coastal Health (BEACH) Act of 2000 was to bring consistency to beach assessments; however, differences between the states continue based on the various approaches for setting standards and their use in closing impaired beaches.

Several factors limit the effectiveness of current recreational water warning systems, the most prominent of which is the delay in warnings caused by long laboratory sample processing time. Current laboratory measurement methods used to enumerate indicator bacteria (multiple tube fermentation, membrane filtration, and chromogenic substrate) require an 18- to 96-hour incubation period. By the time results are obtained, exposure has already occurred for a day or more. This inadequacy in the notification system is exacerbated because most contamination events are intermittent and indicator levels typically return below thresholds within 24 hours (Boehm et al., 2002; Leecaster and Weisberg, 2001). Thus, contaminated beaches remain open during the laboratory incubation period, but often return to acceptable levels by the time laboratory results are available and warning signs are posted.

BOX 4-5
California's AB411 Beach Standards

The State of California has the most rigorous beach water quality monitoring requirements and standards in the country. Regulations implemented in response to a 1998 state law (AB411) require that three indicator species (enterococci, fecal coliforms, and total coliforms) be measured at least weekly at beaches with more than 50,000 annual visitors. State regulations also define daily and monthly average standards for each indicator, as well as a daily standard for the ratio of total to fecal coliforms. These thresholds were established based on a California-specific epidemiologic study (Haile et al., 1999), and the law requires that public warning signs be posted whenever any of the thresholds are exceeded. Implementation of AB411 requirements resulted in an eightfold increase in the number of public warnings issued. Most of the increase was due to inclusion of an enterococci standard that did not previously exist in California. More than 90 percent of the public warnings are associated with enterococci violations, which are several times higher than warnings associated with either of the other indicators (Noble et al., 2003b).

TABLE 4-1 Range of Bacterial Standards Values Used Among States[a]

Indicator	Marine		Freshwater	
	Instantaneous	Average	Instantaneous	Average
Enterococci	61-104	7-155	33-360	33-193
Escherichia coli	125-1,000	100-235	77-1,000	47-130
Fecal coliforms	50-1,500	50-400	200-1,000	200-500
Total coliforms	200-10,000	1,000-2,400	200-5,000	130-2,400

[a]All values are units per 100mL.
SOURCE: EPA, 2002a.

Another shortcoming is the poorly established relationship between presently used indicators and health risk. Recent reviews of beachgoer epidemiology studies (Prüss, 1998; Wade et al., 2003) found that enterococci had the best relationship to health risk among presently used indicators for marine water, but less than half of the studies found a significant health relationship and the dose-response curves establishing the relationship between increased illness and indicator density were highly variable. This inconsistency among epidemiologic study results may be due to geographic variability and differences in the sources of contamination from study to study, and may be one of the reasons for the differences in recreational water indicators and standards among states (see Tables 1-4 and 4-1).

The use of indicators is based on the presumption that they co-occur at a constant ratio with illness-causing pathogens. This premise is flawed because indicator levels in the gastrointestinal tract may vary within a narrow range, but pathogen concentration is highly variable and dependent on which pathogens are in the population at what levels at specific times. Furthermore, upon leaving the intestinal tract, microbial indicators and pathogens degrade at different rates that are mediated by factors such as their resistance to aerobic conditions, ultraviolet radiation, temperature changes, and salinity. As a result, the epidemiological relationship between indicator density and illness patterns can differ depending on the age of the source material, as well as local meteorological and other environmental conditions. Several studies also have found that some indicator bacteria can grow outside the human or animal intestinal system (Desmarais et al., 2002; Fujioka, 2001; Hardina and Fujioka, 1991; Solo-Gabrielle et al., 2000; see also Chapter 3), further confounding the correlation between pathogens and indicators.

The underlying epidemiologic studies are also limited because many reported failures of beach water quality standards are associated with nonpoint source contamination (Lipp et al., 2001a; Noble et al., 2000; Schiff et al., 2003), but the epidemiologic studies used to establish recreational bathing water standards (EPA, 1986) have been based primarily on exposure to human fecal-dominated point source contamination (Haile et al., 1999). Since nonpoint sources generally have a higher percentage of animal fecal contributions, and animals shed bacterial indicators without some of the accompanying human pathogens, there is considerable uncertainty in extrapolating present standards to nonpoint source situations. A poor correlation between bacterial indicators and virus concentrations has been found in the study of nonpoint sources and water quality (Jiang et al., 2001; Noble and Fuhrman, 2001). However, when a human source, such as septic systems, has been present, enterococci have been significantly correlated with viruses (Lipp et al., 2001a).

A major problem with present water contact warning systems is that bacterial indicator concentrations are spatially and temporally variable and most sampling is too infrequent to transcend this granularity.[1] Taggart (2002) found that sequential samples collected at the same location typically varied by a factor of two and samples 100 meters apart typically differed tenfold. Cheung et al. (1990) found

---

[1]For the purpose of this report, the term "granularity" refers to both the natural spatial and temporal variability of pathogens and indicator organisms that occur (and can be measured) in the environment and the level of coarseness or detail that is used in obtaining such measurements. As such, the term has a more specific meaning than "variability," which is more commonly used throughout the report.

that indicator concentration at a site varied fifteenfold within a day, and Boehm et al. (2002) found that elevated indicator counts typically lasted less than two hours as water masses moved past their sampling site. Most beach monitoring occurs only weekly, and more than one-third of beaches nationally are monitored only monthly. Most of this monitoring is based on collection of a single water sample, the interpretation of which is further compromised by measurement variability. For multiple-tube fermentation, laboratory measurement error based on the 95 percent confidence interval exceeds 50 percent of the mean; more than half of the beach warnings issued in Los Angeles are within measurement error of the standard (Noble et al., 2003a). A guidance document to address these issues is needed from EPA.

Granularity and measurement error concerns are exacerbated by the all-or-none paradigm that is pervasive for beach warnings. Most water quality managers choose from only two options in response to high bacterial indicator counts: (1) close a beach because of a perceived health risk or (2) do nothing. Beach closures are usually reserved for sewage spills, with indicator measurements used primarily to help identify the likelihood that a spill has occurred. No action is typically taken based on indicator measurements alone, particularly when high counts are intermittent. Thus, efforts to inform and protect the public are supported only partially through the current use of indicator measurements.

Some locales are beginning to change this dual-action paradigm by adding additional management options. For instance, California now issues beach advisories when a sample exceeds state bacterial indicator standards and there is no apparent evidence of a sewage spill. Advisories differ from closures in that swimmers are not required to exit the water. California's approach, though, is limited because it requires advisories based on comparison to a single-sample bacterial standard. Temporal and spatial granularity of bacterial counts, combined with the day or longer laboratory processing time, leads to frequent misinformation when warnings are based on a single sample.

Several environmental advocacy groups, such as Heal the Bay (http://www.healthebay.org) and the National Resources Defense Council (http://www.nrdc.org/water/oceans/ttw/titinx.asp), are also beginning to transcend the all-or-none and single-sample difficulties by using the magnitude and frequency of standard failures to develop "letter grades" to describe water quality of recreational beaches. Letter grades have been used successfully in some parts of the country to provide the public with information about the health quality of restaurants and are readily understandable to the public. Such grades can effectively address the granularity issue by integrating data over a longer time period but to be effective they require more frequent monitoring than the monthly sampling that is conducted in many parts of the country.

Source Identification

When a public health risk or water quality impairment is identified through measurement-based systems, the next step is often to conduct investigations to identify the source of contamination. There are two primary purposes of source identification. The first is to decide whether a health warning should be issued because a recreational water body closure is typically issued only after determining that a human fecal source is associated with the high bacterial indicator levels. The second is to identify the most promising approach for fixing the problem. For example, should a local agency be looking for a leaking sewage pipe or for a flock of birds as the source of the problem? From a regulatory point of view, source tracking also feeds directly into the total maximum daily load (TMDL; see also Chapter 1) requirement of the Clean Water Act (CWA) for problem characterization in impaired waters.

Four basic approaches have been used for source identification of microbial contamination, commonly referred to as microbial source tracking. The first involves spatially intensive sampling to identify the source through gradients in indicator density. The second enhances typical indicator measurement with genotypic/phenotypic examination, based on the presumption that certain strains of microorganisms have coevolved with their host and demonstrate specificity to that type of animal. The third is direct measurement of alternative indicators or pathogens that are more closely linked to human intestinal tracts than bacteria such as *E. coli* and enterococci. The fourth is measurement of chemical compounds that are specific to human waste streams, such as caffeine and coprostanol. Some of these methods are only able to discriminate between human and nonhuman sources, while others are able to distinguish specific animal sources contributing to fecal pollution (e.g., dogs, cattle, birds). The following sections describe and evaluate the source tracking methodologies that are currently being used.

Intensive Sampling Approaches

The most frequent practice when routine monitoring identifies a persistent bacteriological water quality problem of unknown origin is to conduct spatially and temporally intensive sampling with standard bacterial indicators, along with efforts to visually identify waste sources. These types of "sanitary surveys" are often preferred because they can be performed using existing equipment and manpower. Although an extensive discussion of sanitary surveys and criteria for their use is beyond the scope of this report, the committee believes these surveys are generally effective when the problem is a leaking pipe in which concentrations are linearly related to location. This approach has been used successfully to identify sewage and manure leaks (Burkholder et al., 1997; Mallin et al., 1997), but it is less effective when the problem is a nonpoint source. Intensive surveys are mostly limited by the long time necessary for laboratory analysis, requiring them to proceed multidirectionally, rather than following a contamination trail in a single direction based on differential concentrations. This makes them expensive and highly impractical when there are multiple tributaries in a system.

Phenotypic and Genotypic Indicator Approaches

Phenotypic and genotypic methods are being used increasingly to track fecal contamination sources (Scott et al., 2003; Simpson et al., 2002). These approaches are based on the presumption that fecal bacteria in the intestines of animals co-evolve to become host specific. Over time, this process produces both identifiable phenotypic traits and changes in gene sequences. Phenotypic and genotypic methods exploit these changes to link fecal bacteria with hosts. This coevolution may be enhanced by differences in food sources among animals. In domesticated animals and humans, it may also be enhanced by the introduction of different types of antibiotics among species.

Both phenotypic and genotypic methods can be divided further into those that require a library of bacterial isolates of known origin and those that do not (Table 4-2). A "library-dependent" method requires cataloging a large number of phenotypic or genotypic patterns from fecal bacteria of known origin. Source identification is then achieved through matching the patterns produced by bacteria obtained in ambient water samples to those in the database.

The most frequently used phenotypic method is multiple antibiotic resistance profiling (MAR; Cooke, 1976; Hagedorn et al., 1999; Harwood et al., 2000; Parveen et al., 1997). This method is based on growing isolates on selective media containing a suite of antibiotics of varying types and concentrations. The resulting resistance "fingerprints" comprise a library. Resistance patterns of indicator bacteria isolated from ambient samples are then matched against the library to determine probable sources. Carbon source utilization (CSU) is similar to MAR except that the fingerprints are based on differential growth in various carbon source growth media (Hagedorn et al., 2003). F+ RNA coliphages can also be used in this way because they belong to four groups or serotypes that differ in

TABLE 4-2 Classification of Genotypic and Phenotypic Methods

Method Type	Library Dependent	Library Independent

Genotypic	Ribotyping
	Repetitive intergenic DNA
	  sequences (rep-PCR)
	Pulsed-field gel electrophoresis
	  (PFGE)
		Host-specific molecular markers (PCR)
		Enterotoxin biomarkers (PCR)
		Terminal restriction fragment length
		  polymorphism (t-RFLP) analysis of
		  total bacterial community

Phenotypic	Multiple antibiotic resistance Carbon source profiling	F+ RNA coliphage serotyping

their occurrence in humans and animals. F+RNA coliphage isolates can be typed by a simple serological test based on prevention of phage replication (absence of host cell lysis, called "neutralization") in the presence of its specific antiserum (Hsu et al., 1995). F+ RNA coliphage grouping also can be done by nucleic acid hybridization as an alternative genotypic method.

The phenotypic source identification methods have the advantage of rapid processing of multiple samples, relatively low cost per sample, and use of equipment already present in most microbiological laboratories. Recently, the detection of specific antibiotic resistance traits in enteric bacteria found in environmental media and fecal waste sources has been used to identify bacterial sources without the development of an extensive reference database of isolates. In these studies, enteric bacteria impacted by anthropogenic sources, such as human sewage or animal manure, were more likely to be antibiotic resistant than the same species of bacteria from ambient sources (Chee-Sanford et al., 2001). Mathew et al. (1998, 1999) also determined that resistance patterns differed between farm types and between pigs of differing ages.

Currently, the most widely used library-dependent genotypic methods include pulsed-field gel electrophoresis (PFGE), ribotyping, and repetitive-intergenic DNA sequence polymerase chain reaction (rep-PCR), although other methods such as denaturing gradient gel electrophoresis (DGGE) are also being investigated (Simpson et al., 2002). In the future, microarray technology should lend itself to the detection of a wide variety of gene sequences in support of microbial source tracking (see Appendix C for further information). All of these methods require characterization of some bacterial genetic sequence to create a reference library and are summarized below:

- PFGE involves generating a DNA "fingerprint" by digesting the complete genomic DNA from a pure culture of bacteria using restriction endonucleases (enzymes that cut DNA at specific sequences of the genetic code), with the resulting banding pattern constituting the fingerprint. Assessments of PFGE's effectiveness as a microbial source tracking tool are contradictory and limited (Parveen et al., 2001; Scott et al., 2002).
- Ribotyping is similar to PFGE, except that only DNA fragments containing ribosomal genes are identified. Ribotyping has been used for bacterial pathogens (*Salmonella*, *Vibrio*), but source tracking applications to date have focused on *E. coli* (Carson et al., 2001; Parveen et al., 1999). This genomic method has been used mostly to distinguish between human and other animal sources, but it also has been used to discriminate between various animal hosts (Hartel et al., 2002; Scott et al., 2003). Although it is the most widely published of the genetic methods, no standard approach has yet been developed with regard to the numbers of isolates measured, the necessary library size, the most effective restriction enzyme, or even which bacterial indicator species has the greatest level of host specificity.

• Rep-PCR amplifies specific repetitive elements distributed throughout the bacterial genes to produce a complex banding pattern. This method has been shown to differentiate between human and various animal sources, but only a few studies of its use have been published (Carson et al., 2003; Dombek et al., 2000).

Non-library-dependent genotypic source tracking methods currently in use include host-specific molecular markers (PCR), terminal restriction fragment length polymorphism (t-RFLP) analysis of total prokaryotic community, and enterotoxin biomarkers (PCR) and are summarized below:

• Host-specific molecular markers focusing on t-RFLP to members of *Bacteriodes-Prevotella* and *Bifidobacterium* have been used successfully for source identification in a few cases (Bernhard and Field, 2000a,b; Bernhard et al., 2003). A related method, total prokaryotic community profiling using t-RFLP, has been used to differentiate between source waters at their ocean terminus (Patricia Holden, University of California at Santa Barbara, personal communication, 2003). This latter method is limited to local applications because a unique source sample is necessary for comparison with receiving waters.
• Enterotoxin biomarkers using species-specific PCR primers to amplify toxin genes found in *E. coli* have been found to differentiate cow, human, and pig waste (Khatib et al., 2002), but geographic and temporal stability of this method is unknown.

Although many genotypic and phenotypic methods are promising, none have been thoroughly tested or are yet widely accepted in the regulatory community. Of greatest concern is the poor understanding of the temporal and geographic stability of traits and genetic sequences (Hartel et al., 2002; Jenkins et al., 2003). High variability in either parameter could restrict use of many methods to local venues due to the time and cost constraints inherent in constructing suitable libraries. It is also unclear which bacterial species show the greatest host specificity, and different bacterial species could show greater or lesser specificity in different hosts depending on the method employed. Finally, the lack of standardized protocols for performing these methods may preclude non-research-oriented laboratories from adopting them and inhibit sharing of libraries.

Direct Measurement Approach

Direct measurement of certain bacteria or viruses that are exclusive to human waste or are rarely found in animals is another approach that has been used to distinguish human from animal sources of fecal contamination (though see Box 4-6). The advantage of direct measurement methods is that there is no need for a large database to match a "fingerprint" against. Thus, these methods can be directly incorporated into microbiological monitoring schemes with the presence of

---

BOX 4-6
Source Specificity of Pathogens

The relationship between fecal indicators and human enteric pathogens varies with the fecal contamination source (human, animal, domestic sewage from septic systems, municipal sewage from cities, combined animal feeding operation waste, etc.). Relatively few pathogens, such as hepatitis A virus, *Shigella* bacteria, and the amoebic protozoan *Entamoeba histolytica*, are unique to humans. Most enteric bacterial and protozoan parasites of humans are harbored by a variety of mammals. Many enteric viruses were once considered unique to humans, with other animals harboring taxonomically similar viruses that do not infect humans. However, evidence for the distinctiveness of human and nonhuman enteric viruses is beginning to decline as the likelihood of cross-species transmission between human and nonhuman strains of some enteric viruses is becoming apparent. A recent example is hepatitis E virus (HEV), which is found in humans, swine, rats, and other animals. The human and swine strains of HEV are very similar at the genetic and protein levels; the human virus has been found to infect swine and the swine virus to infect nonhuman primates (Emerson and Purcell, 2003). Given the current available data, the relative risks of exposure to human pathogens from either human or nonhuman animal sources of fecal contamination are difficult to quantify for any particular source of fecal contamination.

---

the microbe indicative of human waste input. A number of direct measurement indicators have been used for microbial source tracking as described below and summarized in Table 4-3:

- *Bifidobacterium* spp. are obligate anaerobic bacteria. Mara and Oragui (1983) developed a human bifid sorbitol agar that can be used with membrane filtration (Clesceri et al., 1998). There is no indication that these bacteria reproduce in the environment; however, the survival is highly variable and methods for their recovery and detection in water and other environmental samples are inefficient.

- *Bacteroides fragilis* bacteriophage is a virus that infects anaerobic bacteria found in the intestinal tract. Application of this phage has been used primarily in Europe, where it was determined that a bacterial host-specific strain, HSP40, was found in 10 percent of humans but not in animals (Tartera and Jofre, 1987). The phage does not replicate in the environment and has good stability, but it may be found in relatively low numbers and be undetectable in areas with large dilution (McLaughlin, 2001). Furthermore, some studies have determined that the

TABLE 4-3 Advantages and Disadvantages of Select Microorganisms Used for Microbial Source Tracking

Microorganism	Advantages	Disadvantages
Bifidobacterium spp.	Sorbitol-fermenters may be human specific	Low numbers present in environment Variable survival rates Culture methods not well developed
B. fragilis HSP40 bacteriophage	Possibly human specific	Not always present in sewage or present in low numbers Not always human specific, depending on bacterial strain and geographic location
F+ RNA bacteriophage	Groups are correlated with sources (humans versus animals and livestock) with small % of overlap	Lower numbers in warm marine and tropical waters due to variable survival rates
Human enteric viruses	Human specific if cell cultures and genome targets to amplify are chosen carefully Addresses hazard identification component of risk assessment paradigm	Low concentrations in water and other samples Better cultivation and molecular methods needed
Rhodococcus coprophilis	Specific indicator of grazing animal fecal sources	Detected by culture for a lengthy period (weeks)

SOURCE: Adapted from Scott et al., 2002.

host bacterium also detected *B. fragilis* phages from animals (swine) as well as from humans, leading to questions about the human specificity of the system (Chung, 1993).

- F-specific coliphages are viruses that infect male (F+) strains of *Escherichia coli* bacteria that produce F pili. The F pili possess the receptor that allows for detection and distinction of the F+ groups of coliphages. There are two taxonomically distinct F-specific phage groups, one containing RNA (Leviviridae) and another containing DNA (Inoviridae). Members of the two groups can easily be separated and identified by determining if they are resistant (Inoviridae) or sensitive (Leviviridae) to RNase. The F+ RNA coliphages are morphologically similar to many human enteric viruses. By means of plaque isolation techniques, phages have long been used as virus indicators due to their similar transport and fate (Havelaar, 1993; IAWPRC, 1991; Kott et al., 1974). Serotyping or genotyping the F+ RNA coliphages into their four main groups can further distinguish the origin as human or animal (Hsu et al., 1995), although there is some overlap of the groups. However, Schaper et al. (2002) found that serotypes II and

III were associated with human sewage, but human samples also contained serotypes I and IV. Animal samples contained all four serotypes, with the majority of the F+ coliphage being serotypes I and IV. They found statistical significance in the assignment of serotypes to specific human or animal sources, but the distinction may not be as definitive as previously thought since there was some overlap between the serotypes and their expected animal sources.

• Human enteric viruses are potentially definitive for human fecal contamination and have been monitored directly in water since the 1960s. Originally cell culture and serology were used to isolate and identify the viruses, but PCR and reverse transcription PCR (RT-PCR) methods are now more common (Jiang et al., 2001; Lee and Kim, 2002). The specificity of these methods to human enteric viruses depends on careful choice of the cell cultures for isolation and target genomic region for PCR or RT-PCR amplification.

• *Rhodococcus coprophilus* is a fecally excreted bacterium associated with grazing animals, and its presence in water has been suggested as an animal-specific fecal indicator (Long, 2002; Mara and Oragui, 1985). Unfortunately, methods to culture this bacterium require long incubation periods, and molecular methods for its detection in water have not yet been developed.

Although several species are specific enough to human sources of fecal contamination that they are suitable candidates for direct measurement, two challenges remain with the direct measurement approach. The candidate species generally require specialized methods, such as molecular techniques, that are available only in research laboratories. Also, direct measurement approaches are capable only of defining the presence of human material, not the percentage of fecal contamination that human source material encompasses. Without the ability to distinguish whether human material comprises a small or large fraction of the fecal contamination, managers do not have all the information necessary to determine the most effective course of corrective action.

Chemical Approaches

A number of organic compounds are primarily specific to human fecal contamination. These chemical indicators can be either natural products found in human feces (e.g., fecal steroids, aminopropanone) or synthetic chemicals found in products, such as detergents, that are specific to household waste streams (e.g., linear alkylbenzenes, trialkylamines, nonylphenols, whitening agents) (Takada and Eganhouse, 1998).

Fecal steroids (i.e., sterols, stanols, stanones) have been used frequently for differentiating between human and nonhuman sources (Grimalt et al., 1990; Leeming et al., 1996; Standley et al., 2000). The most frequently used steroid is coprostanol, which is produced by catabolism of cholesterol in the intestinal tract and is abundant in human feces (Hatcher and McGillivary, 1979; Maldonado et al., 2000; Takada and Eganhouse, 1998; Venkatesan and Kaplan, 1990; Venkatesan and Mirsadeghi, 1992). While useful, interpretation of fecal steroid data is often confounded because these compounds can be produced via both in vivo biotic and in situ abiotic processes.

Linear alkylbenzenes (LABs), which are residues of anionic surfactants used in detergents and are an example of synthetic industrial chemical indicators. They have been widely used as chemical indicators of human fecal contamination (Eganhouse et al., 1988; Gustafsson et al., 2001; Phillips et al., 1997; Takada and Ishiwatari, 1987; Zeng et al., 1997), but are extremely hydrophobic and susceptible to degradation under aerobic conditions. Thus, LAB concentrations tend to decrease rapidly away from the point of discharge, limiting their usefulness as an indicator for sewage-contaminated waters.

A third widely used chemical indicator of human fecal contamination is caffeine. Although caffeine occurs naturally in more than 60 species of plants, none of these are indigenous to the United States, and any caffeine detected in surface waters must have originated from anthropogenic sources (Seigener and Chen, 2002). While source specific, caffeine is soluble in water and is diluted rapidly once released into the water column. Moreover, it does not sequester in sediments and is rapidly transported long distances from the source, so its concentration is usually low in environmental samples.

Although chemical indicators can provide some useful information regarding fecal contamination, they are generally less promising than genotypic or phenotypic methods because they are more costly and time consuming. Chemical methods are also typically less sensitive than biological methods and generally require large sample volumes to achieve adequate detection limits. Perhaps the biggest drawback to using chemical indicators is that so little is known about their fate in aquatic systems. To achieve their full potential utility, more research is needed to gain a better understanding of the transport, transformations, and persistence of these compounds under various environmental conditions. Despite their shortcomings, chemical indicator-based methods can be effective when used to complement other methods of microbial pollution monitoring.

Status and Trends Assessment

Bacterial indicators are frequently used in ambient water quality monitoring to assess whether water bodies meet state-designated water quality standards for beneficial use(s), such as drinking water supply, water contact recreation, and shellfish harvesting (see also footnote 3, Chapter 1). Ambient water quality monitoring can also focus on trends, to determine whether conditions are degrading or whether mitigation activities are leading to improvements. These beneficial use assessments generally involve integration of data over longer time periods and larger spatial scales than public health warning systems, which typically focus on short-term decisions about individual sites. As a result, the indicator properties that are desirable for warning systems, such as rapidity of laboratory processing and results, differ for this application. Examples of indicator use in various types of ambient water monitoring are described below.

Surface Water Assessments

As described in Chapter 1 (see also Table 1-2), EPA is required under Sections 305(b) and 303(d) of the Clean Water Act to provide a national water quality assessment and to identify those water bodies that are failing to achieve their designated ambient water quality standards and beneficial uses. Water bodies not in compliance are subject to EPA's Total Maximum Daily Load Program, which requires that discharges to that water body be reduced to the level necessary for achieving beneficial uses. TMDLs are initiated through an assessment of multiyear data and are typically accompanied by monitoring programs to assess trends in both source inputs and receiving water quality. Coliform (total, fecal, and *E. coli*) bacteria have been among the most frequent indicators used for identifying microbial water body impairment. Although coliforms have been useful for the status assessment portion of these activities, their use in trends assessment is often compromised because sample processing typically focuses on quantifying only a small part of the possible response range. Whereas health managers are interested in whether indicator concentrations exceed a risk-based threshold, trend detection requires measurement of indicator concentrations at levels below public health thresholds, since trend detection is not well achieved when data at the low end of the range are classified as non-detects. Trend detection is also hampered when values are censored at the high end of the range. This problem can be addressed within the context of existing bacterial indicator methods by processing multiple dilutions of a sample, but this adds significantly to the processing costs.

Although current microbial indicators meet most trend assessment needs, they can be improved. For instance, most TMDL listing decisions are based on indicator bacteria measurements without assessment of whether the source is human, even though TMDLs are intended to limit anthropogenic discharges to the water body. Similarly, trends assessments could be improved by determining whether ambient water quality changes were due to human or nonhuman inputs. The source tracking techniques discussed above are appropriate to these applications. In addition, most marine assessments are still based on coliforms, even though enterococci have been found to be more closely associated with health risk and significantly more marine water bodies would be detected as impaired if enterococci were used instead of coliforms (Noble et al., 2003b).

Shellfish Water Assessments

As noted in Chapter 1, extensive discussion of the use of indicators for shellfish microbial water quality monitoring is beyond the committee's statement of task. However, it is appropriate to note that comparing coliform counts to standards based on biweekly or monthly average concentration is typical for monitoring shellfishing waters. Longer-term averages are used because shellfish tissue concentrations reflect this longer exposure period. The critical factor in these programs is ensuring that collections are made either in an unbiased manner or at times when concentrations are likely to be highest. Numerous studies have demonstrated the adverse effects of rainfall on bacteriological water quality (e.g., Ackerman and Weisberg, 2003; Boehm et al., 2002, Lipp et al., 2001a; Noble et al., 2003c), and these periods need to be included in the sampling effort, despite inconveniences associated with sampling during rain events.

While shellfish monitoring programs have generally been effective at protecting public health, they are most effective when complemented with a sanitary inspection of the surrounding shoreline. These surveys describe the location of human and animal populations in the shellfish growing area, land uses in the watershed that could impact water quality, and the location and magnitude of fecal waste sources (wastewater treatment plants and their effluents, on-site septic systems, etc.). Understanding these activities can be used to better focus the monitoring and aid in the interpretation of trends.

Drinking Water Supplies

Microbiological monitoring has historically played only a minor role in the evaluation of surface and groundwater sources of drinking water. Water utilities have relied on treatment (filtration and disinfection) and focused their monitoring on treatment effectiveness, rather than monitoring indicators of waterborne pathogens in their source water. However, recent waterborne outbreaks of cryptosporidiosis have led EPA to revise the Interim and Long Term 2 Enhanced Surface Water Treatment Rule (LT2ESWTR) that will require monitoring for *Escherichia coli* by utilities with surface source drinking water supplies (EPA, 2002b, 2003; see also Chapter 1 and Table 1-1). Depending on *E. coli* levels, some utilities will also be required to monitor for *Cryptosporidium*. The annual average concentration of these indicators will be used to define monitoring and treatment requirements that will provide increased control of *Cryptosporidium*.

Although the proposed requirements of the LT2ESWTR are a step in the right direction, surface water monitoring will remain a low-frequency activity (either monthly or every two weeks) and will miss transient impairments in microbial water quality. More specifically, the proposed sampling schedule does not directly consider the impact of precipitation events on source water quality, even though the microbial quality of both surface and groundwater becomes appreciably worse following precipitation events, and this is the period of greatest vulnerability to waterborne outbreaks (Curriero et al., 2001; Rose et al., 2000).

Groundwater quality monitoring is rare, despite data that show the majority of drinking water outbreaks of disease in the United States result from groundwater systems (see Chapter 1 and Figure 1-2). Although there was no final national regulation for groundwater quality at the time this report was prepared, some states have wellhead protection programs for drinking water supplies using groundwater sources. The Ground Water Rule is expected to be promulgated sometime in late 2004, and will define disinfection needs for source water based on the vulnerability of the aquifer according to its hydrogeological characteristics and bacteriological quality (EPA, 2000). More specifically, coliform bacteria have been recommended as the indicator of choice for groundwater, with an option for including coliphage or direct virus monitoring. The known risks from viruses in fecally contaminated groundwater, combined with evidence that coliphages are better indicators of viruses than are indicator bacteria, and that human enteric viruses are detectable in fecally contaminated groundwater using current technologies, suggest that coliphage or direct virus monitoring would enhance the assessment of groundwater microbiological quality and would make better indicators of human health risk (see Chapter 6 for further information).

Prediction-based Warning Systems

The typical application of indicators for public health warning systems involves measuring bacterial indicators to assess recent water quality conditions. One shortcoming of this approach is that it does not prevent exposure, since people swim in (or drink) the water prior to sampling, during sample processing, and while mitigative or warning actions are being taken. An alternative approach is to develop predictive models that prevent exposure.

One example is the use of rainfall as a predictive indicator. Rainfall is associated with elevated bacterial indicator levels on both daily (Curriero et al., 2001; Kistemann et al., 2002; Schiff et al., 2003) and seasonal (Boehm et al., 2002; Lipp et al., 2001b) time scales. These elevated levels typically result from urban runoff and combined sewage-stormwater system overflows.

Several states issue swimmer warnings based on rainfall. For example, five county health departments in southern California routinely issue warnings not to swim in the ocean for three days following a rainstorm of 0.1 inch or more (Ackerman and Weisberg, 2003). Although California's warnings are only advisory, Monmouth County in New Jersey routinely closes two beaches that typically have elevated bacterial concentrations following runoff events for 24 hours following 0.1 inch or more rain; the closure is extended to 48 hours following 2.8 inches or more of rain (David Rosenblatt, New Jersey Department of Environmental Protection, personal communication, 2003).

While rainfall-based warnings are valuable, they are based on limited em-

---

BOX 4-7
Advanced Predictive Modeling

Great Lakes scientists are using multiple regression techniques to develop more sophisticated models for predicting beach water quality in Chicago and Milwaukee (Olyphant and Whitman, in press). These models include rainfall during the previous 24 hours, wind, solar radiation, water temperature, lake stage, water turbidity, and pH. Rainfall, wind, and turbidity are indicative of the strong influence that storms have on *E. coli* concentrations. At the Milwaukee beach, storm effects result primarily from sewage overflows into tributary rivers that get pushed shoreward by easterly winds. The Chicago beach is not directly influenced by stream inflows, but storms stir up *E. coli* laden sand in the breaker zone. Solar radiation is a negative term in the model that reflects UV-mediated bacterial die-off during bright sunshine. Water temperature and lake stage represent conditions that lead to high bacterial concentrations during non-storm periods. Bacterial populations grow faster in warm water and bacteria become more concentrated when lake levels fall at the beach in Chicago. These models were evaluated by comparing predictions of *E. coli* concentration exceeding EPA's recommended threshold of 235 CFU/100mL with measured concentrations. The model correctly predicted 66 of 90 events at the Milwaukee beach and 50 of 57 events at the Chicago beach. Model errors were evenly split between false negatives and false positives for the Milwaukee beach, but six of the seven incorrect predictions for Chicago are ones that would have led to overprotective actions.

---

pirical evidence. Ackerman and Weisberg (2003) found that 91 percent of storms with precipitation greater than 0.25 inches led to an increase in the number of Los Angeles beaches failing bacterial water quality standards. However, the response was more equivocal for storms with precipitation between 0.1 and 0.25 inches, when factors such as spatial coverage of the storm, antecedent rainfall, and size and type of watershed become potentially more important in determining the need for warnings. More complex models that incorporate these factors, as well as similar studies conducted in other parts of the country (see Box 4-7), will have to be developed before predictive models become widely accepted tools for public health warnings.

Another predictive-based warning system, which operates on a longer time scale, involves land use as an indicator of fecal contamination. Many recreational bathing areas, drinking water sources, and shellfishing areas are located in drainage basins that are undergoing development pressure. Changing land use, such as increased urbanization or conversion of rangelands to agricultural lands, can affect pathogen contributions within the drainage area. Mallin et al. (2000, 2001) have demonstrated a statistical relationship between the amount of development in a watershed and downstream bacterial concentrations, but these results are likely to be site specific. The theoretical framework for more generalizable models that predict receiving water contaminant concentrations based on land use, such as HSPF (Hydrologic Simulation Program—Fortran; Bicknell et al., 1997), are available, but the runoff relationships necessary to parameterize these models are not well developed. Further work on these models is needed before managers can use them to define the level of development at which increased mitigation activities will be necessary to ensure acceptable water quality.

Lobitz et al. (2000) have suggested that remote sensing data can also serve as a predictive tool for bacterial waterborne outbreaks. They indicate that *Vibrio cholerae* occurs commensally with species of phytoplankton, the density of which can be tracked through satellite imagery. Moreover, satellite imagery of circulation and sea surface temperature can be used to predict future blooms. While such modeling approaches need more empirical testing, rapid advances in remote sensing technology (e.g., Isern and Clark, 2003) will provide new opportunities for developing such models.

SUMMARY: CONCLUSIONS AND RECOMMENDATIONS

Microbial water quality indicators are used in a variety of ways within public health risk assessment frameworks, and the most desirable indicator attributes—and therefore the most appropriate indicators—naturally depend on their manner of use. Despite their importance and longevity, Bonde's attributes of an ideal public health indicator need to be refined. These historic definitions of indicators have been tied to the methods used to measure them, but the development of new measurement methods necessitates separate criteria for evaluating the biological and method attributes of indicators. Separate criteria allow one to choose the indicator with the most desirable biological attribute for a given application and to match this with a measurement method that best meets the need of the application.

The most important biological attribute is a strong quantitative relationship between indicator concentration and the degree of public health risk. One of the most important method attributes is its specificity, or ability to measure the target indicator organism in an unbiased manner. Speed of the method (processing time and rapidity of results) is also an important characteristic in many cases, particularly when warning systems are involved and human exposure occurs during the laboratory analysis period.

Many public health applications use microbiological indicators, including public health warning systems, source identification, and status or trends assessments. No single indicator or analytical method (or even a small set of indicators or analytical methods) is appropriate to all applications. A suite of indicators and indicator approaches is required for different applications and different geographies.

Several factors limit the effectiveness of current recreational water warning systems, the most prominent of which is the delay in warnings caused by long laboratory sample processing times. Current laboratory measurement methods used to enumerate indicator bacteria (multiple tube fermentation, membrane filtration, and chromogenic substrate) are too time consuming. They require an 18- to 96-hour incubation period, during which the public is exposed to potential health risks. One approach that is increasingly being used to address this problem is predictive models intended to prevent exposure.

Another shortcoming of present warning systems is the poorly established relationship between presently used indicators and health risk. Current studies do not address all sources of contamination, have not identified the etiological agents of illness, have not been conducted in enough geographical locations, and do not address chronic exposure. Many reported failures of beach water quality standards are associated with nonpoint source contamination, but the epidemiologic studies used to establish recreational bathing water standards have been based primarily on exposure to point source contamination dominated by human fecal material.

A major problem with present water contact warning systems is that bacterial indicator concentrations are spatiotemporally variable and most sampling is too infrequent to transcend this granularity. The predominant all-or-none decision framework of either closing the beach or taking no action at all, sometimes on the basis of a single sample, magnifies the errors associated with this temporal and spatial granularity.

There are many promising microbial source identification techniques that can help in deciding whether a health warning should be issued or in identifying the best approach for fixing the problem. However, these techniques are not yet standardized or fully tested.

Groundwater quality monitoring is rare, despite data showing that the majority of waterborne outbreaks of disease in the United States result from groundwater systems. Viral contamination of groundwater is a particular concern because the small size and considerable environmental persistence of viruses make it more likely they will reach and contaminate groundwater. The known risks from viruses in fecally contaminated groundwater, and evidence that human enteric viruses are detectable in fecally contaminated groundwater, suggest that coliphage or direct virus monitoring would enhance the assessment of groundwater microbiological quality and would make a better indicator of human health risk.

The discussion in this chapter and the preceding conclusions support the following recommendations:

- Since it is not possible to identify a single, unique indicator or small set of indicators capable of identifying all classes of waterborne microbial pathogens, priority should be given to developing a phased monitoring approach that relies on a flexible "tool box" of indicators and indicator approaches that are used according to strategies appropriate to the specific applications (see Chapter 6).

- The link between indicators and pathogens, and among indicators, pathogens, and adverse health outcomes, would be strengthened by including measurements of both indicators and pathogens in comprehensive epidemiologic studies. In particular, studies to better assess the role of nonpoint sources in occurrence of human pathogens and indicator organisms, disease outbreaks, and endemic health risks in recreational waters should be conducted. Use of alternative indicators need to be included in these studies.

- Improved indicators for viruses in groundwater sources of drinking water need to be developed.

- New paradigms for reporting water contact health risk, such as "letter grades" for public beaches, need to be developed. The present all-or-none closure decisions can misinform the public because of large spatiotemporal heterogeneity in indicator concentrations. Letter grades—which have been used successfully in some parts of the country to provide the public information about the health quality of restaurants—are one option that would effectively address the granularity issue by integrating data over a longer time period and are readily understandable.

- Investment should be made in developing rapid analytical methods. The most commonly used warning systems involve laboratory methods that are too time consuming to achieve the best possible public health protection. New molecular methods, which do not have the long incubation time requirements of present culture-based methods, are on the near term horizon (see Chapters 5 and 6).

- There are several promising source identification (i.e., microbial source tracking) techniques on the horizon that should be incorporated into monitoring systems when they have been adequately validated. Public health risk from exposure to fecally contaminated water is likely to vary depending on whether high indicator concentrations resulted from animal or human sources, and microbial source tracking tools will allow public health managers to incorporate that distinction into their decision making.

- No matter how rapid measurement techniques become, they will always be retrospective. Models that predict future water quality conditions, based on factors such as rainfall, are potentially valuable tools for warning the public before exposure occurs, but the scientific foundation for these models has to be enhanced before they can be widely used.

REFERENCES

Ackerman, D., and S.B. Weisberg. 2003. Relationship between rainfall and beach bacterial concentration on Santa Monica Bay beaches. Journal of Water and Health 1: 85-89.

Allen, M. 1977. The impact of excessive bacterial populations in coliform methodology. American Society for Microbiology Annual Conference.

Bernhard, A.E., and K.G. Field. 2000a. Identification of nonpoint sources of fecal pollution in coastal waters by using host-specific 16S ribosomal DNA markers from fecal anaerobes. Applied and Environmental Microbiology 66: 1587-1594.

Bernhard, A.E., and K.G. Field. 2000b. A PCR assay to discriminate human and ruminant feces based on host differences in *Bacteroides prevotella* 16S ribosomal DNA. Applied and Environmental Microbiology 66: 4571-4574.

Bernhard, A. E., T. Goyard, M. Simonich, and K.G. Field. 2003. A rapid method for identifying fecal pollution sources in coastal waters. Water Research 37: 909-913.

Bicknell, B.R., J.C. Imhoff, J.L. Kittle, Jr., A.S. Donigian, Jr., and R.C. Johanson. 1997. Hydrological Simulation Program - Fortran Users Manual for Version 11. Athens, Georgia: U.S. Environmental Protection Agency, National Exposure Research Laboratory.

Boehm, A.B., S.B. Grant, J.H. Kim, S.L. Mowbray, C.D. McGee, C.D. Clark, D.M. Foley, and D.E. Wellman. 2002. Decadal and shorter period variability and surf zone water quality at Huntington Beach, California. Environmental Science and Technology 36: 3885-3892.

Bonde, G. 1966. Bacteriological methods for estimation of water pollution. Health Laboratory Science 3: 124-128.

Burkholder, J.M., M.A. Mallin, H.B. Glasgow, Jr., L.M. Larsen, M.R. McIver, G.C. Shank, N. Deamer-Melia, D.S. Briley, J. Springer, B.W. Touchette, and E.K. Hannon. 1997. Impacts to a coastal river and estuary from rupture of a large swine waste holding lagoon. Journal of Environmental Quality 26: 1451-1466.

Cabelli, V.J., A.P. Dufour, M.A. Levin, L.J. McCabe, and P.W. Haberman. 1979. Relationship of microbial indicators to health effects at marine bathing beaches. American Journal of Public Health 69: 690-696.

Camper, A.K., and G.A. McFeters. 1979. Chlorine injury and the enumeration of waterborne coliform bacteria. Applied and Environmental Microbiology 3: 633-641.

Carson, C.A., B.L. Shear, M.R. Ellersieck, and A. Asfaw. 2001. Identification of fecal *Escherichia coli* from humans and animals by ribotyping. Applied and Environmental Microbiology 67: 1503-1507.

Carson, C.A., B.L. Shear, M.R. Ellersieck, and J.D. Schnell. 2003. Comparison of ribotyping and repetitive extragenic palindromic-PCR for identification of fecal *Escherichia coli* from humans and animals. Applied and Environmental Microbiology 69: 1836-1839.

Chee-Sanford, J.C., R.I. Aminov, I.J. Krapac, N. Garrigues-Jeanjean, and R.I. Mackie. 2001. Occurrence and diversity of tetracycline resistance genes in lagoons and groundwater underlying two swine production facilities. Applied and Environmental Microbiology 67: 1494-1502.

Cheung, W.H.S., R.P.S. Hung, K.C.K. Chang, and J.W.L. Kleevens. 1990. Epidemiological study of bathing beach water pollution and health related bathing water standards in Hong Kong. Water Science and Technology 23: 243-252.

Chung, H. 1993. F-specific coliphages and their serogroups and *Bacteroides fragilis* phages as indicators of estuarine water and shellfish quality. Ph.D. dissertation, University of North Carolina, Chapel Hill.

Clesceri, L.S., A.E. Greenberg, and A.D. Eaton, eds. 1998. Standard Methods for the Examination of Water and Wastewater, 20th Edition. Washington, D.C.: American Public Health Association.

Cooke, M.D. 1976. Antibiotic resistance among coliform and fecal coliform bacteria isolated from sewage, seawater, and marine shellfish. Antimicrobial Agents and Chemotherapy 9: 879-884.

Curriero, F.C., J.A. Patz, J.R. Rose, and S. Lele. 2001. The association between extreme precipitation and waterborne disease outbreaks in the United States, 1948-1994. American Journal of Public Health 91: 1194-1199.

Desmarais, T.R., H.M. Solo-Gabriele, and C.J. Palmer. 2002. Influence of soil on fecal indicator organisms in a tidally influenced subtropical environment. Applied and Environmental Microbiology 68: 1165-1172.

Dombek, P.E., L.K. Johnson, S.T. Zimmerly, and M.J. Sadowsky. 2000. Use of repetitive DNA sequences and the PCR to differentiate *Escherichia coli* isolates from human and animal sources. Applied and Environmental Microbiology 66: 2572-2577.

Edberg, S.C., and D.B. Smith. 1989. Absence of association between total heterotrophic and total coliform bacteria from a public water supply. Applied and Environmental Microbiology 55(2): 380-384.

Eganhouse, R.P., D.P. Olaguer, B.R. Gould, and C.S. Phinney. 1988. Use of molecular markers for the detection of municipal sewage sludge at sea. Marine Environmental Research 25: 1-22.

Emerson, S.U., and R.H. Purcell. 2003. Hepatitis E virus. Reviews in Medical Virology 13(3): 145-154.

EPA (U.S. Environmental Protection Agency). 1986. Ambient Water Quality Criteria for Bacteria - 1986. Washington, D.C.: Office of Water. EPA 440-5-84-002.

EPA. 2000. National Primary Drinking Water Regulations: Ground Water Rule; Proposed Rule. Federal Register 65(91): 30193-30274.

EPA. 2002a. EPA's Beachwatch Program: 2001 Swimming Season. Washington, D.C.: Office of Water. EPA 823-F-02-008.

EPA. 2002b. National Primary Drinking Water Regulations: Long Term 1 Enhanced Surface Water Treatment Rule. Federal Register 76: 1812-1844.

EPA. 2003. National Primary Drinking Water Regulations: Long Term 2 Enhanced Surface Water Treatment Rule; Proposed Rule. Federal Register 68(154): 47640-47795.

Feachem, R.G., D.J. Bradley, H. Garelick, and D.D. Mara, eds. 1983. Sanitation and Disease, Health Aspects of Excreta and Wastewater Management. New York: John Wiley & Sons.

Fujioka, R.S. 2001. Monitoring coastal marine waters for spore forming bacteria of faecal and soil origin to determine point from non-point source pollution. Water Science and Technology 44: 181-188.

Geldreich, E.E. 1978. Interferences to coliform detection in potable water supplies. Pp. 13-19 in Evaluation of the Microbiology Standards for Drinking Water. C.W. Hendricks, ed. Washington, D.C.: EPA-570-9-78-00C.

Gerba, C.P., S.M. Goyal, R.L. LaBelle, I. Cech, and G.F. Bogdan. 1979. Failure of indicator bacteria to reflect the occurrence of enteroviruses in marine waters. American Journal of Public Health 69: 1116-1119.

Gerba, C.P. 2001. Assessment of enteric pathogen shedding by bathers during recreational activity and its impact on water quality. Quantitative Microbiology 2: 55-68.

Griffin, D.W., E.K. Lipp, M.R. McLaughlin, and J.B. Rose. 2001. Marine recreation and public health microbiology: Quest for the ideal indicator. BioScience 51: 817-825.

Grimalt, J.O., P. Fernandez, J.M. Bayona, and J. Albages. 1990. Assessment of fecal sterols and ketones as indicators of urban sewage inputs to coastal waters. Environmental Science and Technology 24(3): 357-363.

Gustafsson, Ö., C.M. Long, J. Macfarlane, and P.M. Gschwend. 2001. Fate of linear alkylbenzenes released to the coastal environment near Boston Harbor. Environmental Science and Technology 35: 2040-2048.

Hagedorn, C.S., S.L. Robinson, J.R. Filtz, S.M. Grubbs, T.A. Angier, and R.B. Reneau, Jr. 1999. Using antibiotic resistance patterns in the fecal streptococci to determine sources of fecal pollution in a rural Virginia watershed. Applied and Environmental Microbiology 65: 5522-5531.

Hagedorn, C.S., J.B. Crozier, K.A. Mentz, A.M. Booth, A.K. Graves, N.J. Nelson, and R.B. Reneau. 2003. Carbon source utilization profiles as a method to identify sources of faecal pollution in water. Journal of Applied Microbiology 94: 792-799.

Haile, R.W., J.S. Witte, M. Gold, R. Cressey, C.D. McGee, R.C. Millikan, A. Glasser, N. Harawa, C. Ervin, P. Harmon, J. Harper, J. Dermand, J. Alamillo, K. Barrett, M. Nides, and G. Wang. 1999. The health effects of swimming in ocean water contaminated by storm drain runoff. Journal of Epidemiology 104: 355-363.

Harden, H.S., J.P. Chanton, J.B. Rose, D.E. John, and M.E. Hooks. 2003. Comparison of sulfur hexafluroride, fluorescein and rhodamine dyes and the bacteriophage PDR-1 in tracing subsurface flow. Journal of Hydrology 277(1-2): 100-115.

Hardina, C.M., and R.S. Fujioka. 1991. Soil, the environmental source of *Escherichia coli* and enterococci in Hawaii's streams. Environmental Toxicology 6: 185-195.

Hartel, P.G., J.D. Summer, J.L. Hill, J.V. Collins, J.A. Entry, and W.I. Segars. 2002. Geographic variability of *Escherichia coli* ribotypes from animals in Idaho and Georgia. Journal of Environmental Quality 31: 1273-1278.

Harwood, V.J., J. Whitlock, and V.H. Withington. 2000. Classification of the antibiotic resistance patterns of indicator bacteria by discriminant analysis: Use in predicting the source of fecal contamination in subtropical Florida waters. Applied and Environmental Microbiology 66: 3698-3704.

Hatcher, P.G., and P.A. McGillivary. 1979. Sewage contamination in the New York Bight: Coprostanol as an indicator. Environmental Science and Technology 13: 1225-1229.

Havelaar, A.H. 1993. Bacteriophages as models of human enteric viruses in the environment. American Society for Microbiology News 59: 614-619.

Hsu, F-C., Y-S Shieh, J. van Duin, M.J. Beekwilder, and M.D. Sobsey. 1995. Genotyping male-specific RNA coliphages by hybridization with oligonucleotide probes. Applied and Environmental Microbiology 61: 3960-3966.

IAWPRC (International Association on Water Pollution Research and Control) Study Group. 1991. Bacteriophages as model organisms in water quality control. Water Research 25: 529-545.

Isern, A.R., and H.L. Clark. 2003. The ocean observatories initiative: A continued presence for interactive ocean research. Marine Technology Society Journal 37:26-41.

Jenkins, M.B., P.G. Hartel, T.J. Olexa, and J.A. Stuedemann. 2003. Putative temporal variability of *Escherichia coli* ribotypes from yearling steers. Journal of Environmental Quality 32: 305-309.

Jiang, S., R. Noble, and W. Chu. 2001. Human adenoviruses and coliphages in urban-runoff impacted coastal waters of southern California. Applied and Environmental Microbiology 67: 179-184.

Khatib, L.A., Y.L. Tsai, and B.H. Olson. 2002. A biomarker for the identification of cattle fecal pollution in water using the LYIIa toxin gene from the enterotoxigenic *Escherichia coli*. Applied Microbiology and Biotechnology 59: 97-104.

Kistemann, T.C., C. Koch, F. Dangendorf, R. Fischeder, J. Gebel, V. Vacata, and M. Exner. 2002. Microbial load of drinking water reservoir tributaries during extreme rainfall and runoff. Applied and Environmental Microbiology 68: 2188-2197.

Kott, Y., N. Roze, S. Sperber, and N. Betzer. 1974. Bacteriophages as viral pollution indicators. Water Research 8: 165-171.

Labelle, R.L., C.P. Gerba, S.M. Goyal, J.L. Melnick, I. Cech, and G.F. Bogdan. 1980. Relationships between environmental factors, bacterial indicators and the occurrence of enteric viruses in estuarine sediments. Applied and Environmental Microbiology 39: 586-596.

LeChevallier, M.W., M. Abbaszdegan, A.K. Camper, G. Izaguirre, M. Stewart, D. Naumovitz, M. Mardhall, C.R. Sterling, P. Payment, E.W. Rice, C.J. Hurst, S. Schaub, T.R. Slifko, J.B. Rose, H.V. Smith, and D.B. Smith. 1999a. Emerging pathogens - bacteria. Journal of the American Water Works Association 91: 136-172.

LeChevallier, M.W., M. Abbaszdegan, A.K. Camper, G. Izaguirre, M. Stewart, D. Naumovitz, M. Mardhall, C.R. Sterling, P. Payment, E.W. Rice, C.J. Hurst, S. Schaub, T.R. Slifko, J.B. Rose, H.V. Smith, and D.B. Smith. 1999b. Emerging pathogens - viruses, protozoa, and algal toxins. Journal of the American Water Works Association 91: 110-121.

Lee, S.-H., and S.-J. Kim. 2002. Detection of infectious enteroviruses and adenoviruses in tap water in urban areas in Korea. Water Research 36: 248-256.

Leecaster, M.K., and S.B. Weisberg. 2001. Effect of sampling frequency on shoreline microbiology assessments. Marine Pollution Bulletin 42: 1150-1154.

Leeming, R., A. Ball, N. Ashbolt, and P. Nichols. 1996. Using fecal sterols from humans and animals to distinguish fecal pollution in receiving waters. Water Research 30: 2893-2900.

Lipp, E.K., R. Kurz, R. Vincent, C. Rodriguez-Palacios, S.R. Farrah, and J.B. Rose. 2001a. The effects of seasonal variability and weather on microbial fecal pollution and enteric pathogens in a subtropical estuary. Estuaries 24: 266-276.

Lipp, E.K., N. Schmidt, M.E. Luther, and J.B. Rose. 2001b. Determining the effects of El Niño-Southern Oscillation events on coastal water quality. Estuaries 24: 491-497.

Lobitz, B., L. Beck, A. Huq, B. Wood, G. Fuchs, A.S. Faruque, and R. Colwell. 2000. Climate and infectious disease: Use of remote sensing for detection of *Vibrio cholerae* by indirect measurement. Proceedings of the National Academy of Sciences 97(4):1438-1443.

Long, S. 2002. Development of Source-Specific Indicator Organisms for Drinking Water (Project #2645). Report Order Number 90911. Denver, Colorado: American Water Works Association Research Foundation.

Maldonado, C., M.I. Venkatesan, C.R. Philips, and J.M. Bayona. 2000. Distribution of trialkylamines and coprostanol in San Pedro shelf sediments adjacent to a sewage outfall. Marine Pollution Bulletin 40(8): 680-687.

Mallin, M.A., J.A.M. Burkholder, and J. Springer. 1997. Comparative effects of poultry and swine waste lagoon spills on the quality of receiving streamwaters. Journal of Environmental Quality 26: 1622-1631.

Mallin, M.A., K.E. Williams, E.C. Esham, and R.P. Lowe. 2000. Effect of human development on bacteriological water quality in coastal watersheds. Ecological Applications 10: 1047-1056.

Mallin, M.A., S.H. Ensign, M.R. McIver, G.C. Shank, and P.K. Fowler. 2001. Demographic, landscape, and meteorological factors controlling the microbial pollution of coastal waters. Hydrobiologia 460: 185-193.

Mara, D.D., and J.I. Oragui. 1983. Sorbitol-fermenting bifidobacteria as specific indicators of human faecal pollution. Journal of Applied Bacteriology 55: 349-357.

Mara, D.D., and J.I. Oragui. 1985. Bacteriological methods for distinguishing between human and animal faecal pollution of water: Results of fieldwork in Nigeria and Zimbabwe. Bulletin of the World Health Organization 63(4): 773-783.

Mathew, A.G., W.G. Upchurch, and S.E. Chattin. 1998. Incidence of antibiotic resistance in fecal *Escherichia coli* isolated from commercial swine farms. Journal of Animal Science 76: 429-434.

Mathew, A.G., A.M. Saxton, W.G. Upchurch, and S.E. Chattin. 1999. Multiple antibiotic resistance patterns of *Escherichia coli* isolates from swine farms. Applied and Environmental Microbiology 65: 2770-2772.

McFeters, G.A., J.S. Kippin, and M.W. LeChevallier. 1986. Injured coliforms in drinking water. Applied and Environmental Microbiology 5: 1-5.

McLaughlin, M.R. 2001. Application of *Bacteroides fragilis* phage as an alternative indicator of sewage pollution in Tampa Bay, Florida. M.S. thesis. University of South Florida, St. Petersburg.

Noble, R.T., J.H. Dorsey, M.K. Leecaster, V. Orozco-Borbon, D. Reid, K.C. Schiff, and S.B. Weisberg. 2000. A regional survey of the microbiological water quality along the shoreline of the Southern California Bight. Environmental Monitoring and Assessment 64: 435-447.

Noble, R.T., and J.A. Fuhrman. 2001. Enterovirsuses detected by reverse transcriptase polymerase chain reaction from the coastal waters of Santa Monica Bay, California: Low correlation to bacterial indicator levels. Hydrobiologia 460: 175-184.

Noble, R.T., S.B. Weisberg, M.K. Leecaster, C.D. McGee, K. Ritter, K.O. Walker and P.M. Vainik. 2003a. Comparison of beach bacterial water quality indicator measurement methods. Environmental Monitoring and Assessment 81: 301-312.

Noble, R.T., D.F. Moore, M.K. Leecaster, C.D. McGee, and S.B. Weisberg. 2003b. Comparison of total coliform, fecal coliform, and enterococcus bacterial indicator response for ocean recreational water quality testing. Water Research 37: 1637-1643.

Noble, R.T., S.B. Weisberg, M.K. Leecaster, C.D. McGee, J.H. Dorsey, P. Vainik, and V. Orozco-Borbón. 2003c. Storm effects on regional beach water quality along the southern California shoreline. Journal of Water and Health 1: 23-31.

Olyphant, G.A., and R.L. Whitman. 2004. Elements of a predictive model for determining beach closures on a real time basis: The case of 63rd Street Beach, Chicago. Environmental Monitoring and Assessment. (in press).

Parveen, S., R.L. Murphree, L. Edminston, C.W. Kaspar, K.M. Portier, and M.L. Tamplin. 1997. Association of multiple-antibiotic-resistance profiles with point and nonpoint sources of *Escherichia coli* in Apalachicola Bay. Applied and Environmental Microbiology 63: 2607-2612.

Parveen, S., K.M. Portier, K. Robinson, L. Edminston, and M.L. Tamplin. 1999. Discriminant analysis of ribotype profiles of *Escherichia coli* for differentiating human and nonhuman sources of fecal pollution. Applied and Environmental Microbiology 65: 3142-3147.

Parveen, S., N.C. Hodge, R.E. Stall, S.R. Farrah, and M.L. Tamplin. 2001. Genotypic and phenotypic characterization of human and nonhuman *Escherichia coli*. Water Research 35: 379-386.

Phillips, C.R., M.I. Venkatesan, and R. Bowen. 1997. Interpretations of contaminant sources to San Pedro shelf sediments using molecular markers and principal components analysis. Pp. 242-260 in Molecular Markers in Environmental Geochemistry, R.P. Eganhouse, ed. Washington, D.C.: American Chemical Society.

Pisciotta, J.M., D.F. Rath, P.A. Stanek, D.M. Flanery, and V.J. Harwood. 2002. Marine bacteria cause false-positive results in the Colilert-18 rapid identification test for *Escherichia coli* in Florida waters. Applied and Environmental Microbiology 68: 539-544.

Prüss, A. 1998. Review of epidemiological studies on health effects from exposure to recreational water. International Journal of Epidemiology 27: 1-9.

Reynolds, K.A., C.P. Gerba, and I.L. Pepper. 1996. Detection of infectious enteroviruses by an integrated cell culture-PCR procedure. Applied and Environmental Microbiology 62: 1424-1427.

Robertson, W.J. 1984. Pollution indicators and potential pathogen microorganisms in estuarine recreational waters. Canadian Journal of Public Health 75: 19-24.

Rose, J.B., S. Daeschner, D.R. Deasterling, F.C. Curriero, S. Lele, and J. Patz. 2000. Climate and waterborne disease outbreaks. Journal of the American Water Works Association 92: 77-87.

Rose, J.B., D.E. Huffman, K. Riley, S.R. Farrah, J.O. Lukasik, and C.L. Harman. 2001. Reduction of enteric microorganisms at the Upper Occoquan Sewage Authority water reclamation plant. Water Environment Research 73: 711-720.

Schaper, M., J. Jofre, M. Uys, and W.O.K. Grabow. 2002. Distribution of genotypes of F-specific RNA bacteriophages in human and non-human sources of faecal pollution in South Africa and Spain. Journal of Applied Microbiology 92: 657-667.

Schiff, K.C., J. Morton, and S.B. Weisberg. 2003. Retrospective evaluation of shoreline water quality along Santa Monica Bay beaches. Marine Environmental Research 56: 245-254.

Scott, T.M., J.B. Rose, T.M. Jenkins, S.R. Farrah, and J. Lukasik. 2002. Microbial source tracking: Current methodology and future directions. Applied and Environmental Microbiology 68: 5796-5803.

Scott, T.M., S. Parveen, K.M. Portier, J.B. Rose, M.L. Tamplin, S.R. Farrah, and J. Lukasik. 2003. Geographical variation in ribotype profiles of *Escherichia coli* isolated from humans, swine, poultry, beef, and dairy cattle in Florida. Applied and Environmental Microbiology 69(2): 1089-1092.

Seigener, R., and R.F. Chen. 2002. Caffeine in Boston Harbor seawater. Marine Pollution Bulletin 44: 383-387.

Seyfried, P.L., N.E. Brown, C.L. Cherwinsky, G.D. Jenkins, D.A. Cotter, J.M. Winner, and R.S. Tobin. 1984. Impact of sewage treatment plants on surface waters. Canadian Journal of Public Health 75: 25-31.

Seyfried, P.L., R.S. Tobin, N.E. Brown, and P.F. Ness. 1985a. A prospective study of swimming-related illness: I. Swimming-associated health risk. American Journal of Public Health 75: 1068-1070.

Seyfried, P.L., R.S. Tobin, N.E. Brown, and P.F. Ness. 1985b. A prospective study of swimming-related illness: II. Morbidity and the microbiological quality of water. American Journal of Public Health 75: 1071-1075.

Simpson, J.M., J.W. Santo-Domingo, and D.J. Reasoner. 2002. Microbial source tracking: State of the science. Environmental Science and Technology 36: 5729-5289.

Solo-Gabrielle, H.M., M.A. Wolfert, T.R. Desmarais, and C.J. Palmer. 2000. Sources of *Escherichia coli* in a coastal subtropical environment. Applied and Environmental Microbiology 66: 230-237.

Standley, L.J., L.A. Kaplan, and D. Smith. 2000. Molecular tracer of organic matter sources to surface water resources. Environmental Science and Technology 34: 3124-3130.

Taggart, M. 2002. Factors affecting shoreline fecal bacteria densities around freshwater outlets at two marine beaches. Ph.D. dissertation, University of California, Los Angeles.

Takada, H., and R. Ishiwatari. 1987. Linear alkylbenzenes in urban riverine environments in Tokyo: Distribution, source, and behavior. Environmental Science and Technology 21: 875-883.

Takada, H., and R.P. Eganhouse. 1998. Molecular markers of anthropogenic waste. Pp. 2883-2940 in Encyclopedia of Environmental Analysis and Remediation, R.A. Meyers, ed. New York: John Wiley & Sons.

Tartera, C., and J. Jofre. 1987. Bacteriophages active against *Bacteroides fragilis* in sewage-polluted waters. Applied and Environmental Microbiology 53: 1632-1637.

Venkatesan, M.I., and I.R. Kaplan. 1990. Sedimentary coprostanol as an index of sewage addition in Santa Monica Basin, Southern California. Environmental Science and Technology 24: 208-214.

Venkatesan, M.I., and F.H. Mirsadeghi. 1992. Coprostanol as sewage tracer in McMurdo Sound, Antarctica. Marine Pollution Bulletin 25: 328-333.

Wade, T.J., N. Pai, J.N.S. Eisenberg, and J.M. Colford, Jr. 2003. Do U.S. Environmental Protection Agency water quality guidelines for recreational waters prevent gastrointestinal illness? A systematic review and meta-analysis. Environmental Health Perspectives 111(8): 1102-1109.

Zeng, E.Y., A.R. Khan, and K. Tran. 1997. Organic pollutants in the coastal marine environment off San Diego, California. 3. Using linear alkylbenzenes to trace sewage-derived organic materials. Environmental Toxicology and Chemistry 16: 196-201.

Zmirou, D., J.P. Ferley, J.F. Collin, M. Charrel, and J. Berlin. 1987. A follow-up study of gastrointestinal diseases related to bacteriologically substandard drinking water. American Journal of Public Health 77: 582-584.

5

New Biological Measurement Opportunities

INTRODUCTION

Recent and forecasted advances in microbiology, molecular biology, and analytical chemistry make it timely to reassess the long-standing paradigm of relying primarily or exclusively on traditional microbial (predominantly bacterial) indicators for waterborne pathogens in order to make public health decisions regarding the microbiological quality of water. This chapter provides an overview and discusses various issues and methods for making biological measurements. It underscores some of the key issues in making measurements both generically and specifically for pathogens and indicators of waterborne pathogens. The methods are evaluated critically in terms of their attributes, including potential applicability for measuring indicators and pathogens, as well as their limitations. The issues of standardization and validation of methods are then discussed, followed by a look toward the future that describes how new and emerging technologies and science will facilitate waterborne pathogen and indicator measurements. The chapter closes with a summary of its conclusions and recommendations.

Spatial and Temporal Granularity

As discussed in Chapter 4 and illustrated in Figure 5-1, the spatial and temporal scales (i.e., the "granularity") at which indicators and indicator organisms are employed may differ widely among applications. Small spatial and short temporal scales (area A) are of particular interest in beach monitoring programs and,

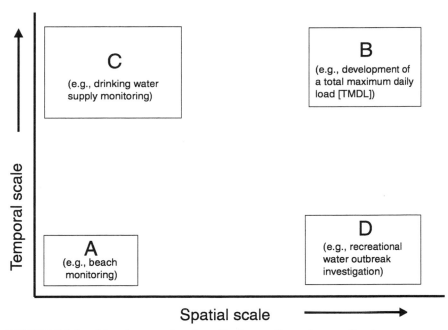

FIGURE 5-1 Spatial and temporal scales of indicators for various applications.

potentially, to transient contamination of groundwater. Larger spatial scales and longer temporal scales (area B) are of importance in understanding overall sources of microbial loadings to a watershed (that may serve as a water supply) or in studying the contamination of an aquifer or well. Small spatial scales but long temporal scales (area C) may be useful in understanding "typical" conditions at a water supply intake on a river system for the purposes of developing treatment configurations to meet drinking water standards for finished water. Large spatial scales (area D) but short temporal scales may be useful in understanding the occurrence of contamination over a large recreational area under outbreak conditions or from a storm event.

The temporal and spatial requirements for each particular application largely dictate the types of indicators or indicator approaches employed and the methods for measuring these indicators. As discussed throughout this report, particularly in Chapter 6, what is needed is a phased monitoring approach that makes use of a flexible "tool box" in which a variety of indicator methods and approaches are available for measuring a given indicator or pathogen for differing applications and circumstances. In many indicator applications, the level of perceived public health threat will determine the method or methods employed, as well as the spatiotemporal granularity. The indicator method, frequency, and spatial coverage of sampling will have to be "adaptive" in the sense that more frequent samples taken over larger areas with more sensitive methods will be required when the threat level is high (e.g., following high rainfall events) than when the threat is low. In some cases, the number and type of indicators measured may also differ with changing environmental conditions.

Classical Methods and Their Limitations

Most of the indicator applications described in previous chapters rely on biological measurements of bacteria. The classical laboratory techniques presently used for those measurements are primarily culture based, involving quantification of a metabolic or growth response after a suitable incubation period in an appropriate substrate. As reviewed in Chapter 1, culture based methods have been used for more than 100 years in water and related areas of environmental microbiology and have been considered adequate to provide quantification of indicator organism (predominantly bacteria) concentrations. Culture methods may be limited by their incubation period since most require 24 hours or longer, during which time the public is potentially exposed to a health risk (see Chapter 4 for further information).

The current choices of detection methods for indicator bacterial species or groups were motivated by the associated technical difficulties in culturing many types of waterborne pathogens. However, it is now possible to detect the growth of some specific pathogenic as well as indicator bacteria and also some viruses and parasites in as little as a few hours. For example, in clinical diagnostic and food microbiology bacteriology, automated bacteria culture detection and identification can be achieved in four to six hours (Fung, 2002; Lammerding et al., 2001; Murray et al., 1999); however, these and other advanced methods for rapid culture detection have not been well developed for or adapted to the rapid detection of indicator or pathogenic bacteria in water and other environmental samples. One reason why rapid culture-based detection works well in clinical diagnostic microbiology is that clinical specimens often contain high concentrations of the bacteria of interest, thereby allowing them to be cultured to even higher concentrations in only a few generations. In contrast, water and other environmental samples often contain very few bacteria of interest and therefore, many generations of bacterial growth are needed before these bacteria are readily detected by culture methods. Besides bacteria, coliphages—which are bacterial viruses infecting *Escherichia coli* (*E. coli*) that have been shown to be useful microbial indicators of fecal contamination and predictors of human health effects from recreational water exposures (see also Chapters 3 and 4)—can be cultured and detected in as little as six to eight hours by some methods (Lee et al., 1997; Sobsey et al., 1990).

As discussed in Chapter 3, many types of pathogenic and indicator bacteria present in the environment are in various states of physiological health and fitness, depending on their origin, properties, and how long they have been in the environment. The state of the microbes is influenced by the extent to which they have been exposed to various environmental stresses such as extreme temperatures and pH levels, hypo- or hypertonic salts, aerobic or anaerobic conditions, UV radiation, heavy metals, and various other antimicrobial chemicals, including chemical disinfectants such as chlorine (Hurst, 1977; McFeters and Camper, 1983; McFeters et al., 1986a,b). Therefore, enteric bacteria and many other bacteria in aquatic environments that are stressed, injured, and physiologically altered, may or may not be detected by various culture methods (Edwards, 2000).

Typical culture methods for pathogen and indicator bacteria in water and other environmental samples greatly underestimate the true concentrations of viable and potentially infectious cells—sometimes by as much as a thousandfold (Colwell and Grimes, 2000; Ray, 1989). For example, the anaerobic enteric bacteria that are so plentiful in the human and animal gastrointestinal tract, such as *Bifidobacteria* and *Bacteroides* (see also Chapter 4), are very difficult to culture from water and other environmental media because they are highly sensitive to very low concentrations of oxygen. While these bacteria would appear to be attractive candidate indicators of fecal contamination, the inability to culture them efficiently from water and other environmental media has been a major impediment to their potential use as fecal indicator microbes. However, the advent of nucleic acid based molecular methods to detect these bacteria now makes it more plausible and practical to consider them as fecal indicators (Barnhard and Field, 2000).

The underestimation of bacteria concentrations also results in part because the differential and selective media used to culture many types of waterborne pathogens and indicators contain inhibitory agents intended to suppress the growth of nontarget bacteria. Such agents also suppress the growth of injured or stressed target bacteria. In addition, other culture conditions, such as elevated incubation temperatures, may contribute to the lack of growth of target bacteria. Because bacteria injury is induced by the chemical disinfection and other treatment processes applied to water and wastewater, McFeters and colleagues (1986a,b) greatly improved the detection of injured coliform bacteria in water (by more than 10-fold) by the use of a medium that contained fewer inhibitory ingredients. According to some authorities, such bacteria can become viable but non-culturable (VBNC), as discussed in Chapter 3 and below.

Whether the VBNC pathogenic and indicator bacteria in water are infectious for human and other hosts and, in the case of the pathogens, pose health risks, remains uncertain and is quite controversial (Bogosian and Bourneuf, 2001; Bogosian et al., 1998; Kell et al., 1998). Some studies have reported that bacteria in the VBNC state have the ability to infect humans or animals (Colwell et al., 1996; Jones et al., 1991). Other investigators have not been able to infect animal hosts with so-called VBNC bacteria or have reported evidence that a few culturable bacteria within a large population of non-culturable bacteria could be responsible for the observed infections (Hald et al., 1991; Medema et al., 1992;

Smith et al., 2002). Because of the lack of scientific agreement of the public health significance of VBNC bacteria and the objections of some authorities even to the use of this terminology, this report does not attempt to address the VBNC issue in the context of microbial indicators of pathogens and human health risks from waterborne pathogens. However, the report does address issues related to the detection of stressed, injured, and otherwise physiologically compromised bacteria in water, the roles and appropriateness of both culture and non-culture methods to detect and quantify bacteria and other waterborne microbes, and the quantitative relationships between bacteria concentrations in water and the human health effects from exposure to water by ingestion and other routes.

The advent of increasingly sophisticated and powerful molecular biology techniques provide new opportunities and alternative approaches to improve upon present indicators and pathogens by both culture and non-culture methods. Molecular methods do not require incubation to culture bacteria because they can directly quantify existing cellular or subcellular structural properties. Therefore, these methods have the potential to be more rapid than culture methods, providing results in as little as minutes to a few hours rather than the typical overnight incubation time for culture methods. Some of these nucleic acid-based methods employ amplification schemes in which a small amount of indicator genetic material is replicated up to a billionfold for easy detection. They also have the potential to be less expensive, making direct measurement of pathogens more economically feasible. Much of the rest of this chapter is devoted to describing the types of molecular methods that are presently under development and have the potential to replace, supplement, or greatly improve the quality of information of classical (largely bacterial) culture-based methods in the future. It is important to mention that Appendix C (Detection Technologies) supplements the discussion (both generally and specifically) of these and other methods by describing them in more detail. Furthermore, molecular methods can be coupled with or linked to microbial culture methods in ways that can increase sensitivity, decrease detection time, and provide conclusive and rapid confirmation of identity and infectivity (e.g., Reynolds et al., 1996).

Targets and Opportunities

Several different analytes can be measured in microorganisms. For purposes of this discussion, microbes can be divided broadly into cells and viruses. Cells can be detected by the following categories of analytes, as summarized in Figure 5-2.

Nucleic Acids

Deoxyribonucleic acids and ribonucleic acids have unique sequences of nucleotide bases (adenine, thymidine [uracil in RNA], cytosine, and guanine) that

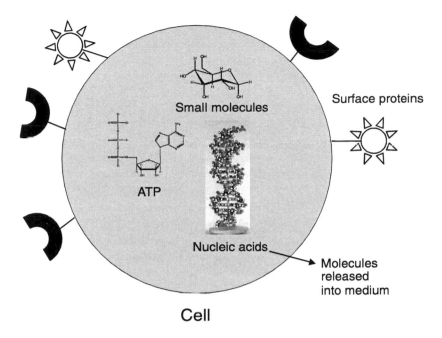

FIGURE 5-2 Targets to measure on or in a cell. Note: ATP = adenosine 5′-triphosphate.

enable the unequivocal identification of a particular organism. DNA and/or RNA is present in all cells and viruses. Cells contain both DNA and RNA, whereas viruses contain either DNA or RNA but never both. The choice of nucleic acids and the ways in which they are measured in microorganisms can provide different kinds of information with regard to microbial identification, viability, and infectivity or culturability. For example, some nucleic acid targets and the methods for their detection can provide very broad identification of a family or genus of microorganism, while other targets can provide very specific identification of species, strain, or subtype. Some nucleic acid targets can be taken as measures of viability or infectivity, such as messenger RNA (mRNA) of cellular microbes or mRNA production by viruses in infected cells. In some cases, mRNA targets are evidence of culturability or infectivity. In general, RNA correlates with viability because nucleases present in most biological samples destroy RNA rapidly. Therefore, both the presence and quality of RNA and the specific sequences present can provide a reasonable indication of viability (see more below). In developing methods to detect and quantify waterborne microorganisms and microbial indicators of pathogens, it is important to consider both the targets for detection and methods of detection with consideration of the value and interpretation of resulting data.

Surface Proteins

Proteins present on the surface of a microbe and to a lesser extent those located within a microbe are often unique and offer a means to definitively identify a microorganism of interest. The most common method of analyzing such proteins is the use of immunoassays in which specific antibodies are raised against the proteins and used as binding reagents. Both monoclonal and polyclonal antibodies can be used. Polyclonal antibodies tend to be more broadly reactive, which makes them useful in detecting microbes as broad groups, such as genera. Monoclonal antibodies have greater specificity because they recognize and bind to a very specific epitope or functional group on or in the target microorganisms. The uniqueness of the epitope depends on its function within the microbe. Some epitopes are common to all members of a microorganism family, genus, or species (group or "common" antigens); others can be highly specific, appearing only in an individual strain, subtype, or variant.

Other approaches to microbe identification based on proteins can employ non-antibody ligands, such as aptamers or phage display libraries, that will specifically recognize and bind to a particular protein or an epitope on it (Breaker, 2002). Such ligand binding probes to identify microorganisms, including bacterial spores (e.g., *Bacillus anthracis*; Zhen et al., 2002), are becoming more accessible because of the advances made in protein identification and mapping within microbes and the advances made in the synthesis of in vitro proteins, oligonucleotides, or oligopeptides. Certain proteins on the surface or in the interior of microbes can be detected by ligand binding assays. The presence of these markers on or in the cell can be evidence of microbe viability or infectivity. Certain proteins in cells and viruses may be present in a native state only when the microbe is intact and infectious. Therefore, the ability to specifically detect that molecule by a ligand-binding assay can be taken as a measure of viability or infectivity.

Carbohydrates (Polysaccharides)

Carbohydrates or polysaccharides present on the surface of a microbe or within a microbe also can offer a unique way to definitively identify a microorganism of interest. Many of these specific carbohydrates are oligosaccharides covalently bound to proteins to create glycoproteins. Such molecules on the surfaces of cells and viruses often have high specificity or uniqueness in identifying a microorganism. Immunoassays can be used to detect, identify, and quantify such polysaccharides or glycoproteins, again using specific polyclonal or monoclonal antibodies raised against the microbe or the specific target molecules. Like proteins, the specificity of polysaccharide epitopes depends on their function within the microbe, with some antigens common to all members of a microorganism family, genus, or species and others being highly specific for individual strains, subtypes, or variants. Non-antibody ligands also can be used to detect, identify, and quantify specific polysaccharide epitopes. As with microbial proteins, ligand-binding probes to microbial polysaccharides are becoming more accessible because of the advances made in functional polysaccharide identification and mapping and the advances made in ligand-binding chemistry. As is the case for certain nucleic acids and proteins, the detection of certain polysaccharides on or in microbes by ligand binding can be evidence of a microorganism's viability or infectivity (Feng and Woo, 2001). Certain polysaccharides in cells and viruses are active receptors for attachment and infection and are present in the native state only when the microbe is intact and infectious. Therefore, detecting such molecules by a ligand-binding assay is a measure of viability or infectivity.

Other Small Molecules

Some microorganisms contain or release characteristic metabolites or products, such as sugars, polysaccharides, antibiotics, alkaloids, lipids, and (protein-based) enzymes and toxins into their environment or growth medium. These compounds may be products of either primary or secondary metabolism and can provide a distinct signature for the microorganism of interest. Many methods are available for analyzing such compounds including mass spectrometry, colorimetric assays, enzymatic assays, and various chromatographic methods. For example, adenosine 5'-triphosphate (ATP) is often measured as an indicator of viable and possibly infectious cells, because it is degraded rapidly when the cell dies (e.g., bioluminescence assays; Deininger and Lee, 2001).

Special Considerations for Viruses

Viruses are typically detected either by their DNA or RNA (for RNA viruses) and their surface proteins (either the capsid or the envelope; see Figure 5-3). Although many viruses do not contain small molecules or detectable amounts of internal protein, most animal viruses do. When present, these internal proteins can also be targets for detection, although often they are less accessible than surface proteins. Because viruses are inert outside their host cells, determining the infectivity of a virus often depends on culturing it in host cells. When they do infect host cells, viruses begin to produce new, virus-specific molecules that can be targeted for detection by molecular and other chemical methods as evidence of their presence, infectivity, and concentration. Virus-specific nucleic acids, such as mRNA and proteins, including both structural and nonstructural proteins, can be targeted for detection by chemical, immunochemical, and molecular methods. In addition, all viruses have specific functional groups or epitopes on their surfaces that are used for attachment to host cells. If the cell receptor or its functional ligand constituent can be identified, such a molecule can be used to detect and quantify viruses through a ligand-binding assay.

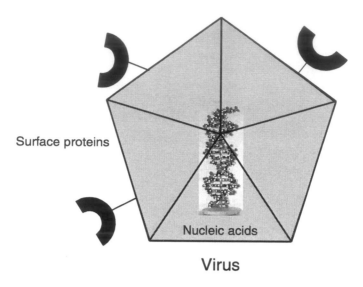

Virus

FIGURE 5-3 Targets to measure on or in a virus.

Special Consideration for Protozoa

Like bacteria, waterborne protozoa are single-celled organisms and consist of many of the same components. Unique to the enteric protozoa, however, is the formation of an (oo)cyst as part of its environmental and infectious stage (see Chapter 3 for further information). In most cases, this structure is currently detected by microscopy through the aid of stains and antibodies against the (oo)cyst cell wall. Enteric protozoa are obligate parasites and are similar to viruses in that they need a host organism to reproduce. Thus, determination of the potential viability of protozoa has been studied using vital dye inclusion-exclusion as a measure of the integrity of an (oo)cyst's outer wall as well as its inner cytoplasmic and nuclear membranes. Huffman et al. (2000) showed, however, that vital dyes grossly overpredicted infectivity of *Cryptosporidium* under some circumstances. Cell culture methods are now being used and have been found to be statistically comparable to animal infectivity for the determination of infectious oocysts (Slifko et al., 2002). Methods continue to evolve, and as with other microorganisms, polymerase chain reaction (PCR) techniques to target the nucleic acid components as well as methods that combine cell culture and PCR are now being used for detecting protozoa in water (Quintero-Betancourt et al., 2002; see also Appendix C). For example, the free-living amoeba (e.g., *Naegleria*) can be isolated from water using a culture technique (i.e., their growth in the trophozoite stage is responsive to a bacterial culture). In addition, PCR, probes, and culture methods are now being combined to identify those species and subtypes that are particularly lethal to humans (Kilvington and Beeching, 1995).

ISSUES IN SAMPLING AND ANALYSIS

The process of making a measurement consists of the four steps shown in Figure 5-4. A common misunderstanding is that measurement is the only critical step in the analysis process.

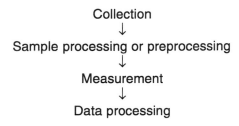

Collection
↓
Sample processing or preprocessing
↓
Measurement
↓
Data processing

FIGURE 5-4 Four steps involved in performing a measurement.

However, as discussed below, all four components of the process must be considered to ensure accurate analysis of microbial water quality.

Collection and Sampling Issues

The first step in performing a measurement is collecting a sufficiently representative sample, and this remains one of the most challenging problems in water quality monitoring. By representative, it is meant that the sample will reliably portray the presence and concentrations of the analyte of interest (e.g., a microorganism or a chemical) in the water being evaluated or analyzed for its quality. Furthermore, it is important that the sample also be representative of human exposures that may lead to pathogen ingestion and any resulting infection and illness. As noted previously, it is important to recognize that the presence and concentrations of microorganisms and chemicals in water and other environmental media can be highly variable over time (at different times) and space (at different locations within the same body of water). Therefore, as described in Chapter 4, obtaining representative samples often requires taking multiple samples over an extended period (e.g., daily, weekly, monthly), sometimes from different locations within a body of water during the same time period. The importance of addressing variability in microorganism concentrations in water as related to human exposures to pathogens has been well documented in recreational water epidemiologic studies (Fleisher et al., 1993; Kay et al., 1994). The temporal variability of microbial occurrence in groundwater has also been documented (EPA, 2000).

Collecting representative samples requires careful consideration of the objectives or purpose of sampling in the context of the need to obtain a reliable estimate of microbial exposure in a timely fashion. Unfortunately, sample collection often involves simply "grabbing" a volume of water and placing it in a storage vessel. For many samples, it is important to preserve the sample, by refrigeration or chemical preservatives, to avoid degradation. All or a fraction of the sample is then taken to the analysis site for further processing. Typically, a sufficient sample volume is taken either to determine whether a microorganism or other analyte is present (i.e., presence-absence) or to estimate the concentration of microbes or other analytes in the water being analyzed (e.g., number of microbes per unit volume).

For microorganisms of public health concern in water, both types of analysis (presence-absence and concentrations estimates) are now used for estimating exposures and making decisions regarding the acceptability of the water for beneficial use under the Clean Water Act (CWA; see also Chapter 1), such as drinking water supply. In some cases, the goal of the analysis is to document that samples of a certain volume (e.g., 100 mL) do not contain a particular microorganism the vast majority of the time (e.g., absence of total coliforms in 95 percent of successive 100-mL drinking water samples) or ever (e.g., absence of *Escherichia coli* in successive 100-mL volumes of drinking water all of the time). In other cases, the goal of the analysis is to document that samples of a certain volume contain a particular microorganism at concentrations below a threshold level considered indicative of an unacceptable health risk (e.g., maximum allowable concentrations of fecal indicator bacteria in recreational bathing waters). In water analysis based on either presence-absence or estimates of concentration, the variability of microbial concentrations is typically addressed by taking repeated samples from the body of water over time and determining both central tendency (e.g., mean or median) and dispersion (e.g., minimum-maximum values, interquartile range, 95 percent confidence limits).

The focus of data analysis and interpretation is often on typical exposures that are portrayed by central tendencies and dampened extremes, such as 95 percent confidence limits, that are based on logarithmically transformed data. Recent evidence from food microbiology and foodborne disease outbreaks indicates that measures of central tendency and the use of logarithmic transformations of microbial concentration data for the purposes of calculating geometric means and corresponding logarithmic measures of dispersions may be inappropriate for extrapolating to higher exposures and estimating corresponding health risks (Paoli, 2002). Such transformations tend to suppress the effects of extreme values, including the high values on the upper end of a frequency distribution that represent the greatest levels of exposure and health risk. Characterizing the extremes of exposure is necessary because illnesses can result from combinations of rare events that lead to high levels of exposure. Therefore, data on the magnitude and probability of deviations at the high extremes are needed and must be taken into consideration. The widespread use of logarithmic transformations and measures of central tendency and dispersion of log-transformed data to estimate exposures and health risks needs to be reconsidered in water microbiology, epidemiology, and health risk assessment.

An important characteristic of sampling when there is the likelihood of only low-level detection is that, although the species (microorganism or chemical) of interest may be present in the water being sampled, it may be present at such a low concentration that a given (typical) sample will not contain it. In such situations, the term "Poissonian sampling" comes into play. Simply put, Poissonian sampling aims to determine how many samples of a given volume, or what volume of sample, must be analyzed to ensure that the species of interest is present or not at the prescribed threshold level.

Given a random sample, Figure 5-5 below illustrates the typical numbers involved. The key parameter is s, which is essentially the "average" number of target microbes or molecules that one finds in the sample volume being analyzed. For example, if there are 100 target microbes or DNA genomes per milliliter in a sample, then a sample volume of 10 µL will, "on average," contain one microbe or genome per sample volume. For lower concentrations of targets, the volume required to ensure a representative sample increases accordingly.

A variety of factors also must be considered in devising and using sample plans when estimating microbial concentrations in water. Sampling may be intended to observe long-term trends in the concentration of microbes in a body of water in which the emphasis is on determining if a measure of central tendency, such as a geometric mean concentration based on replicate samples over a monthly period, is below a specified value. In this situation, the emphasis is on estimating the typical concentrations of microorganisms in water over a long

PCR counting statistics for low concentrations
(assuming a Poissonian distribution)

Probability (p) of finding $(n$ targets$) = \dfrac{(s^n)^* \exp(-s)}{n!}$

Where s = sample concentration $\times$ sample volume
If $s = 1$, $p(0) = 0.37$
If $s = 3$, $p(0) = 0.05$
If $s = 10$, $p(0) < 10^{-4}$

FIGURE 5-5 PCR statistics for low concentrations. SOURCE: Raymond Mariella, Lawrence Livermore National Laboratories, personal communication, 2002.

period of time as a measure of the acceptability of the water for a beneficial use. In such applications, microbial sampling plans still have to address sources of variability in microbial concentrations and the detection of extreme high concentrations of microorganisms from events such as precipitation and other increases in microbial loads. Additionally or alternatively, sampling plans may be intended to estimate the concentration of microorganisms at a single point in time when populations are exposed (e.g., swimmers on a particular day at a bathing beach). In this case, emphasis must be placed on obtaining temporally and spatially representative samples of water to determine whether the concentration of microbes is below that producing an acceptable health risk from a single exposure event (see Chapter 4 for further information). Sampling plans and procedures to estimate the risks from such short-term exposures may have to be quite different from sampling plans and procedures intended to estimate long-term trends and typical concentrations (see also Figure 5-1).

Whether sampling is intended to estimate concentrations for determination of immediate or short-term exposure risks or longer-term trends, it is clear that little information can be obtained from analysis of a single sample of water for a microbial indicator or pathogen. Statistically based sampling methods must be used in conjunction with analyses of multiple samples in order to estimate how microbial levels and human exposures change with varying water quality conditions. Sampling plans must be able to identify when and where microbial concentrations in water are at their highest levels since this is when human health risks are greatest.

Because of the issues associated with collecting representative samples, additional research to develop improved methods for rapid sample concentration and effective, reproducible sample recovery should be supported.

Preprocessing

Once a sample has been collected, it may be subjected to several steps designed to prepare it for analysis of the target microorganisms. For example, in the case of bacterial analysis by culture, a water sample may be filtered to collect the bacteria on a membrane filter that is then placed on a culture medium for incubation and the development of bacterial colonies (see also Appendix C). In this case, the bacteria are separated and concentrated from the sample water prior to culture and enumeration. Similarly, the physical preprocessing steps of filtration or sedimentation by centrifugation have been used to recover microbes from water samples, while at the same time both concentrating them and separating them from other constituents. Several preprocessing steps are available to help purify the sample so that the desired components can be measured without potential interference from other sample constituents. For example, for analysis of nucleic acids it is necessary to remove organic matter (e.g., humic acids) and cellular debris and metals (e.g., iron, aluminum, heavy metals) because they can inhibit the reactions employed for analysis (Kreader, 1996; Reynolds et al., 1997).

Preprocessing to separate microorganisms or molecular targets in microorganisms from matrix constituents includes chemical methods such as precipitation (with inorganic salts, polyethylene glycol, or acids); solvent extraction (e.g., chloroform); adsorption (to charged surfaces of filters, minerals, or synthetic polymers); chelation (of heavy metals and multivalent cations); chromatography (ion exchange and size exclusion); multiple aqueous phase separation using soluble polymers; treatment with detergents; and ligand binding (e.g., immunocapture, immunomagnetic separation).

A second aspect of processing or preprocessing in biological measurement often involves amplifying the desired microbe or other target analytical component. In some cases, amplification of the target microbe is an essential feature of the measurement method, such as culturing bacteria in liquid (broth) or on solid (agar or membrane filter) media for their quantification or enumeration. Another increasingly used example of processing or preprocessing is PCR, which amplifies a specific DNA sequence that may be present in a sample. More specifically, the DNA sequence of interest is amplified 2^n where n = number of PCR cycles. In this manner, the DNA sequence of interest is amplified exponentially, and the resulting sample contains a high concentration of the sequence of interest and can be measured and detected easily. RNA targets, such as the genomic RNA of enteric viruses and mRNA, also can be amplified as a processing or preprocessing method. Usually this amplification is done first by synthesizing a complementary nucleic acid strand (cDNA) to the target RNA sequence by reverse transcription (RT) and then applying PCR to the resulting double-stranded molecule. This method is referred to as RT-PCR.

Quantitative Versus Qualitative Measurements

It is important to recognize that while methods aimed particularly at treated drinking water or groundwater have focused on detecting presence versus absence of a particular indicator in a given volume, in most ambient water applications, obtaining quantitative information has been the ultimate goal. Thus, it is important to obtain reliable estimates of concentrations of target microorganisms or the indicator being measured. The Most Probable Number (MPN) statistical approach has long been used in environmental water microbiology along with quantal (i.e., presence-absence) assays and has been automated to the extent that labor associated with the dilutions and replicate assays is now less tedious and costly (e.g., semiautomated quantification, liquid-based methods for *E. coli* such as Quanti-Tray®). Furthermore, to establish or study risk estimates associated with a given water sample, the concentration of the pathogen or indicator may be required. Threshold concentrations of certain pathogens or indicators must often be determined to assess whether the water is in compliance with regulations. A simple presence-absence measurement without quantification is insufficient in such cases. In other cases, such as the deliberate introduction of a toxic chemical or biological threat agent nominally not found in water, a qualitative presence-absence determination is generally sufficient because it indicates that a problem exists. As noted previously, it is important that quantitative measurements consider extreme events (high concentrations) and reliably represent the frequency distribution of these events and their temporal and spatial variability.

Measurement

Once a sample has been suitably processed or preprocessed, a measurement is made. This step may involve injecting the sample into an instrument; titrating a sample; scoring cultures as positive or negative for microbial growth; or enumerating colonies, plaques, or foci present on agar or membrane filters or in infected cells after a requisite incubation period. The end result is to collect data in the form of spectra, counts, volume, optical density, and so on. These data are simply values that correspond to some parameter being detected by the instrument or by the individual taking the readings. Various methodologies widely employed for making measurements of microorganisms in water samples are summarized later in this chapter and in Appendix C.

Data Collection and Processing

Once measurement data have been scored, they are collected (compiled) and processed. Processing involves manipulating or analyzing the data based on the presence, absence, or concentration of the analyte being analyzed. In simple cases, data processing is straightforward—for example, scoring the presence or absence of a particular analyte (e.g., virus or bacterium) by simply observing and recording the positive or negative result obtained during the measurement phase of analysis. In other cases—for example, estimating the concentrations of microbes cultured in different and replicate volumes of broth media—the numbers of positive and negative cultures of the total cultures inoculated per sample volume have to be processed through calculation of an MPN or a 50 percent infectious dose (ID_{50}). Some measurements generate complex or large amounts of data that must be subjected to detailed analysis before a result for the presence, absence, or concentration of the target analyte can be determined. For example, amplified nucleic acid from PCR may have to be subjected to nucleic acid hybridization or nucleotide sequencing before the sample can be confirmed as positive. Mass spectral data must be processed to correlate the measured spectrum to the spectra of various compounds stored in a database. With the increasing amount of data coming from high-density arrays, mass spectra, and long-term time series with high spatial coverage, there is a need to devote additional resources and effort to data storage and processing.

ASSESSMENT OF METHODS AND THEIR ATTRIBUTES

In analyzing water and other environmental samples for indicators of pathogens or for pathogens themselves, three main options are available: (1) analyze for live or infectious microorganisms (pathogens or indicators); (2) analyze for microorganisms without conclusively determining their infectivity or viability; or (3) analyze for another constituent in the sample (a surrogate) that is indicative and predictive of the presence and concentration of the pathogen or microbial indicator (e.g., a chemical associated with fecal contamination; see Chapter 4 for further information).

Direct Analysis of Microorganisms by Infectivity, Culturability, or Viability

On the basis of indicating public health risk of infection or disease from exposure to microbially contaminated water, the direct measurement of a pathogen or a reliable microbial indicator of pathogens by culture or infectivity is generally considered the "gold standard," and both should be the goal of any new measurement technique. That is, if a pathogenic microorganism can be cultured and shown to be infectious, it indicates that the organism is viable and potentially able to cause infection and disease given sufficient exposure and a susceptible host. Analyzing for a microbial indicator by culture or infectivity can also be predictive of such a health risk, provided the indicator is otherwise a reliable predictor of infectious pathogens. Various methods of analyzing for pathogens by culture or infectivity are available and have been reviewed and described in detail elsewhere (Hurst et al., 2002; Sobsey, 1999, 2001).

Briefly, the most commonly used culture methods for bacteria are colony counts on membranes or agar medium plates and liquid broth cultures. In either format for culture, the target bacteria are detected by and distinguished from other bacteria by use of differential and selective media that have specific ingredients for multiplication of the target bacteria, such as chemical inhibitors against the growth of non-target bacteria, and constituents (specific growth substrates or indicator chemicals such as oxidizing or reducing agents) that cause the growth of the target bacteria to be unique in appearance and distinguishable from non-target bacteria (differential ingredients).

Because viruses are obligate intracellular parasites, culturing them requires the use of susceptible host cells. The viruses will attach to and penetrate the host cell, where they will multiply (replicate), usually with subsequent release of progeny viruses and death and lysis of the host cell. This process of infection can be detected by death or lysis, as manifested by the disappearance of the cells ("clearing") from a broth culture; the development of virus-induced morphological changes in the appearance of the cell (cytopathogenic effects, or CPEs); or the development of discrete circular, cleared areas in a layer or lawn of cells in an agar medium, which are referred to as plaques. Some viruses do not produce visible lysis or CPEs in host cells although the viruses have multiplied within the cells. In such cases, the presence of the viruses in the infected cells can be detected by molecular methods (nucleic acid, protein, or enzyme) or by immunochemical and immunohistochemical methods, as described later in this chapter.

The application of cell culture for *Cryptosporidium parvum* infectivity began in the early 1990s and was reviewed recently by Rose et al. (2002). A variety of end points are currently being utilized to determine the concentration of infectious oocysts in water samples. Immuno-based assays utilizing antibodies to *C. parvum* sporozoites and other life-cycle stages, coupled with a secondary antibody conjugated to a fluorescent dye or enzyme, have been employed. Molecular-based assays using either PCR or RT-PCR methodologies to amplify DNA or RNA targets extracted from infected cells or oligonucleotide probes that can detect nucleic acids in situ have also been developed for speciation and genotyping. In a study of surface waters and filter backwash waters, infectious oocysts of a variety of strains were detected in 4.9 and 7.4 percent of the samples, respectively, using cell culture methods (DiGiovanni et al., 1999).

Another way to analyze for pathogens or microbial indicators of pathogens is by direct observation of their viability. Viability can be analyzed on the basis of several different measures or end points, such as physical movement (e.g., of larvae in ova), hatching (e.g., excystation of protozoan cysts or oocysts), enzyme activity, oxidation-reduction, synthesis of macromolecules, and uptake or exclusion of dyes. In the case of some microorganisms, viability measurements are likely to be good predictors of infectivity because the end point is actually detecting the activity of a living organism. For example, the detection of viable helminth ova by microscopic examination for the movement of mature larvae within the ova is likely to be strongly associated with human or animal infectivity. However, some viability assays, such a excystation of protozoan (oo)cysts or dye exclusion (or uptake), are poor predictors of infectivity for human or animal hosts.

Analysis of Microorganisms by Measurement of Their Constituents or Components

As noted previously, measuring components of microbes is often used for their analysis in water and other environmental media, and some of these approaches are summarized below (for further information see Hurst et al., 2002; Sobsey, 1999, 2001). Although the techniques described below are designed to measure specific components of the microorganism as discussed elsewhere in this report (see also Box 4-2), the most important biological attribute of an indicator is a strong quantitative relationship between indicator concentration and the degree of public health risk. That analysis of the constituent should itself be a reliable predictor of human health risk from exposure to microbiologically contaminated water.

Strategies

For all the non-culture based methods of microbial analysis, there are two general approaches that can be used to identify the presence of a specific microorganism: (1) targeting a single *specific* component of the organism that is unique and characteristic of that organism and (2) using *fingerprinting* in which a pattern of components signifies the presence of the microorganism. In the latter case, the individual components may not be unique to the particular microorganism of interest, but their concentration, co-occurrence, or sequence generates an overall response profile characteristic of the microorganism.

Nucleic Acid Analysis

DNA sequences can be present at only one copy per cell, which poses an extremely difficult detection challenge. Some target genes, however, such as unique intragenic sequences in DNA and certain forms of RNA, can be present in multiple copies in a cell, making these sequences easier to detect. DNA is typically amplified first using PCR (described earlier and more extensively in Appendix C). PCR is sequence specific, although sometimes in the absence of a specific target sequence, nonspecific amplification of non-target sequences may occur; in addition, under low stringency conditions non-specific binding of the primers also occurs. These nonspecific products will usually give negative results in subsequent analysis of the PCR products. PCR followed by analysis such as hybridization provides two levels of discrimination.

Another method based on hybridization is fluorescent in situ hybridization (FISH). In this approach, fluorescent probes specific to different regions on the chromosome containing different labels are hybridized to the intact microorganism and the pattern of colors on the chromosome are viewed by microscopic examination under a fluorescent microscope. The unique banding pattern corresponds to the microorganism of interest.

RNA is present in bacterial cells or protozoa as messenger RNA (mRNA), ribosomal RNA (rRNA), or transfer RNA (tRNA). In RNA viruses, the RNA is present as the viral genetic material, as either a single- or double-stranded molecule, or as either one continuous strand or multiple unique strands. Messenger RNA is present in many copies per cell and, as described above, is typically first converted into complementary DNA by reverse transcription and then PCR-amplified for analysis (RT-PCR).

Nucleic acid analysis can be conducted using several strategies. Specific unique sequences can be selected that have no counterparts in any other microorganism. In this manner, the presence of the amplified DNA (the amplicon) signals the occurrence of the specific sequence of interest. In this type of analysis, prior sequence information about the specific organism of interest is required. It is important to note that the uniqueness of nucleic acid sequences can vary—from sequences that are common or shared among closely related microbes and can therefore detect families, genera, or other groups of microbes, to highly specific sequences that can identify a single species, strain, or variant of a microbe. Alternatively, DNA or RNA fingerprinting is performed in which the pattern of nucleic acid sequences is correlated with a particular microorganism. In this case, sequence information is not essential as long as the pattern is known a priori. In this approach, the identifying pattern is based on a reference microbe or microbes. In most environmental applications, a collection of reference microorganisms, called a library, is created against which to compare the pattern observed in an environmental isolate obtained from a sample (see Chapter 4 for further information).

After amplification, DNA is analyzed either by sequencing or by hybridization to a unique complementary genetic sequence (a gene probe) or to an array containing the complementary genetic sequence. For example, Dombek and colleagues (2000) used membrane filtration to first concentrate *E. coli* samples from a variety of sources (humans, chicken, cows, ducks, geese, pigs, and sheep) and then microarray technology (DNA fingerprinting; see Appendix C for a detailed discussion of microarrays in detecting waterborne and foodborne pathogens) to identify their sources with success rates ranging from 89.5 to 100 percent. If only one or a limited number of sequences are required for identification, rapid or real time PCR can be employed in which a fluorescence signal appears only when the sequence of interest is present. The advantages of rapid (real time) PCR over traditional PCR methods include faster results and fewer handling steps (see Appendix C for further information). Rapid PCR methods are amenable to field use, and several commercial vendors have instruments available for bio-warfare agent field detection. Their use to detect enteric microbes in field samples has already been reported (Donaldson et al., 2002). Therefore, as these methods improve and become more widely available, there is considerable promise for the expanded application of this rapid PCR technology to detect microbial pathogens and indicators in environmental waters.

Although these molecular biology methods directed at nucleic acids were developed primarily for the Human Genome Project[1] with application in clinical medicine, these powerful techniques have direct applicability to waterborne pathogen or indicator detection is clearly feasible and has already been done (Cook, 2003; Griffin et al., 2003; Keer and Birch, 2003). So far, much of the application of nucleic acid amplification, detection, and characterization is by

---

[1]See http://www.ornl.gov/TechResources/Human_Genome/home.html for further information.

hybridization using macro-scale methods (e.g., various forms of conventional PCR and RT-PCR and filter or other hybridization to detect amplicons). However, microarray technology is becoming increasingly available for use in applied and environmental microbiology, primarily directed at gene expression of living microorganisms under different environmental conditions. In these applications, the identities of the microorganisms may already be known, many cells are exposed to the conditions under study, and numerous copies of the target nucleic acid are available for detection. DNA microarrays have been applied successfully to the detection of pathogenic microorganisms in environmental samples. Direct detection of extracted or accessible nucleic acids is possible when present at high concentrations (>10 cells per mL; Chandler et al., 2001), and for low numbers of target organisms, nucleic acid amplification is applied prior to hybridization (including in microarrays), nucleic acid sequencing, or other detection and characterization technologies.

One of the important issues to address in the application of nucleic acid technologies to the detection of pathogens and indicators in water and other environmental samples is to what extent and how such technologies can distinguish infectious and viable microbes from those that are noninfectious or inactivated, including the detection of nucleic acids from dead and degraded microorganisms. It is important to recognize that the detection of waterborne pathogens or indicators whether viable and potentially infectious or not, may provide sufficient information to assess vulnerability to contamination or to make decisions about public health risk (see also Chapters 4 and 6). For example, under circumstances where the basis of acceptable risk is the absence or not exceeding a specified maximum concentration of the nucleic acid of a pathogen or indicator in a specified sample size. Such a management approach has been proposed for Noroviruses or F+ coliphages as Norovirus indicators in raw bivalve molluskan shellfish (Dore et al., 2000, 2003).

Under some other conditions, the detection of nucleic acids from inactivated microorganisms would not necessarily be indicative of a health risk and would not be considered useful information in support of management decisions. For example, water subjected to physical or chemical disinfection that inactivates viruses and other microbes can still contain detectable nucleic acids of these inactivated viruses (Sobsey et al., 1998). One approach to overcoming the problem of detecting the nucleic acid of inactivated viruses is to couple nucleic acid detection methods such as PCR with microbial culture methods. Initial amplification of the microbes through culturing is then followed by methods to detect their nucleic acids. This has been done successfully for enteric viruses and is referred to as integrated cell culture-PCR (ICCPCR; Blackmer et al., 2000; Reynolds et al., 1996).

Another nucleic acid approach to detecting viable and potentially infectious microorganisms is to target only those nucleic acids found in organisms in this state, such as messenger RNA. For viruses, another approach is to detect only fully intact and potentially infectious RNA viruses by first exposing them to proteases and then ribonucleases, to which intact and potentially infectious viruses are resistant. These enzyme pre-treatments degrade damaged and noninfectious viruses, leaving only the intact and potentially infectious viruses to be detected by subsequent nucleic acid amplification methods (Nuanualsuwan and Cliver, 2002). Yet another approach to the detection of infectious viruses is to amplify only full length viral genomic nucleic acid. If the nucleic acid has been degraded to less than full length fragments or contains lesions causing inactivation, then nucleic acid amplification does not occur. Thus, several promising methods to the detection of the nucleic acids of only intact, viable, and infectious microorganisms are in development and being validated.

As the methods for recovery, concentration, and purification of target microbes and their nucleic acids are further improved for the application of various nucleic acid methods, including microarray technology, it is likely that these approaches will become more widely applicable to the detection, quantification, and identification of microbes in water and other environmental media. The committee concludes that the introduction of molecular techniques for nucleic acid analysis is a growth opportunity for the field of waterborne pathogen detection and recommends that U.S. Environmental Protection Agency (EPA) resources be invested to accelerate the introduction and further development of these techniques. Lastly, it should be noted that microbial toxins, which are proteins, cannot be detected by PCR or other nucleic acid analysis-based methods.

Immunological Methods

For surface proteins, large peptides, and their glycosylated derivatives, immunological methods offer a high degree of specificity and sensitivity. Although small molecules such as toxins generally cannot be detected using immunological methods, surface proteins on bacteria, protozoa, and viruses can be unique to the microbe and detected by immunological methods of analysis. In this approach, an antibody is raised against the microorganism or the purified protein to be detected. Both polyclonal and monoclonal antibodies can be obtained, and an immunoassay is developed around these antibodies. Typically, the antibodies are employed to capture the analyte, carry a label to the analyte, or both. The most common immunological method used is the enzyme linked immunosorbent assay (ELISA) in which a capture antibody bound to a surface is used to bind and concentrate the analyte. A second antibody, labeled with an enzyme, is then bound to a second recognition site on the analyte. Finally, a chromogenic or fluorogenic substrate is added that is converted to an observable product, which can be detected. A detailed description of the ELISA method appears in Appendix C.

Another immunological method now becoming widely used in environmental microbiology is immunomagnetic separation (e.g., Gehring et al., 1996; Mitchell et al., 1994). This method can be employed as both a processing (recovery, concentration, and purification) method and a detection method (or at least part of a detection method). The typical application of immunomagnetic separation is to have the antibodies bound to a solid phase, such as paramagnetic beads, and then react those beads with the sample for target microbe recovery by its binding to the antibody on the solid phase. Typically, another method, such as culture, immunofluorescence microscopy, or nucleic acid hybridization or amplification, is then used for detection and confirmation of the target microbe or its components. In some applications, reaction of the target microbe antigen with the antibody on the bead is a sufficient basis for detection using an electronic sensor.

For example, an electrochemiluminescence (ECL) technology for detecting *Cryptosporidium parvum* oocysts in environmental water samples has been recently developed (Lee et al., 2001). The method is reported to be quantitative and reproducible, and requires only minimal sample processing. Currently, the ECL assay detects as few as 1 oocyst in 1 mL of concentrated test sample with sample turbidity of up to 10,000 nephelometric turbidity units (NTUs). In this study, water and sewage samples collected during a cryptosporidiosis outbreak were tested by ECL assay. *Cryptosporidium parvum* oocysts were found in the source water at the time of outbreak, and a sharply decreasing level of oocysts in sewage samples was observed over a three-month period following the outbreak. The use of immunocapture technologies in conjunction with electrochemical detectors is one of several approaches to rapid and improved immunological detection of waterborne pathogens and indicators.

Another direction for further advancement of immunodetection is for viable waterborne microorganisms. For example, a quantitative immunoassay capable of detecting low numbers of excystable, sporozoite-releasing *C. parvum* oocysts in turbid water samples has been developed (Call et al., 2001). Monoclonal and polyclonal antibodies have been developed against a sporozoite antigen released only during excystation or when the oocyst is mechanically disrupted. In this assay, oocysts in the test sample are first excysted and then centrifuged. The soluble sporozoite antigen is captured by monoclonal antibodies attached to a magnetic bead. The captured antigen is then detected by ruthenium-labeled polyclonal antibodies via electrochemiluminescence. This viability assay can detect as few as 50 viable oocysts in a 1-mL assay sample with a turbidity as high as 200 NTUs. With further development, refinement, and validation, immunoassays may eventually be able to detect a variety of different viable microbes in water and other environmental samples.

Another immunologically based method is flow cytometry. In this method, microbial cells are labeled with a fluorescently-labeled antibody. Multiple antibodies are employed, with each antibody specific to a particular microorganism. The labeled mixture is then passed through the flow cytometer, which interrogates the solution and determines the numbers of each microorganism based on the occurrence of each label (e.g., Collier and Campbell, 1999; Veal et al., 2000).

Miscellaneous Methods

A wide variety of other available methods can be employed to measure constituents or components of cells. One approach to molecular detection of waterborne pathogens and indicators is based on the use of ligand-binding assays to recover and detect target microbes. Many microbes possess specific surface receptor molecules or epitopes that bind to specific molecular targets and have various functions, such as cell attachment, transport of molecules for nutrition, or molecular processing for immune response or other biological activities. As these molecules and their corresponding targets are elucidated, the molecules to which these epitopes bind can be used for microbe capture and detection—analogous to the use of antibodies for such purposes, as described earlier. Because some microbe surface receptors are used for initiation of infection in host cells, the ability of the microbes to bind to their specific target ligand can be used to detect intact, chemically functional, and potentially infectious or viable microorganisms. Such assays for viability based on the ability to bind to specific receptors are now under development for waterborne microorganisms and are likely to be developed further.

Other chemical and biochemical constituent analysis also can be used to ascertain the presence of viable organisms (e.g., ATP detection with luminescence detection; Deininger and Lee, 2001), the presence of specific toxins, or the organism's protein profile. One of the more powerful and increasingly used analytical methods is mass spectrometry, which is employed for whole organism analysis as well as small molecule analysis. In the former case, bacteria or bacterial spores can be injected directly into the mass spectrometer and their lipid and/ or protein fragmentation profiles can be used to identify them (Ishida et al., 2002; Madonna et al., 2001; see Appendix C for an example). Alternatively, using preconcentration followed by front-end separation such as gas or liquid chromatography coupled with mass spectrometery detection, small molecular components can be analyzed. Significant advances will be required before mass spectrometry can be used in the field because the instruments are generally large, and require significant amounts of power.

Attributes of Methods

All of the methods described thus far have both positive and negative aspects associated with their use in the detection of waterborne pathogens or microbial indicators of pathogens. Table 5-1 provides a qualitative description, based on the collective expertise of the committee, of how each major grouping of detection methods (i.e., culture, immunological, nucleic acids, cell components) currently performs relative to each desirable method attribute described in Chapter 4 (see Box 4-3). In some cases, there is a wide range of performance for each attribute within a given method.

Multi Parameter Measurements

Microbial methods can be designed to measure a single parameter or multiple parameters to detect and quantify microorganisms. For example, some culture methods detect and quantify microorganisms by the ability to display several parameters such as growth in a lactose medium at 44.5°C with acid and gas production as the basis for fecal coliforms analysis. Other methods are based on the ability of the target microbe to utilize a specific array of organic substrates as a basis for its identification, whereas still others— such as immunological detection with a specific monoclonal antibody or PCR amplification with a specific primer set—may detect only a single organism or closely related group of microorganisms. As noted throughout this report, at present it is impossible at present to completely capture and characterize the microbial quality of water for all pathogens by any of the currently available methods. With new and emerging technologies, it is likely to become possible to simultaneously measure multiple analytes in a water sample, thus providing a better basis for judging the microbial quality of the water from which it is taken and any associated health risks. Although any of these methods can be used for pathogen detection and identification, the simultaneous use of multiple capture methods based on orthogonal detection principles (e.g., antibodies and nucleic acid probes) can significantly increase detection specificity of waterborne pathogens and their indicators.

Three strategies can be used to obtain multiparameter measurements: (1) integrate data from many different measurement technologies, (2) integrate several measurement techniques into a single system, and (3) develop instruments that have the intrinsic ability to make multi-analyte measurements (e.g., arrays). The committee recommends multiparameter approaches in which many technologies and methods are integrated to obtain the best possible information from available samples.

More consideration should also be given to "broad range" survey methods, for example, broad range ribosomal RNA or DNA PCR with high throughput sequencing, DNA microarray-based analysis, or mass spectrometry-based analysis of PCR products. While these approaches are not currently ready for widespread routine use, they are critical for building databases dedicated to background characterization and identification of predictive patterns for waterborne pathogens and indicator organisms.

STANDARDIZATION AND VALIDATION OF METHODS

Whatever indicators or pathogens are ultimately selected as the best suited for measuring microbial water quality to achieve specific applications, several methods are already available or on the horizon. Such methods may include more conventional methods based on cultivation and membrane filtration, perhaps using new formulations of biochemical media, or they may be novel and use inno-

TABLE 5-1 Comparison of Major Categories of Microbial Detection Methods by Desirable Attributes

Method Attributes	Culture	Immunological	Nucleic Acid	Cell Component
Specificity to desired target organism	Low to moderate	High	High	Moderate to high
Broad applicability	High	High	High	Low to moderate
Precision	Moderate	Low to high	High	Moderate to high
Adequate sensitivity	Moderate to high	Low to moderate	High	Not applicable
Rapidity of results	Low to moderate	Moderate to rapid	Moderate to rapid	Low to moderate
Quantifiable	Moderate	Low	Low to moderate	Not applicable
Measures viability or infectivity	Yes	No	No, but possible	No, but possible
Logistical Feasibility				
Training and personnel requirements	Low to high	Moderate	Moderate	Low to high
Utility in field	Low to high	Moderate to high	Moderate to high	Low to high
Cost	Low to high	Moderate	Moderate	Low to high
Volume requirements	Low to high	Low	Low	Low to high

vative and emerging technologies such as biosensors based on antibodies or microarrays based on nucleic acid hybridization. Building on these advances, it is possible to divide measurement methods into two broad, but not mutually exclusive, categories:

1. Research methodologies: methods that have been published but are used primarily in academic, industry, and government research laboratories

2. Conventional methodologies: methods that have been assessed with some model of standardization, are widely accepted and used, and are applicable to industry and private laboratories

The data, research, and information needs (both short and long term) to advance "research methods" into those considered "conventional" are central to the following discussion and the committee's statement of task (see Executive Summary and Box ES-1). In this regard, the ability to provide timely, accurate, and reliable data is central to the goals of water quality monitoring, testing, and reporting. Thus, the process of method development and validation is directly linked to the quality of data. Standardizing and then validating prospective methods can follow several models, but most have in common a prescription and terminology regarding their specified application (see Box 5-1). The approach also focuses on a number of the methods attributes described in Chapter 4 (especially Box 4-3) and Table 5-1. Thus, a guide for those who are examining new methods or modifying existing methods for new applications should include a description and study of these key elements in the published literature to enhance the method's potential to be improved or to benefit from wider acceptability and use.

Organizations Involved in Developing Standards

Several organizations and associations are involved in the development of standard methods for evaluating water quality or microorganisms in water (see Box 5-2 and Table 5-2). These methods have been developed primarily in response to industry needs and in some cases to the needs of local, state, or even the U.S. government. Although many standards are developed according to a consensus process in which the views of all stakeholders on the scientific basis of the approach are taken into account, many standards are also voluntary and focus on international standardization that is based on voluntary involvement of all interests. In general, the need for a standard is first articulated and then defined by a group of technical experts, the details are negotiated, and there is finally an approval by the organization and its members. It is important to note that the defined technical aspect for a standard may or may not involve testing and the explicit consideration of the attributes of a method through a validation process.

For the purposes of this report, the primary groups involved in standard development for the microbiological assessment of water include the EPA, Interna-

---

BOX 5-1
Terms Frequently Used in the Development of
Biological Measurement Methods

Standards involve the development of a common language for something established by an authority or by general consent that can be established and used for the measure of quantity, weight, extent, value, or quality; they are a means for determining what a thing or process should be. Standardization in the field of water quality includes definition of terms, sampling of waters, measurement, and reporting of water characteristics.

Validation is the process of demonstrating that a method is acceptable for its intended purpose.

The **accuracy** of a measurement is defined by how close a result comes to the true value and determining the accuracy often involves calibration of the analytical method with a known standard.

Precision is a measure of the reproducibility of measurements and is usually described by the standard deviation, standard error, or confidence interval. Precision relates to the quality of an operation by which a result is obtained.

Specificity is a measure of the ability of a method to discriminate the desired target accurately (e.g., the microorganism, protein, genetic sequence) in the presence of all potential sample components, including other microorganisms. The response in test mixtures is compared with the response in solution containing only the target.

The **sensitivity** or **detection limit** of a method is the lowest concentration that produces a response detectable above background or noise level of the system.

---

tional Organization for Standardization (ISO), American Public Health Association (APHA), American Water Works Association (AWWA), Water Environment Federation (WEF), and American Society for Testing and Materials International (ASTM).

As noted throughout this report, EPA currently has published methods for various bacterial, protozoan, and viral indicators and pathogens in response to specific rules and programs under the Safe Drinking Water Act (SDWA) and the CWA. EPA's current validation process for microbiological method is discussed

---

BOX 5-2
Standards in the United States

In the United States, many organizations comprise the U.S. standardization system; these include government and nongovernmental organizations involved in the development of both mandatory and voluntary (consensus) standards. Mandatory standards are set by the government, and regulatory standards focus generally on health, safety, environmental, or other criteria. Regulatory agencies such as the EPA regularly reference hundreds of existing voluntary consensus standards, in lieu of developing their own, that have the force of law once they are referenced in a government regulation. In this regard, the U.S. National Technology Transfer and Advancement Act of 1996 (Public Law 104-113) requires federal agencies to adopt private sector standards, particularly those of standards-developing organizations (see also Table 5-2), wherever possible rather than creating proprietary, nonconsensus standards (see http://ts.nist.gov/ts/htdocs/210/nttaa/nttaa.htm or http://astm.org for further information).

---

later. The ISO methods that are relevant to this report are found under TC 147/SC 4 (TC refers to the technical committee; SC refers to the subcommittee) that developed the microbiological methods. There are 19 published ISO standards including standards for coliforms, *Escherichia coli*, *Clostridium*, bacteriophage, *Salmonella*, *Pseudomonas* (*fluorescens* and *aeruginosa*), *Legionella*, *Campylobacter*, *Cryptosporidium*, and *Giardia*.

The APHA, AWWA, and WEF regularly publish *Standard Methods for the Examination of Water and Wastewater*, which is currently in its twentieth edition (APHA, 1998) and remains one of the oldest publications used worldwide for water testing methods. The current edition includes more than 400 methods detailed in a step-by-step format; each method describes the applications and potential uncertainties associated with its use. The work of identification, selection, and ultimate inclusion of prospective methods is conducted by volunteer review committees that utilize the published literature to produce a consensus-based standard method. At present, there are standard methods for bacteria, enteric viruses, and enteric protozoa; however, only the coliform bacteria, *E. coli*, and heterotrophic bacteria methods are approved by EPA for use.

To date, the standardization of microbiological methods in the United States has generally followed one or more models, but most have in common consideration of the intended application, and one typical route to standardization is inclusion in the aforementioned *Standard Methods for the Examination of Water and Wastewater*. Another typical route to the standardization of a microbiological

TABLE 5-2 Select Organizations Associated with Standards Development

Organization(s) and URL	Type
American National Standards Institute (ANSI; www.ansi.org)	Not-for-profit, nongovernmental organization
Association of Analytical Communities (AOAC International; www.aoac.org)	Not-for-profit organization with ties to federal government funding
American Society for Testing and Materials International (ASTM; www.astm.org)	Not-for-profit organization
American Public Health Association (APHA; www.apha.org); American Water Works Association (www.awwa.org); Water Environment Federation (WEF; http://www.wef.org)	Not-for-profit health, drinking water, and wastewater associations (respectively)
U.S. Environmental Protection Agency (www.epa.gov)	U.S. government regulatory agency
International Organization for Standardization (ISO; www.iso.ch)	Private agency headquartered in Geneva, Switzerland
National Institute of Standards and Technology (NIST; www.nist.gov)	U.S. government agency

[a]Multilaboratory testing entails the evaluation of a method with a highly specified protocol using multiple tests to evaluate inter-laboratory (generally 5 to 11 laboratories) precision and accuracy of the method that requires coordination of reagents and matrix spikes.

method is through the D-19 ASTM group for water. To be published, each method must follow a prescribed interlaboratory testing protocol, with 11 laboratories participating, defining accuracy and precision. Similar to publication in *Standard Methods*, the application for each method evaluated is described to include the specification of limitations. ASTM has developed microbiological methods for water, including coliphage, enteric protozoa, enteric viruses, and heterotrophic bacteria. EPA had worked closely with ASTM and provided appreciable funding

Role and Activities
Does not develop standards; serves as the U.S. member body to the ISO in the development of ISO standards; accredits standards-developing organizations according to their consensus processes and accredits standards developed by others as American National Standards
Previously known as the Association of Official Analytical Chemists. An international provider and facilitator in the development, use, and harmonization of validated analytical methods and laboratory quality assurance programs and services; AOAC provides three methods validation programs (described elsewhere) that require multilaboratory assessment[a]
Develops and publishes voluntary consensus standards for materials, products, systems, and services nationally and internationally that require multilaboratory assessment
Develops committees and editorial board for the publication of *Standard Methods for the Examination of Water and Wastewater*; its 20th edition (APHA, 1998) is approved for use by the EPA
EPA's Office of Science and Technology is responsible for preparing standards to be used in support of government regulations. EPA publishes laboratory analytical methods that are used by industrial and municipal facilities in analyzing the chemical and biological components of wastewater, drinking water, and other environmental samples required by EPA regulations under the authority of the CWA and SDWA (see also Tables 1-1 and 1-2). Almost all such standards are published by EPA as regulations under Title 40 of the Code of Federal Regulations (CFR) and require multilaboratory assessment
Develops voluntary standards; its membership is comprised of the recognized national standard setting bodies of 140 nations. ISO has more than 180 technical committees devoted to almost all areas of standardization. Final publication of an ISO standard requires the majority consensus of technical committee members and two-thirds of the ISO voting membership
Assists U.S. industry in the development and application of technology, with leading expertise in the area of technology standards and industry standardization issues; is also actively involved in voluntary consensus standards development activities

for interlaboratory testing; however, no major funding has been provided for this mechanism in the last 10 years.

Validation Process

The development of a standard method and implementation for a biological measurement through validation can be a long, tedious, and expensive proposition. The validation process itself, however, can focus on assessing the method attributes in a systematic manner. Thus, as certain attributes of a method are evaluated and ranked as most important for a given application, this can drive the validation process as well as the criteria that are established for acceptability. Although methods used for many regulatory actions must undergo more rigorous testing, those used for gathering information that may be employed in more adaptive management strategies might not have to meet such stringent criteria.

Validation of a measurement method includes studies on specificity, linearity, accuracy, precision, range, detection limit (sensitivity), and consideration of robustness and related issues. Early stages may focus primarily on specificity, linearity, accuracy, and precision; however, the validation process should be considered iterative, particularly as a variety of water matrices are tested.

Specificity Studies

Whether using whole cell, cultivation, genetic, or antibody methods, the detection of the specific target will have to be assessed against a variety of other targets. Some methods may try to capture a broad group of organisms (e.g., all enteroviruses, whether animal or human); others may focus more narrowly (e.g., coxsackieviruses from humans only). The specificity is described as the ability of the method to correctly classify organisms (in groups or as specific species or subtypes).

Linearity Studies

Linearity can be evaluated by preparing standard solutions at various concentration levels (five to six have generally been recommended), thus demonstrating the performance of a method over a range of organism densities. Interestingly, nucleic acid-based methods generally have been designed and tested with varying concentrations after extraction (i.e., using dissolved DNA or RNA as the target). It is recommended that new methods also be tested using whole microorganisms, rather than just extracted DNA or RNA targets for linearity and specificity studies.

Accuracy Studies

There are generally three approaches for determining the accuracy of a biological measurement method. First, a sample with a known concentration is tested; however, reference standards for microorganisms may often be difficult to obtain so this approach is not used very often. Second, and more commonly, the new method is compared to an existing method that is accepted as accurate. This approach makes it difficult, however, to test a new molecular method against an accepted cultivation method because they are measuring different targets. The third approach focuses on calculating the recovery of known numbers of a microorganism spiked into a water matrix. Both blank matrix spikes and various types of water are typically used. Triplicate tests at a minimum of three levels over a range of 10 to 1,000 times the target concentration are often used. As discussed previously, measuring aquatic microorganisms may involve several steps or a series of methods added together for concentration, purification, and detection. Thus, a series of recovery experiments may be required at each stage of the process, or it may prove more expeditious to evaluate the entire method. The range and detection limits of a method are determined from the linearity and accuracy studies.

Precision Studies

There are three levels to the evaluation of precision: (1) intralaboratory tests performed by one individual, (2) intralaboratory tests performed by multiple analysts, and finally (3) tests performed through interlaboratory testing (i.e., multiple analysts from multiple laboratories). The precision of a method can then be determined through analysis of the amount of scatter in the results obtained from multiple analyses of a homogeneous sample. The exact sample and standard preparation procedures that will be used in the final method should be tested. Statistical equivalence is often used to evaluate the results from different laboratories, or the range of results is used to evaluate acceptability.

Robustness and Other Considerations

It is important to note that interferences in the use of a method may affect the results because of constituents in the water matrix (e.g., concentration of organic material, pH). Thus, the sensitivity to such variables helps define the robustness of a prospective method or its ability to remain unaffected by changes. In addition, maintenance of instruments, stability of reagents, and types of controls to be used will have to be described in the validation of a microbiological method.

The level of false positives and false negatives associated with the use of a method can be assigned through the validation process. These help to understand a method's specificity. Depending on the nature of the method and its applications, criteria for what constitutes "acceptable" performance can be discussed. Three major questions arise when a new method is tested and its precision and accuracy are defined: (1) Is the performance acceptable for each desired application? (2) Can the method's performance be improved? (3) What will be the performance of the method if it is used in a wide variety of laboratories?

EPA and Association of Analytical Communities
(AOAC International) Methods Validation Approaches

The EPA uses a number of different procedures for microbiological analysis of water and other environmental samples. Internally, EPA has established a methods approval streamlining process, called the 1600 Series, to develop, evaluate, and standardize methods designed to detect microorganisms in water and other environmental media. Through this process, Series 1600 methods have been developed and published for *Giardia, Cryptosporidium*, coliphages, enterococci, *Aeromonas,* and other microorganisms. In addition, through recent legislation, EPA has attempted to identify, approve, and accept microbiological methods developed and evaluated by entities other than the agency (EPA, 2002). In a recent publication, EPA (2002) approved updated versions of analytical methods developed by the ASTM, those included in *Standard Methods*, the U.S. Geological Survey, and the U.S. Department of Energy for use in various CWA and SDWA compliance monitoring programs. These latest approvals included multiple editions of the same method, which the agency believes will benefit the regulatory and regulated communities by increasing method selection flexibility and by allowing the continued use of time-tested procedures.

In 1996, EPA proposed a regulation to streamline the program for approving laboratory test procedures and quality control measures that are used to gather data and monitor compliance under the CWA and the SDWA. This effort was geared toward reducing the regulatory burden imposed on industries and municipalities, and the technology development and laboratory services communities. It is also proposed to lower the barriers to innovative technology. A draft *Guide to Method Flexibility and Approval of EPA Water Methods* was released in 1996 as a result of this effort. However, this report was never finalized; it has not been updated since its original release, and there are no plans to do so in the near future (Lisa Almodovar, EPA, personal communication, 2003). While this proposed program placed the burden of the cost and time for implementation of a new measurement method on industry and laboratories, it provided little incentive for development of new methods or modification of current methods for application to a new matrix. It also focused on compliance monitoring and thus (for example) did not provide incentives for testing methods to address microorganisms on the 1998 Drinking Water Contaminant Candidate List (EPA, 1998; see also Chapter 6).

The development and standardization of new chemical measurement methods has had AOAC International behind it for more than 100 years, with significant funding from several governmental agencies, though primarily the U.S. Department of Agriculture and the U.S. Food and Drug Administration. There are three main programs within AOAC (see Table 5-2 and www.aoac.org for further information):

1. The Official Methods of Analysis[SM] program of AOAC provides multilaboratory validation for nonproprietary and commercial proprietary methods where the highest degree of confidence in performance is required to generate credible, defensible, and reproducible results.

2. The Peer-Verified[SM] methods program provides independent laboratory validation for nonproprietary methods where rapid validation and some degree of confidence in performance are needed.

3. The Performance Tested[SM] methods program provides similar independent laboratory validation of performance claims but for commercial proprietary methods where rapid validation and some degree of confidence are also needed.

Common elements of these three AOAC programs include the following:

- typical validation time of 12 months minimum;
- minimum of 8-10 independent labs;
- in-house method validation data review;
- publication of methods in the *Official Methods of Analysis of AOAC INTERNATIONAL* and the *Journal of AOAC INTERNATIONAL*;
- citing of methods in U.S. Code of Federal Regulations; and
- permission of proprietary methods or rapid methods.

The committee concludes that the AOAC's Peer-Verified Methods program may provide a good approach for the assessment and validation of new microbiological methods in an efficient and defensible manner since its intent is to categorize a group of tested methods that have not yet been subjected to a full collaborative study.

The codification of a method sufficient to reach widespread acceptance and use requires acceptable reproducibility among laboratories and acceptable (and known) levels of false positives and false negatives. In addition, it has long been known that the analysis of truly replicate samples (e.g., aliquots from a well mixed homogenous suspension of microorganisms) should result in observations that are distributed in accordance with Poisson statistics (Armitage and Spicer, 1956; Ziegler and Halvorson, 1935). It is likely that any new method that achieves wide acceptance will need to go through one of the official standardization processes. In this regard, the testing (number of labs, samples, etc.) and evaluation process is likely to be similar (or even more rigorous) than that undertaken to include coliform presence-absence tests in an earlier edition of *Standard Methods* (Clark and el-Shaarawi, 1993; Pipes et al., 1986).

Recent and ongoing major developments in new methods in microbiology with applications for public health-related water quality have necessitated a new approach for their rapid assessment, standardization, and validation. For example, a recent review on pollution of coastal waters by Griffin and colleagues (2003) found more than nine articles published in the last few years that all used PCR-based methods for rapid detection of viruses with a variety of specificity and sensitivity but little evidence of validation. Water utilities are purchasing PCR or quantitative PCR machines with funds associated with the new "water security" measures, yet the methods and applications for their effective use are not easily obtainable. It is clear that a major effort is needed in the area of methods for the examination of microbial water quality and health. To move new methods into those deemed conventional, a process is required that not only allows for standardization but also for implementation of the methods so that widespread acceptance and use can result. In addition, there is a need to develop standard approaches for interpreting results, particularly results coming from non-culture-based methods.

While the committee concludes that an AOAC Peer-Verified approach, or its equivalent, may be the best way forward, a major program on methods development must be established with water research laboratories in academic institutions in collaboration with industry research and government research laboratories. The committee recommends that such a program for the development and validation of microbiological analysis of water contain several elements:

1. EPA should strengthen its current role and appropriate interactions and partnerships with standard-setting organizations, including ASTM, AOAC, and ISO that are largely individually driven, to facilitate microbial methods development and help focus on new and innovative methods. In addition, regular and ongoing involvement of professional organizations such as the American Society for Microbiology can bring credible independent third party input.

2. A nationwide database should be created that compiles and serves as a clearinghouse for all microbiological methods that have been utilized and published for studying water quality. Research methods, particularly those that have great potential for evolving to conventional methods, will need to be documented. The funding of methods development has been limited, especially for new and emerging methods and innovative indicators. Rather, the development of new and improved methods has been substantially funded (largely by EPA and the American Water Works Association Research Foundation; AWWARF) for only a few pathogens, specifically those targeted for regulations in drinking water such as *Cryptosporidium* and *Giardia*. Greater efforts are required to support methods development for new and emerging microbial detection technologies, for more pathogens, and for new and improved candidate indicators of waterborne pathogens. Approaches for the development and maintenance of an on-line database of new microbiological methods for the analysis of water should be investigated. Guidance on the appropriate data needed for methods studies should be included in this database, and a method for iterative development of consensus methods on-line should be provided.

3. A specific program on promising research methodologies should be sup-

ported by EPA. These methodologies need not be microorganism specific, but should be application specific, focusing on the desirable attributes of the method.

CHALLENGES AND PITFALLS OF NEW TECHNOLOGY

A variety of challenges and potential pitfalls will be associated with developing and using improved and rapid methods for microbiological water quality compared with continued use and reliance on time-tested, widely accepted, traditional microbiological methods. These are discussed below.

Scientific Principles in Identifying a Culture

Many new methods employing current and innovative technologies may be more sensitive than the classical methods. However, when there are discrepancies between new (e.g., PCR) and classical methods in identifying a culture (e.g., *Salmonella* in food and water), the decision as to which method is correct will likely center around determining the identity of the unknown culture. In such situations, the issue of phenotypic expression of cells versus genotypic composition of cells becomes very crucial. Another important comparison issue revolves around the statistical treatment of information. For classical methods, a few attributes will be used to pinpoint the identity of an unknown culture; however, in modern diagnostic kits that utilize many method attributes, whether an unknown culture is identified as *Salmonella* 90 or 85 percent accurately makes the identification rather ambiguous.

As noted previously, an important question concerns the viability and infectivity of the cell being monitored. Some molecular techniques such as PCR can amplify the DNA of a dead cell and give a positive response concerning the presence of a pathogen in a sample that might be perfectly safe. The different scientific principles involved in determining the identity of a culture deserves more research and debate.

Physical and Separation Issues

Many new biological measurement methods are becoming increasingly miniaturized, computerized, and automated. As the technology for making biological measurements continues to scale down to micro-scale dimensions, the corresponding sample sizes required for such analyses similarly decrease (e.g., submicroliter samples are easily assessed using many of today's technologies) and the introduction of a target pathogen or indicator organism from a water sample to interact with these extremely small entities (e.g., a million dots of DNA on a microscope slide) becomes a challenge. Furthermore, the use of such small volumes itself poses a serious water sampling problem since the samples presented to the instrument may not contain the species to be measured simply because they are not adequately representative of the bulk sample or because the detection method is not sensitive enough. This sampling problem is of particular concern for pathogens that can pose unacceptable health risks at very low concentrations.

Thus, there is a need to address the sensitivity of miniaturized detection methods and ensure that sample collection, preprocessing or processing, concentration, and purification are given adequate attention. In other words, microorganisms of concern must either be removed from large-volume samples and presented to these miniaturized methods in small aliquots, or larger sample volumes must be passed by the sensor. Currently, this area represents one of the most important technological challenges to the analysis of pathogens and indicators in water and other environmental samples. Separation of target microorganisms from water during sample preparation before application to modern and sophisticated detection systems is an important area for further research. Specifically, elimination, reduction, and destruction of inhibitors, debris, food particles, lipids, proteins, organic and inorganic particles, cellular matter, and so forth, in samples are all important issues to be resolved.

Cost and Technology Transfer

Many modern diagnostic and detection systems utilize sophisticated instrumentation that may be excessively costly for most potential users. In fact, the average cost of an automated instrument can easily reach or surpass $30,000 and perhaps even $250,000 for a mass spectrometer (e.g., Fourier transform infrared spectrometer). Of course, if one performs a large number of tests regularly, the average cost per test will be low, but in many instances, smaller laboratories may find that the volume of tests does not justify the cost of the instrument. For example, a laboratory that routinely conducts less than 100 *Salmonella* tests per week has little or no need for a sophisticated, automated, and very expensive instrument that can perform thousands of tests per week. The committee recognizes the lack of technical, infrastructure, and financial resources required to implement water monitoring in many parts of the United States and recommends that efforts be made to support the development of inexpensive and rapid fieldable methods for testing microbial water quality.

Finally, while many detection technologies exist that are applicable to the detection of waterborne pathogens and indicator organisms, they are primarily laboratory based. The need to develop rapid fieldable methods will require the concurrent development of reagents, methods, and the attendant portable instruments that can survive repeated transport and use in the less stringently controlled environments of the field.

Unrealistic Expectations and Resistance to Change

New users of an automated measurement device often expect the system to operate perfectly and perform all their necessary tests immediately. Usually, such expectations are too high. Once it is discovered that a particular system does not satisfy all immediate measurement needs, some users will either discard the system totally or develop a negative impression of the system. Users have to understand that an instrument is designed and marketed after extensive testing for specific applications and that even a slight deviation from the specified protocol (e.g., putting acidic water into a sensor not designed to handle a low-pH sample) may result in unsatisfactory performance. No system is 100 percent perfect all the time. At present, there is no sensor that can be placed into a water sample and left alone to make an autonomous measurement without some level of attention.

There is also an intrinsic resistance to change that pervades virtually every analytical community where certain well-established methods have been employed successfully for long periods. This innate conservatism is well founded in some cases where new methods have not been validated. A specific application can sometimes lead to errors or compromise existing long-term data sets. In such cases, it is important that new methods be tested side by side with well-established methods so that the user can acquire a degree of comfort with the new method. The ideal situation is to design "foolproof" systems so that no human error can interfere with the operation of the system from the point of sample application to the end results. Although the microbiological community is moving ever closer to that reality, it is not yet achievable.

LOOK TO THE FUTURE

Today's measurement techniques are aimed at detecting viable organisms or specific components present in the organism of interest and correlating their presence to human health risk assessment.

Sensors and biosensors are beginning to play a role in several application areas including clinical medicine, environmental science, and process control fields. Analyzers are designed to integrate the steps of sampling, preprocessing, and measurement into a single, functional device. In some cases, sampling is determined by sensor placement; no preprocessing is required due to the sensor's specificity; signal transduction-detection is an integral function; and sensors offer the potential for real-time monitoring capability because they measure continuously. Further advances in sensors, including making them sufficiently robust for field deployment, will enable them to address some of the measurements discussed throughout this chapter.

Another area that will have an important impact on microbial analysis is molecular recognition. The use of combinatorial methods such as phage display (Sidhu, 2000), aptamers (O'Sullivan, 2002), and combinatorial chemistry has expanded the ability to rapidly and efficiently generate and screen new molecular entities that may be helpful in producing new recognition elements that can be used as labels, purification reagents, and sensors.

A technology area that will enable significant reductions in sample preparation and separation times is in the field of microfluidics and microelectromechanical systems (MEMS). Complete "lab-on-a-chip" devices are being created out of inexpensive materials such as glass or plastic. These chips contain fully integrated analytical systems with the ability to concentrate, separate, and detect a multitude of analytes including nucleic acids, proteins, and small molecules. Because the overall device sizes are small compared to most benchtop analytical systems, they can perform analysis in second-to-minute time frames. Further advances in chip design and detection schemes should facilitate more complex and sensitive analysis.

One of the most exciting fields of current research in science and technology is the area of nanotechnology. In this area, defined as systems with features on the nanometer scale, functional devices and materials are being developed at an increasing rate. While the material costs associated with the technology may be high, the number of devices one can prepare from a small amount of material is enormous. For example, a gram of nanoparticles contains trillions of individual particles, each of which can, in principle, serve a particular function.

With the advent of nanotechnology, and even microtechnology, materials costs will therefore actually decrease. The ability to pack functions, such as communications hardware, on-board processing, and signal transduction, into ever-smaller devices suggests that in the not-too-distant future it should be possible to create sensors with a high degree of measurement capability in an extremely small device.

One of the more recent trends in sensing systems is array technology (see also Appendix C). In these systems, tens to thousands of sensor elements can be placed on a single substrate with overall dimensions of several square millimeters. The burgeoning area of DNA microarrays for genomics is driving advances in this area. Developments in protein and carbohydrate arrays will further advance the applicability of arrays to microbial analysis. Nanotechnology will cause feature sizes to shrink even further. The ability to place so many sensors on a single device raises the prospect of what has been referred to as a "universal sensor"—a system able to detect virtually anything of interest. Such systems can be built on chips in which a sensor is present for every analyte of potential interest. Alternatively, such arrays may be able to measure *patterns of response* in which signatures of various analytes signify the presence of various water quality conditions, organisms, or toxins. In this approach, pattern recognition algorithms, combined with prior training, could be used to assess water quality and identify potential hazards. One of the advantages of such a system is its ability to be anticipatory, such that new or difficult-to-culture pathogens could be detected by presenting a signature that is similar to known pathogens.

As described previously and elaborated further in the Appendix C, mass spectral techniques for performing microbial analysis using entire mass spectral patterns is in its infancy and should also have an impact, with the limitation that it is unlikely to be an inexpensive, portable field analytical tool.

With the prospect for such an enormous amount of data to be collected from the many sensors disposed on arrays, the large numbers of sensor arrays deployed for water monitoring, and the continuous data streams coming from these sensor networks, attention must be paid to data analysis, intelligent decision making, and archiving.

Ongoing research in the micro- and nanotechnology field, combined with efforts in array sensing and intelligent processing, should provide the tools for creating inexpensive, ubiquitous, universal sensing and detection systems beginning now and continuing over the next several decades.

While many of the new and innovative molecular methods discussed in this chapter (and Appendix C) enhance the opportunity for direct measurement of pathogens, more effective use of direct pathogen measurement will require establishment of the correlation between pathogen concentration and health risk. There are presently no standards on which to base health risk decisions for most pathogens. Current epidemiologic studies (as reviewed elsewhere in this report), on which recreational water exposure standards are based, have been conducted almost exclusively for indicator bacteria such as fecal coliforms and enterococci.

Even for presently used indicator bacteria, the relationship to health risk will have to be reestablished for the new molecular-based methods. Existing epidemiology studies have all been based on quantifying exposure using culture-based methods, which measure some aspect of metabolic activity. Some of the new indicator and pathogen methods quantify the presence of cellular structure, but many do not assess the ability to grow or to infect. As such, they have the potential to overestimate health risk relative to present standards.

Consistent with its previous related recommendations, the committee recommends that epidemiologic studies should be designed and performed to both establish the correlation among indicator and pathogen concentrations and health risk, and reestablish the health risks associated with existing and new pathogen indicators for new (non-culture-based) detection methods.

SUMMARY: CONCLUSIONS AND RECOMMENDATIONS

Recent, emerging, and forecasted advances in microbiology, molecular biology, and analytical chemistry make it timely to reassess the long-standing paradigm of relying primarily or exclusively on traditional microbial (primarily bacterial) indicators for waterborne pathogens to support public health decision making regarding the microbiological quality of water. Although classic microbiological culture methods for detection of indicator microorganisms and pathogens have proved effective over many decades, they suffer from a number of limitations that are discussed throughout this report. The advent of increasingly sophisticated and powerful molecular biology techniques provides new opportunities to improve upon present indicators and pathogens by both culture and non-culture methods. What is needed is a phased monitoring approach that makes use of a flexible tool box, in which a variety of indicator methods and approaches are available for measuring a given indicator or pathogen for different applications that considers spatial and temporal scale (granularity) issues. The need for such a phased monitoring approach and examples of its implementation are discussed in detail in Chapter 6.

It is vital that all four components of the process of performing a measurement (i.e., collection, sample processing or preprocessing, measurement, and data processing; not just the measurement itself) be considered in order to make an accurate analysis of microbial water quality. The collection of representative samples requires careful consideration of the objectives or purpose of sampling in the context of the need to obtain a reliable estimate of microbial exposure in a timely fashion. Furthermore, the widespread use of logarithmic transformations and measure of central tendency and dispersion of log-transformed data to estimate exposures and health risks needs to be reconsidered in water microbiology, epidemiology, and health risk assessment.

At present, most water quality measurement methods are single-parameter based. Ongoing research in the micro- and nanotechnology field, combined with efforts in array sensing and intelligent processing, should provide the tools for creating inexpensive, ubiquitous universal sensing and detection systems beginning now and over the next several decades. This development is essential because the committee recognizes the lack of technical, infrastructure, and financial resources required to implement advanced water quality monitoring methods in many parts of the United States. The microbiological community needs to develop and implement multiparameter approaches in which many technologies and methods are integrated to provide the best possible information. Similarly, the water monitoring community needs to be aware of new developments in these areas that can be brought to bear on microbiological water quality monitoring.

Although evolving detection methods will be increasingly able to rapidly detect specific pathogens, the use of well-characterized (conventional) indicator approaches will continue to be necessary because our understanding of existing and emerging pathogens will never be complete. Regardless, more effective use of direct pathogen measurement discussed in this chapter will require establishment of the relationship between pathogen concentration and health risk (see also Chapters 2 and 4). Similarly, the relationship to health risk will have to be reestablished for presently used indicator bacteria and new (non-culture-based) methods.

The funding of methods development has been relatively poor to date for many pathogens, for new and emerging methods, and for new and innovative indicators. Investigations into only a few pathogens, specifically those targeted for regulations in drinking water such as *Cryptosporidium* and *Giardia*, have been substantially funded (largely by EPA and AWWARF) for the development of new and improved methods. Greater and more consistent efforts should be made to support methods development for new and emerging microbial detection technologies, for many more pathogens, and for new and improved candidate indicators of waterborne pathogens.

Newer methods involving immunofluorescence techniques and nucleic acid analysis are proving their value, and novel microtechnologies are evolving rapidly, spurred in part by recent concerns about bioterrorism. Problems associated with sample concentration, purification, and efficient (quantitative) recovery remain and will require significant effort to be resolved. One technology area that will enable significant reductions in sample preparation and separations time is the field of microfluidics and MEMS. Thus, the introduction of molecular techniques for nucleic acid analysis is viewed by the committee as a growth opportunity for waterborne pathogen detection.

With the prospect for such an enormous amount of data to be collected from the many sensors disposed on arrays, the potentially large numbers of sensor arrays deployed for water monitoring, and the continuous data streams coming from these sensor networks, greater attention must be paid to the fields of data analysis, intelligent decision making, and archiving. There is need for a database that compiles and serves as a clearinghouse for all microbiological methods that have been utilized and published for studying water quality. Research methods, in particular those that have great potential for evolving into conventional methods, will have to be documented.

Recent developments in molecular and microbiology methods and their application to public health-related water microbiology have necessitated a new approach for rapid assessment, standardization, and validation of such methods. It is clear that a major effort is needed for accessible methods to examine microbial water quality for health decisions. To move new methods into the mainstream, a process is required that not only allows for standardization and validation but also facilitates widespread acceptance and implementation. In this regard, the committee concludes that the AOAC Peer-Verified approach or its equivalent may be the best way forward. However, a major program on methods development will need to be established with water research laboratories in academic institutions in collaboration with industry research and government research laboratories.

Based on these conclusions, the committee makes the following recommendations:

• A specific program on promising research methodologies for waterborne microorganisms of public health concern should be supported by EPA and other organizations concerned with microbial water quality. Such methodologies need not be microorganism specific, but should be application specific, focusing on the desirable attributes of the method.

• Ongoing research should be supported and expanded to develop and validate rapid, sensitive, and robust methods for detection and measurement of all classes of waterborne pathogens and their indicators. Such expanded research should go beyond pathogenic bacteria and indicators, to include improved methods for the detection of pathogenic viruses and protozoa.

• Additional research is needed to develop improved methods for rapid sample concentration and effective, reproducible microbial recovery. Specifically, elimination, reduction, and destruction of inhibitors, debris, food particles, lipids, proteins, organic and inorganic particles, cellular matter, and so forth, in samples are important issues.

• Research should be funded to develop approaches to the detection of infectious or viable microbes by nucleic acid detection methods, including the use of ligand-binding steps in microbial recovery from samples to select for intact and infectious microbes.

• The adoption of new molecular techniques should be accelerated for waterborne pathogen detection. New methods undergoing validation should be tested using whole microorganisms, rather than just extracted DNA or RNA targets, to perform tests for sensitivity and linearity.

• Focused efforts should be made to support the development of inexpensive and rapid fieldable methods for testing microbial water quality. This will require the concurrent development of reagents, methods, and the attendant portable instruments that can survive repeated transport and use in the field.

• There is a need to address the sensitivity of miniaturized detection methods and ensure that sample collection, preprocessing or processing, concentration, and purification are given adequate attention to achieve representativeness and have the ability to detect microbial concentrations posing unacceptable health risks. This represents one of the most important technological challenges to the analysis of pathogens and indicators in water and other environmental samples and will become more important with the introduction of micro- and nanotechnologies.

• EPA should reinvigorate its role with standard-setting organizations (including ASTM, AOAC International, and ISO) to facilitate microbial methods development that focuses particularly on new and innovative methods. In addition, regular and ongoing involvement of professional organizations such as the American Society for Microbiology will bring credible, independent, third-party input.

• EPA should support the design, development, and maintenance of a nationwide database that compiles and serves as a clearinghouse for all microbiological methods that have been utilized and published for studying water quality. Guidance on the appropriate data needed for methods studies should be included in this database. Finally, a means for iterative development of consensus methods on-line should be provided.

• The committee recommends that epidemiologic studies should be designed and performed to both establish the correlation among indicator and pathogen concentrations and health risk, and reestablish the health risks associated with existing and new pathogen indicators for new (non-culture-based) detection methods.

REFERENCES

APHA (American Public Health Association). 1998. Standard Methods for the Examination of Water and Wastewater, 20th Edition. Washington, D.C.

Armitage, P., and C.C. Spicer. 1956. The detection of variation in host susceptibility in dilution counting experiments. Journal of Hygiene 54: 401-414.

Barnhard, A.E., and K.G. Field. 2000. Identification of nonpoint sources of fecal pollution in coastal waters by using host-specific 16S ribosomal DNA genetic markers from fecal anaerobes. Applied and Environmental Microbiology 66(4): 1587-1594.

Blackmer, F., K.A. Reynolds, C.P. Gerba, and I.L. Pepper. 2000. Use of integrated cell culture-PCR to evaluate the effectiveness of poliovirus inactivation by chlorine. Applied and Environmental Microbiology 66(5): 2267-2268.

Bogosian, G., and E.V. Bourneuf. 2001. A matter of bacteria life and death. EMBO Reports 2(9): 770-774.

Bogosian, G., P.J. Morris, and J.P. O'Neil. 1998. A mixed culture recovery method indicates that enteric bacteria do not enter the viable but nonculturable state. Applied and Environmental Microbiology 64(5): 1736-1742.

Breaker, R.R. 2002. Engineered allosteric ribozymes as biosensor components. Current Opinion in Biotechnology 13(1): 31-39.

Call, J.L., M. Arrowood, L.T. Xie, K. Hancock, and V.C.W. Tsang. 2001. Immunoassay for viable *Cryptosporidium parvum* oocysts in turbid environmental water samples. Journal of Parasitology 87: 203-210.

Chandler, D.P., J. Brown, D.R. Call, J.W. Grate, D.A. Holman, L. Olson, M.S. Stottlmyre, and C.J. Bruckner-Lea. 2001. Continuous, automated immunomagnetic separation and microarray detection of *E. coli* O157:H7 from poultry carcass rinse. International Journal of Food Microbiology 70: 143-154.

Clark, J.A., and A.H. el-Shaarawi. 1993. Evaluation of commercial presence-absence test kits for detection of total coliforms, *Escherichia coli*, and other indicator bacteria. Applied and Environmental Microbiology 59: 380-388.

Collier, J. L., and L. Campbell. 1999. Flow cytometry in molecular aquatic ecology. Hydrobiologia 401: 33-53.

Colwell, R.R., and D.J. Grimes, eds. 2000. Nonculturable Microorganisms in the Environment. Washington, D.C.: American Society for Microbiology Press.

Colwell, R.R., P. Brayton, A. Huq, B. Tall, P. Harrington, and M. Levine. 1996. Viable but nonculturable *Vibrio cholerae* O1 revert to a culturable state in the human intestine. World Journal of Microbiology and Biotechnology 12: 28-31.

Cook, N. 2003. The use of NASBA for the detection of microbial pathogens in food and environmental samples. Journal of Microbiological Methods 53(2): 165-174.

Deininger, R.A., and J. Lee. 2001. Rapid determination of bacteria in drinking water using an ATP assay. Field Analytical Chemistry and Technology 5(4): 185-189.

DiGiovanni, G.D., F.H. Hashemi, N.J. Shaw, F.A. Abrams, M.W. LeChevallier, and M. Abbaszadegan. 1999. Detection of infectious *Cryptosporidium parvum* oocysts in surface and filter backwash water samples by immunomagnetic separation and integrated cell culture-PCR. Applied and Environmental Microbiology 65(8): 3427-3432.

Dombeck, P.E., L.K. Johnson, S.T. Zimmerly, and M.J. Sadowsky. 2000. Use of repetitive DNA sequences and the PCR to differentiate *Escherichia coli* isolates from human and animal sources. Applied and Environmental Microbiology 66(6): 2572-2577.

Donaldson, K.A., D.W. Griffith, and J.H. Paul. 2002. Detection, quantitation, and identification of enteroviruses from surface waters and sponge tissue from the Florida Keys using real-time RT-PCR. Water Research 36(10): 2505-2514.

Dore, W.J., K. Henshilwood, and D.N. Lees. 2000. Evaluation of F-specific RNA bacteriophage as a candidate human enteric virus indicator for bivalve molluscan shellfish. Applied and Environmental Microbiology 66(4): 1280-1285.

Dore, W.J., M. Mackie, and D.N. Lees. 2003. Levels of male-specific RNA bacteriophage and *Escherichia coli* in molluscan bivalve shellfish from commercial harvesting areas. Letters in Applied Microbiology 36(2): 92-96.

Edwards, C. 2000. Problems posed by natural environments for monitoring microorganisms. Molecular Biotechnology 15(3): 211-223.

EPA (U.S. Environmental Protection Agency). 1996. Guide to Method Flexibility and Approval of EPA Water Methods (draft). Washington, D.C.: Office of Water.

EPA. 1998. Announcement of the Drinking Water Contaminant Candidate List. Notice. Federal Register 64(40): 10274-10287.

EPA. 2000. National Primary Drinking Water Regulations: Ground Water Rule; Proposed Rule. Federal Register 65(91): 30193-30274.

EPA. 2002. Guidelines Establishing Test Procedures for the Analysis of Pollutants Under the Clean Water Act; National Primary Drinking Water Regulations; and National Secondary Drinking Water Regulations; Methods Update. Federal Register 67(205): 65219-65253.

Fleischer, J.M., F. Jones, D. Kay, R. Stanwell-Smith, M. Wyer, and R. Morano. 1993. Water and non-water related risk factors for gastroenteritis among bathers exposed to sewage contaminated marine waters. International Journal of Epidemiology 22(4): 698-708.

Feng, S., and P.T.K. Woo. 2001. Cell membrane glycoconjugates on virulent and avirulent strains of the haemoflagellate *Cryptobia salmositica* (Kinetoplastida). Journal of Fish Diseases 24(1): 23-32.

Fung, D.Y.C. 2002. Rapid methods and automation in microbiology. Comprehensive Reviews in Food Science and Food Safety 1: 1-20.

Gehring, A.G., C.G. Crawford, R.S. Mazenko, L.J. Vanhouten, and J.D. Brewster. 1996. Enzyme-linked immunomagnetic electrochemical detection of *Salmonella typhimurium*. Journal of Immunological Methods 195(1-2): 15-25.

Griffin, D.W., K.A. Donaldson, J.H. Paul, and J.B. Rose. 2003. Pathogenic human viruses in coastal waters. Clinical Microbiology Reviews 16(1): 129-143.

Hald, B., K. Knudsen, P. Lind, and M. Madsen. 2001. Study of the infectivity of saline-stored *Campylobacter jejuni* for day-old chicks. Applied and Environmental Microbiology 67(5): 2388-2392.

Huffman, D.E., T.R. Slifko, M.J. Arrowood, and J.B. Rose. 2000. Inactivation of bacteria, virus, and *Cryptosporidium* by a point-of-use device using pulsed broad spectrum white light. Water Resources 34(9): 2491-2498.

Hurst, A. 1977. Bacterial injury: A review. Canadian Journal of Microbiology 23: 936-944.

Hurst, C.J., G.R. Knudsen, M.J. McInerney, L.D. Stetzenbach, and M.V. Walter, eds. 2002. Manual of Environmental Microbiology. Washington, D.C.: American Society for Microbiology Press.

Ishida, Y., A.J. Madonna, J.C. Rees, M.A. Meetani, and K.J. Voorhees. 2002. Rapid analysis of intact phospholipids from whole bacterial cells by matrix-assisted laser desorption/ionization mass spectrometry combined with on-probe sample pretreatment. Rapid Communications in Mass Spectrometry 16(19): 1877-1882.

Jones, D.M., E.M. Sutcliffe, and A. Curry. 1991. Recovery of viable but non-culturable *Campylobacter jejuni*. Journal of General Microbiology 137(Part 10): 2477-2482.

Kay, D., J.M. Fleischer, R.I. Salomon, F. Jones, M.D. Wyer, A.F. Godfree, Z. Zelenauch-Jacquotte, and R. Shore. 1994. Predicting likelihood of gastroenteritis from sea bathing: Results from randomised exposure. Lancet 344: 905-909.

Keer, J.T., and L. Birch. 2003. Molecular methods for the assessment of bacterial viability. Journal of Microbiological Methods 53(2): 175-183.

Kell, D.B., A.S. Kaprelyants, D.H. Weichart, C.R. Harwood, and M.R. Barer. 1998. Viability and activity in readily culturable bacteria: A review and discussion of the practical issues. Antonie Van Leeuwenhoek 73(2): 169-187.

Kilvington, S., and J. Beeching. 1995. Identification and epidemiological typing of *Naegleria fowleri* with DNA probes. Applied and Environmental Microbiology 61(6): 2071-2078.

Kreader, C.A. 1996. Relief of amplification inhibition in PCR with bovine serum albumin or T4 gene 32 protein. Applied and Environmental Microbiology 62(3): 1102-1106.

Lammerding, A.M., A.M. Fazil, and G.M. Paoli. 2001. Microbial food safety risk assessment. In Compendium of Methods for the Microbiological Examination of Foods, F.P. Downes and K. Ito, eds. Washington, D.C.: American Public Health Association.

Lee, J., S. Dawson, S. Ward, S.B. Surman, and K.R. Neal. 1997. Bacteriophages are a better indicator of illness rates than bacteria among users of a white water course fed by a lowland river. Water Science and Technology 35(11-12): 165-170.

Lee, Y.M., P.W. Johnson, J.L. Call, M.J. Arrowood, B.W. Furness, S.C. Pichette, K.K. Grady, P. Reeh, L. Mitchell, D. Bergmire-Sweat, W.R. MacKenzie, and V.C.W. Tsang. 2001. Development and application of a quantitative, specific assay for *Cryptosporidium parvum* oocysts detection in high-turbidity environmental water samples. American Journal of Tropical Medicine and Hygiene 65: 1-9.

Madonna, A.J., K.J. Voorhees, and T.L. Hadfield. 2001. Rapid detection of taxonomically important fatty acid methyl ester and steroid biomarkers using in situ thermal hydrolysis/methylation mass spectrometry (THM-MS): Implications for bioaerosol detection. Journal of Analytical and Applied Pyrolysis 61(1-2): 65-89.

McFeters, G.A., and A.K. Camper. 1983. Enumeration of indicator bacteria exposed to chlorine. Advanced Applied Microbiology 29: 177-193.

McFeters, G.A., M.W. LeChevallier, A. Sing, and J.S. Kippin. 1986a. Health significance and occurrence of injured bacteria in drinking water. Water Science and Technology 18(10): 227-231.

McFeters, G.A., J.S. Kippin, and M.W. LeChevallier. 1986b. Injured coliforms in drinking water. Applied and Environmental Microbiology 51: 1-5.

Medema, G.J., F.M. Schets, A.W. van de Giessen, and A.H. Havelaar. 1992. Lack of colonization of 1 day old chicks by viable, non-culturable *Campylobacter jejuni*. Journal of Applied Bacteriology 72(6): 512-516.

Mitchell, B.A., J.A. Milbury, A.M. Brookins, and B.J. Jackson. 1994. Use of immunomagnetic capture on beads to recover *Listeria* from environmental samples. Journal of Food Protection 57(8): 743-745.

Murray, P.R., E.J. Baron, M.A. Pfaller, F.C. Tenover, and R.H. Yolken, eds. 1999. Manual of Clinical Microbiology, 7th Edition. Washington, D.C.: American Society for Microbiology Press.

Nuanualsuwan, S., and D.O. Cliver. 2002. Pretreatment to avoid positive RT-PCR results with inactivated viruses. Journal of Virological Methods 104(2): 217-225.

O'Sullivan, C.K. 2002. Aptasensors: The future of biosensing. Analytical and Bioanalytical Chemistry 372(1): 44-48.

Paoli, G. 2002. Microbial risk assessment: Lessons learned and future directions. Presented at 1st International Conference on Microbiological Risk Assessment: Foodborne Hazards. College Park, Maryland: University of Maryland.

Pipes, W.O., H.A. Minnigh, B. Moyer, and M.A. Troy. 1986. Comparison of Clark's presence-absence test and the membrane filter method for coliform detection in potable water supplies. Applied and Environmental Microbiology 52: 439-443.

Quintero-Betancourt, W., E.M. Peele, and J.B. Rose. 2002. *Cryptosporidium parvum* and *Cyclospora cayetanensis*: A review of laboratory methods for detection of these waterborne parasites. Journal of Microbiological Methods 49: 209-224.

Ray, B. 1989. Injured Index and Pathogenic Bacteria: Occurrence and Detection in Foods, Water, and Feeds. Boca Raton, Florida: CRC Press.

Reynolds, K.S., C.P. Gerba, and I.L. Pepper. 1996. Detection of infectious enteroviruses by an integrated cell culture-PCR procedure. Applied and Environmental Microbiology 62(4): 1424-1427.

Reynolds, K.S., C.P. Gerba, and I.L. Pepper. 1997. Rapid PCR-based monitoring of infectious enteroviruses in drinking water. Water Science and Technology 35(11-12): 423-427.

Rose, J.B., D.E. Huffman, and A. Gennaccaro. 2002. Risk and control of waterborne cryptosporidiosis. FEMS Microbiology Reviews 26: 113-123.

Sidhu, S.S. 2000. Phage display in pharmaceutical biotechnology. Current Opinion in Biotechnology 11(6): 610-616.

Slifko, T.R., D.E. Huffman, D. Bertrand, J.H. Owens, W. Jakubowski, C.N. Haas, and J.B. Rose. 2002. Comparison of animal infectivity and cell culture systems for evaluation of *Cryptosporidium parvum* oocycts. Experimental Parasitology 101: 97-106.

Smith, R.J., A.T. Newton, C.R. Harwood, and M.R. Barer. 2002. Active but nonculturable cells of *Salmonella enterica* serovar Typhimurium do not infect or colonize mice. Microbiology 148(Part 9): 2717-2726.

Sobsey, M.D., K.J. Schwab, and T.R. Handzel. 1990. A simple membrane filter method to concentrate and enumerate male-specific RNA coliphages. Journal of the American Water Works Association 82(9): 52-59.

Sobsey, M.D., D.A. Battigelli, G.A. Shin, and S. Newland. 1998. RT-PCR amplification detects inactivated viruses in water and wastewater. Water Science and Technology 38(12): 91-94.

Sobsey, M.D. 1999. Methods to identify and detect microbial contaminants in drinking water. Pp. 173-205 in Identifying Future Drinking Water Contaminants. Washington, D.C.: National Academy Press.

Sobsey, M.D. 2001. Microbial detection: Implications for exposure, health effects, and control. Pp 89-113 in Microbial Pathogens and Disinfection By-Products in Drinking Water: Health Effects and Management of Risks, G.F. Craun, F.S. Hauchman, and D.E. Robinson, eds. Washington, D.C.: ILSI Press.

Veal, D.A., D. Deere, B. Ferrari, J. Piper, and P.V. Attfield. 2000. Fluorescence staining and flow cytometry for monitoring microbial cells. Journal of Immunological Methods 243(1-2): 191-210.

Zhen, B., Y.J. Song, Z.B. Guo, J. Wang, M.L. Zhang, S.Y. Yu, and R.F. Yang. 2002. In vitro selection and affinity function of the aptamers to *Bacillus anthracis* spores by SELEX. Acta Biochimica et Biophysica Sinica 34(5): 635-642.

Ziegler, N.R., and H.G. Halvorson. 1935. Application of statistics to problems in bacteriology. IV. Experimental comparison of the dilution method, the plate count, and the direct count for the determination of bacterial populations. Journal of Bacteriology 29: 609-634.

6

A Phased Approach to
Monitoring Microbial Water Quality

INTRODUCTION

Monitoring microbial water quality has been conducted for more than a century by measuring indicator bacteria that occupy human intestinal systems, primarily fecal coliforms, *Escherichia coli*, and some enterococci. The historical origins and premises for the indicators measured are discussed at length in Chapters 1, 2, and 4.

Technological advances described in Chapter 5 provide new opportunities for revising these monitoring procedures. Our increased understanding of microbiology at the molecular level allows existing indicators to be measured using faster and cheaper methods. These advances also provide cost-effective opportunities for measuring new indicators or combinations of indicators, and in some cases, pathogens themselves. There is a strong consensus in the committee that with sufficient support for the necessary research, current indicator systems and their applications will undergo a comprehensive evolution during the coming decade. This evolution will substantially enhance our ability to rapidly and correctly identify when water used for recreational or drinking purposes is contaminated with microorganisms that are pathogenic to humans.

The increasing number and diversity of analytical tools imply that it is timely to reevaluate the appropriateness of currently used indicators. This chapter provides the committee's conclusions and recommendations regarding preferred indicators, in both the short- and the long-term, for a variety of applications. It also provides a monitoring framework within which to make those choices and discusses potential impediments and drivers to implementing the framework. The chapter closes with a summary of its conclusions and recommendations.

PHASED MONITORING APPROACH

Indicators for waterborne pathogens are used to achieve a variety of goals, fulfill various regulations, and meet differing applications under the Safe Drinking Water Act (SDWA) and Clean Water Act (CWA; see Chapters 1 and 4). Often they are used to provide an early warning of potential microbial contamination, an application for which a rapid, simple, broadly applicable technique is appropriate. They are used for health risk confirmation where resulting actions can be costly and time consuming. They are also used to identify the source of a microbial contamination problem which can have terrestrial origins. In both of these latter applications, the time frame and investment in indicators, indicator approaches, and methods must be greater.

Indicator applications also vary according to the media in which they are used. For example, warning systems for groundwater typically focus on the presence or absence of bacterial indicators of fecal contamination because high-quality groundwater does not normally contain fecal bacteria and is often used without disinfection. In contrast, quantitative tests for indicator bacteria are used in monitoring surface drinking water intakes because these waters often show some evidence of fecal contamination and are usually treated with filtration and disinfection. Interpretation of indicator data in recreational water applications is different again because the exposure can be more irregular and involves a more limited population at risk. Furthermore, all of the indicator applications discussed in this report are inextricably linked and to some extent must account for surrounding terrestrial ecosystems (e.g., through fecal loading from agricultural, wild, and domestic animals living in a flood plain) that can affect the microbiological quality of the water being assessed.

A single microbial water quality indicator or small set of indicators cannot meet this diversity of needs and applications. The complexity of issues surrounding microbial water quality assessment requires the use of a "tool box" in which the indicator(s) and method(s) are matched to the requirements of a particular application. Like health investigations, water quality studies may have to proceed through a series of phases with a different suite of tools needed for each phase.

The committee recommends use of a phased, three-level monitoring framework, as illustrated in Figure 6-1, for selecting indicators. The first phase of this framework is screening or routine monitoring (Level A). The objective of this phase is early warning of a health risk or of a change from background condition that could lead to a health risk. This is the most frequent type of monitoring and is routinely conducted throughout the country.

In general, the most important indicator attributes at this level are speed, low cost (logistical feasibility), broad applicability, and sensitivity. Speed is impor-

Phases of investigation

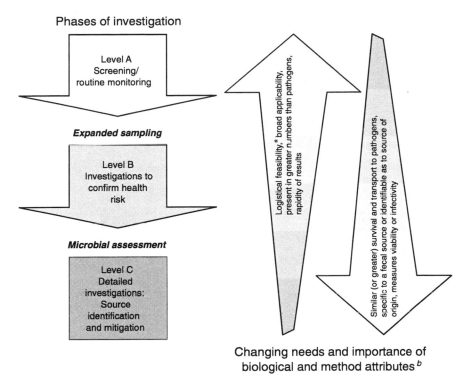

FIGURE 6-1 Recommended three-level phased monitoring framework for selection and use of indicators and indicator approaches for waterborne pathogens.

[a]Includes training and personnel requirements, utility in field, cost, and volume requirements (see Box 4-3).

[b]Not all biological and method attributes discussed in Chapters 4 and 5 are included in this figure nor are they listed in order of importance.

tant because managers have to react quickly to a system change, such as a sewage leak into drinking or recreational waters. Cost is important because the spatial or temporal frequency of monitoring is inversely related to per-sample cost and comprehensive sampling for screening purposes is typically desirable. Methods that have broad applicability to a number of geographic locations, various types of watersheds, and different water matrices are preferred. Sensitivity (i.e., the indicator occurs at more readily measured concentrations than the broad class of pathogens for which it serves as an indicator) is important because managers should not miss potential microbial water quality problems. Method precision and a definitive (quantitative) relation to health risk are less important during this phase. Although these are obviously desirable traits, management decisions should rarely be made at the screening level, unless the indicator concentrations are extreme or supported by ancillary information.

Once screening has identified a potential problem, the second phase involves more detailed studies to confirm a public health risk (Level B). The aim of such investigations is to assess the need for further management actions (e.g., beach closures, boil-water orders) and/or expanded specific data-gathering efforts. There are several ways in which confirmation can be accomplished. A typical approach involves expanded sampling with screening indicators to determine whether the response is repeatable over space and time; Chapter 4 discusses granularity in indicator response and the need for additional sampling to determine whether the signal persists. In some cases, a more reliable processing method is used to confirm that the result is not an artifact.

Ancillary information may be used to help confirm health risk. For instance, visual ("sanitary") surveys for upstream sources of leaking sewage are often conducted to identify point and nonpoint sources of contamination. Changes in water color, odor, or similar parameters may also serve as clues for increased health risk.

The confirmation phase often involves measurement of new indicators, including direct measurement of pathogens. Such studies are not initiated on a routine basis, but would typically be undertaken when screening indicators persist at high levels without a clearly identifiable contamination source. Many of the new and emerging methods described in Chapter 5 and Appendix C—such as quantitative polymerase chain reaction (PCR), microarray technology, and viral cell culture concentrates to screen for a multitude of targets of health concern—will be useful at this stage.

Since confirmation studies focus on assessing health risk, the most important indicator biological attributes during this phase are correlation with contamination sources and transport or survival behavior similar to pathogens. Desirable method attributes include quantifiability and effectiveness at measuring viability or infectiousness, while logistical considerations, rapid turnaround, and broad applicability become less important.

The third phase (Level C) involves studies to determine sources of microbial contamination so that the health risk can be abated through a variety of engineering and policy solutions. However, this chapter and report focuses on identifying the sources of microbial contamination rather than their mitigation. In some cases, source identification is accomplished through expanded spatial sampling to look for gradients. Where recreational waters are concerned, indicator strategies based on molecular signatures are becoming more commonly used in place of screening indicators.

For these detailed investigations, the essential indicator attributes include specificity for a fecal source and quantifiability, while cost considerations and rapidity of results diminish further in importance. Level C studies often overlap in goal with Level B health risk confirmation studies, since identifying the source of contamination facilitates identification of the health risk. Depending on the need to differentiate between source of contamination and public health risk, the ability to measure the infectiousness of the detected indicators or pathogens may also be important.

The standardization and regulation of monitoring methodology generally decrease as indicator use progresses through the three phases. Methods used in the screening phase are typically "standard" and can be accomplished by almost all county health department laboratories. This also holds true for many Level B studies. However, the techniques used during the latter phases may be more specialized and require the expertise of a research laboratory. This is especially true for Level C studies. The responsibility for study costs may also shift through the three phases, with parties potentially responsible for contamination sources typically more involved with the latter two levels of studies.

The committee recommends that the U.S. Environmental Protection Agency's (EPA's) presently available guidance documents and indicator recommendations (focused predominantly on Level A) be expanded to address the latter two phases of selection and use of indicators and indicator approaches for waterborne pathogens. There is an increasing need for national leadership and guidance for the investigation phases that follow screening. Moreover, EPA active support for indicator and method development in the latter two phases is required to fully realize the potential advantages of developing technology. EPA should invest in a long-term research and development program to build a tool box of indicators that will serve as a resource for all three phases of investigation.

The remainder of this chapter describes application of this framework to three typical monitoring situations: marine beaches, surface water sources of drinking water, and groundwater sources of drinking water. The following sections also provide recommendations regarding the most appropriate indicators at present, in the near-term future (including the proposed Ground Water Rule and Interim Enhanced Surface Water Treatment Rule provisions), and in the long-term future at each level of investigation. Lastly, it is important to note that the phased monitoring framework for indicator selection and use described in this chapter is focused on supporting risk management, not risk management actions themselves which are largely beyond the scope of this report. Appropriate risk management decisions depend not only on the results of microbial water quality monitoring but on the application the monitoring is designed to address, such as public health warning systems for recreational beaches (see more below). Under any circumstances, time is of the essence in all three levels of microbial water quality monitoring in order to ensure the public is not exposed to water that is known to be contaminated.

APPLICATION TO MARINE BEACHES

Level A—Screening/Routine Monitoring

The Beaches Environmental Assessment and Coastal Health (BEACH) Act of 2000 requires that EPA develop a consistent national monitoring program for marine and Great Lakes bathing beaches. EPA is creating such a program and recommends enterococci as the preferred indicator for marine systems; it is investigating this group of bacteria for use in freshwater lakes (EPA, 2002; see also EPA, 1999a and Chapter 2). Most states have not yet adopted this recommendation and are still relying on total and fecal coliform bacteria as indicators (see Tables 1-4 and 4-1). EPA has encouraged states to adopt the enterococci standard by requiring its use if their programs are supported by BEACH Act grants.

The committee agrees with EPA's recommendation that enterococci are the best bacterial indicator presently available for screening at marine beaches. While the epidemiological evidence has several deficiencies (see Chapters 2 and 4), enterococci has the best correlation to health risk (EPA, 1986; Prüss, 1998; Wade et al., 2003). It is also the most protective of the standard available indicators. When enterococci have been measured concurrently with fecal coliforms, total coliforms, and *E. coli*, enterococci were the sole indicator failing water quality standards at most sites that failed and exceeded standards at the vast majority of sites where standards failures occurred for other indicators (Noble et al., 2003). It is also relatively inexpensive to measure.

Although enterococci are the best choice for screening at marine beaches, EPA may also want to reexamine whether there is additional/added value in using the total:fecal coliform ratio. Haile et al. (1999) found that enterococci had the best epidemiological relationship with gastrointestinal illness, but the total:fecal ratio had a better relationship with several other illnesses, such as ear infections. The rationale for the ratio is that when fecal coliforms are a high percentage of total coliforms, the presence of a human contamination source is more likely. Haile et al. is the only study to find a strong relationship between the ratio and illness, although it is the only one to have looked for it. The EPA is currently evaluating *Bacteroides* in epidemiological studies (Richard Haugland, EPA, personal communication, 2003) so more information regarding the application of this indicator to assessment of health is forthcoming.

The future holds many possible improvements in beach water quality screening, the most likely of which are enhancements in the speed at which indicators are measured. As detailed in Chapter 4, the greatest shortcoming in present screening methods is the 18- to 96-hour culturing period necessary to obtain results, during which time swimmers are exposed to potentially harmful conditions. Effective screening requires a short quantification period so that water quality managers can react quickly with a public health warning, or conduct further investigations as to whether a public health warning is warranted. A series of new methods that reduce laboratory processing time are in late stages of development (see Chapter 5). Antibody based methods have the greatest potential for shortening the time frame to as little as 20 minutes, but the sensitivity of this technology at present is inadequate to measure bacterial concentrations near water quality standards. Adoption of antibody techniques or other new methods that measure subcellular structures may also require establishment of new standards since the attributes they measure do not assess indicator viability.

Future improvements could also include replacement of the indicator species used, but this is not likely to occur in the near future. While direct pathogen measurement techniques will soon be available, they generally do not fill the need for a broad-spectrum measure. Microarrays that assess the presence of numerous pathogens simultaneously might meet this need, but it is unclear whether microarrays can be used to analyze a large enough water sample to appropriately assess exposure to human pathogens that occur at low density. Cost and operator complexity constraints at the screening level are also an issue. The more likely use of microarrays in the near-term will be at the subsequent phase of investigation, in which microarray technology is used after a short-term incubation, water filtration, PCR, or other concentration or amplification methods to better define the nature of the water quality problem.

Another factor that will limit replacement of present indicators with direct pathogen measurements is the lack of epidemiologic studies to quantify the relationship between concentrations of specific pathogens and health effects (see Chapters 2, 4, and 5 for further information). Previous studies on the relationships between indicators and human health risks have not included efforts to identify the etiological agents causing infection and illness. Such etiological studies are recommended and will be important in developing the pathogen-health effects relationships that will form the basis for water quality standards.

Level B—Investigations to Confirm Health Risk

Confirmation sampling is standard practice in marine beach monitoring. Screening alerts managers to the possibility of health risk from water contact, but additional data collection typically precedes issuance of a beach closure notice. States differ considerably in their approach to confirming health risk. Most states conduct temporal and spatial confirmation sampling prior to issuing a closure notice. This type of confirmation sampling is warranted because several studies have demonstrated considerable variability in indicator bacteria concentrations over small spatial and temporal scales (Boehm et al., 2002; Leecaster and Weisberg, 2001; Taggart, 2002). Many sources are intermittent, and reliance on a single sample may lead to incorrect decisions. While confirmation testing is always appropriate, the need to take immediate action must also be considered to protect public health in situations where other information suggests that a contamination source is, and is likely to continue, discharging to recreational waters.

Many states also take other actions to confirm health risk, including sampling additional indicators, before issuing a beach closure notice. The committee concurs with this since presently measured indicators, such as enterococci, are not specific enough to human waste streams to automatically equate elevated concentrations to a health risk. Hawaii does this by measuring *Clostridium perfringens*, which has been suggested to be more specific to human contamination (Sorenson et al., 1989), although such specificity has not been confirmed more than a decade later. Some states have used the fecal coliform:fecal streptococcus ratio based on a higher prevalence of fecal coliforms in human waste than in animal waste (Feachem, 1975). Other states have used the ratio of fecal coliforms to total coliforms, based on the same concept (Haile et al., 1999). Although these source indicators provide some information and are easily measured by most laboratories, they can be unreliable because of differential survival in the environment (Scott et al., 2002).

The options for improving both spatial/temporal and other forms of health risk confirmation sampling will expand considerably in the near future. An important change will be the availability of rapid methods, allowing confirmation sampling to be conducted in a more cost-effective and timely manner. Methods that provide a response within a few hours will allow the size of the contaminated area to be determined immediately after detecting a standards failure, ensuring that closure decisions are of an appropriate size and are not made on an isolated patch of poor-quality water that dissipates before action is taken.

The future will also include an array of new indicators, including direct pathogen measurements, for confirmation of health risk. Direct pathogen measurements are not presently used for screening largely because no epidemiologic studies have included these measures, which prevents the determination of standards (although such studies are being conducted), and because current techniques are not sensitive enough to generate meaningful data in a screening mode. Pathogen measurements are suitable at the present time for confirmation sampling; however, because the presence of pathogens along with other data on screening indicators (especially the co-occurrence of high concentrations of indicator organisms) suggests a significant public health risk.

Level C—Source Investigations

Confirmation of health risk leads naturally to more detailed investigations that identify and allow mitigation of the microbial contamination source. The most frequently used approach for this has been visual surveys to search for a leaking pipe system. This is often combined with more spatially intensive use of screening indicators to identify gradients in concentration that allow focusing of visual surveys in the proper area.

Source investigation is likely to improve substantially in the near future because of technological developments in two areas. The first is more rapid methods for measuring screening indicators, which will improve the spatially inten-sive surveys. At present, these surveys must sample in multiple directions simul-taneously, because of the long time necessary for laboratory processing, making them impractical when there are multiple tributaries in a given stream system. Development of rapid analysis methods will allow investigations to proceed unidirectionally toward the location of highest bacterial concentration because samples from the confluence of every tributary can be processed quickly.

The other improvement will result from the availability of phenotypic and genotypic source identification (tracking) methods that allow for source differen-tiation by animal species. These methods show great promise, but the field is still new and there is disagreement among practitioners as to which techniques hold the most promise (Griffith et al., 2003; Malakoff, 2002; Simpson et al., 2002). Most methods have been tested in a limited number of locations, often within a single watershed, and with a limited number of possible sources. There have been a few recent attempts at standardized comparative testing (see Box 6-1), but these

---

BOX 6-1
Comparative Testing of Microbial Source Tracking Methods

Microbiological source tracking (MST) methods are potentially pow-erful tools that are increasingly being used to identify sources of fecal contamination in surface waters, but these methods have been subjected to limited comparative testing. The Southern California Coastal Water Research Project recently led an effort (Griffith et al., 2003) in which 22 researchers employing 12 different methods were provided identical sets of blind water samples containing one to three of five possible fecal sources (human, dog, cattle, seagull, or sewage). Researchers were also provided portions of the fecal material used to inoculate the blind water samples for their use as library material. No MST method tested pre-dicted the source material in the blind samples perfectly, although all methods provided useful information. Host-specific PCR performed best at differentiating between human and nonhuman sources, but primers are not yet available for differentiating among nonhuman sources. Hu-man virus and F+ coliphage methods reliably identified the presence of sewage but were not able to identify fecal contamination from individual humans. Library-based isolate methods were able to identify the domi-nant source in most samples but frequently had false positives, identify-ing the presence of fecal sources that were not in the samples. The U.S. Geological Survey is presently conducting a similar type of comparative study using different evaluation criteria (Donald Stoeckel, USGS, per-sonal communication, 2003). Multiple comparative studies are warranted because all desirable attributes of source tracking indicators cannot be tested in a single study.

are limited and have tested only a subset of required attributes. Thus, as stated in Chapter 4, the committee cannot recommend which of these techniques is most suitable at this time, but it does recommend continued support for development and testing of these very promising technologies.

APPLICATION TO SURFACE DRINKING WATER SOURCES

Level A—Screening/Routine Monitoring

Screening for microorganisms in surface water is generally limited to regulatory or voluntary monitoring by water utilities. Routine monitoring for conventional indicators, including total coliforms and either fecal colifoms or *Escherichia coli*, is required under the Total Coliform Rule (EPA, 1989b), but this monitoring applies only to treated water. A requirement for monitoring untreated surface drinking water sources for total coliforms or fecal coliforms is included in the Surface Water Treatment Rule (EPA, 1989a) for systems that wish to avoid filtration. Routine monitoring of all untreated surface drinking water sources for coliforms and either fecal coliforms or *E. coli* is also included in the Interim Enhanced Surface Water Treatment Rule (IESWTR; EPA, 1998c), with monitoring to be conducted at least monthly. At present, there are no other national regulatory requirements for routine monitoring of different microbial indicators or pathogens in surface water sources of drinking water.

The monitoring program at the screening level for surface water supplies should focus not only on the water intake but also on the watershed level. Thus, the indicator systems and attributes will be focused on applications to a large spatial area. It is not clear that total coliform monitoring will be of value; however, water utilities should be encouraged to examine multiple indicators such as *E. coli*, enterococci, coliphage, *Bacteriodes*, and perhaps *Clostridium perfringens* when sampling (sub)tropical waters. Point sources (e.g., sewage discharge points) can be evaluated more readily and may be a constant source of inputs to the watershed. Nonpoint sources will be more difficult to assess because they are likely to be intermittent sources of pollution, often related to meteorological events. Baseline concentrations and changes in levels should be assessed. The integration between Level A and B and/or C studies should be considered.

Turbidity is a reliable indicator of changes in water quality, particularly when those changes occur as a result of meteorological conditions. The U.S. Geological Survey (USGS) and others are currently using flow, rainfall, and turbidity to address potential risk to a system. The availability of technology for on-line and real-time monitoring of turbidity facilitates its inclusion in microbiological monitoring programs.

The committee concludes that better, more reliable drinking water protection can be provided if indicator systems are used to routinely monitor the microbiological quality of untreated surface drinking water supplies. The committee recommends that such routine monitoring be undertaken using the phased approach outlined in Figure 6-1, with greater investments in expanded monitoring and investigations being required as lower-level monitoring indicates the need. Because of an enhanced potential for contamination during wet weather, the committee further recommends that special Level B studies be conducted in each surface water source during at least a half-dozen major wet weather events so that exposure during these periods is well understood. Different intakes from the same waterbody should be considered different sources because contamination is sometimes very local in nature.

Level B—Investigations to Confirm Health Risk

Health risk or Level B assessment for surface drinking water sources has typically taken place via direct pathogen monitoring. Although it is not currently required by federal regulations, some large water utilities have established voluntary routine monitoring programs for selected pathogens, albeit with a relatively low monitoring frequency (typically once per month or once per quarter). Such monitoring is primarily for protozoa and enteric viruses and in many cases is a continuation of the Information Collection Rule (ICR; see Table 1-1) monitoring. The ICR is an example of a special, nationwide study (Level B) conducted as a federal requirement for 18 months from July 1997 to December 1998. An outcome of this special study was the implementation by some of these large water utilities of voluntary long-term monitoring for protozoa and, in more limited cases, for enteric viruses as well. Protozoan monitoring is directed at detection of *Cryptosporidium* and *Giardia*, utilizing the indirect fluorescent assay (IFA) with EPA recommended or approved methods, as described in the ICR or Method 1623, respectively. Monitoring for enteric viruses is conducted with less frequency and by fewer water utilities due, in part, to the complexities and cost of the assay. When voluntary enteric virus monitoring is conducted, the principal method used is the total culturable enteric virus assay in buffalo green monkey kidney (BGMK) cells following the ICR method or its predecessor in *Standard Methods for the Examination of Water and Wastewater* (APHA, 1998; see also Chapter 5).

Level B investigations have been conducted in surface water as part of multistate occurrence studies by independent research groups or in nationwide occurrence studies performed as part of the ICR, the ICR Supplementary Survey, and in the near future, the Long-Term 2 Enhanced Surface Water Treatment Rule (LT2ESWTR; see Table 1-1 and EPA, 2003, for further information). Two large multistate surveys found *Cryptosporidium* in 87 and 51 percent of source waters tested by LeChevallier et al. (1991) and Rose et al. (1991), respectively. The recently proposed LT2ESWTR will require large utilities to monitor for *Cryptosporidium*, *E. coli*, and turbidity for a period of 24 months (EPA, 2003). The results of the two-year monitoring will be used to determine the level of treatment for *Cryptosporidium*, referred to in the rule as "bin classification." Viability or infectivity is of particular interest for *Cryptosporidium* to enable a determination of the potential health risk. In this regard, a recent study by Gennaccaro et al. (2003) demonstrated viable oocycts in reclaimed wastewater effluent using cell culture methods. In addition, genotyping via PCR methods may provide some indication of source. These methods have been widely published, and EPA should develop a consensus method to begin testing in laboratories nationwide as recommended in Chapter 5.

Testing water for other waterborne pathogens is conducted only via special monitoring studies with experimental or non-EPA-approved methods. Such special studies can target a specific pathogen or utilize a representative protozoan, virus group, and/or enteric bacteria for a broader and more complete source water quality assessment. Despite the limited number of special studies conducted to date, they can provide more direct links to the public health issue of concern for the source water under investigation. For certain scenarios, special studies conducted with experimental or research-oriented methods are the most relevant approach, particularly when survival, infectivity, and a high degree of specificity of detection are required. Thus, with special studies, data can be obtained for a specific pathogen (e.g., *Cryptosporidium parvum*), for a group of related pathogens (e.g., enteroviruses), or for a selected suite of pathogens with representative organisms from each of the major waterborne pathogen classes—protozoa, viruses, and bacteria (see also Chapter 3 and Appendix A). Such data can be used more directly to determine their occurrence, survival, and transport in raw surface water.

Analysis of samples for many of the viruses on the 1998 Drinking Water Contaminant Candidate List (CCL; see EPA, 1998a, and NRC, 1999 for further information) will require methods that utilize a combination of cell culture and PCR or PCR directly because viruses such as the noroviruses are not cultivatable. Since the monitoring is of source water, quantitative methods or Most Probable Number (MPN) cell culture methods should be employed, thus permitting an evaluation of the vulnerability of the system and the nature of the challenge to the system by pathogens on a routine or intermittent time frame.

There is often a disincentive for conducting special studies with direct detection of pathogens. Because the nature of the health risk and lack of standards, quality assurance/quality control procedures, or guidelines on acceptable levels, there is a reluctance to address contaminants that the drinking water industry feels it is already controlling (via disinfection) or for which little can be done to prevent contamination of the source water in the first place. Thus, data for most pathogens in untreated water are limited, but as more studies are undertaken, the data will have to be placed into a risk assessment framework (see also Chapter 2), which can be used to make management decisions regarding protection of the watershed or treatment changes in the future (e.g., addition of ozone or UV light). Another disincentive is the lack of guidelines on research methods that would best support the goals for surface water monitoring of microorganisms associated with health risks. The effect of this technology gap is clearly evident in the regulatory arena in both the Unregulated Contaminant Monitoring Rule (UCMR; EPA, 1999b,c) and the CCL (EPA, 1998a), where the listed microbial contaminants are essentially in "regulatory limbo" due to a perceived lack of methods for their detection in water.

To completely understand the potential impacts of the microorganisms listed on the CCL on public health, significant efforts must be made to monitor for their presence in a wide variety of untreated and treated water supplies across the United States. As discussed in Chapter 5, it is also the committee's conclusion that a prerequisite for this monitoring is the development, testing, and discussion of standard methods for their analysis in a free and open environment.

The committee recommends that the EPA (1) prepare a review of published methods for each CCL microorganism (to include groups of related organisms); (2) publish these reviews on the Internet so researchers and practitioners can assess, use, comment, and improve on them; and (3) promote their use in special studies and monitoring efforts in microbial water quality.

Level C—Source Investigations

Monitoring for the presence of fecal contamination is no longer seen as sufficient to establish a priority for public health concerns or as an approach for mitigating or establishing best management practices for reduction of a microbial contaminant in surface water sources of drinking water. Determining the source(s) of microbial contamination is critical for defining the public health issues of concern and establishing priorities for mitigation. The health issues of concern, the risk level, and the dose and mode of exposure will be different if the source of contamination in surface water is from a pasture or from a confined/concentrated animal feeding operation (CAFO), from seagulls or other birds, or from wastewater treatment plant effluent. In current practice, assessments of source of contamination are made via sanitary surveys, reasonable inferences, and with less frequency, chemical tracers (although tracers are used more in groundwater; discussed later). Thus, source tracking studies may be necessary to identify the source or sources of contamination (see Chapter 4 and Tables 4-3 and 4-4 for currently used techniques).

It is clear in this instance that data and information from Level A and B programs could feed into a Level C investigation. The most commonly used microorganisms for source identification in surface water are bacteria, as previously mentioned, although bacteriophage, mammalian viruses, and protozoa may also be used (Scott et al., 2002; Simpson et al., 2002). Because the water industry has been more involved in direct pathogen testing for both *Cryptosporidium* and the enteric viruses (Level B testing), it follows that some percentage of samples could be archived for genotyping.

Mammalian viruses are particularly useful for source identification because they are largely host specific; thus, the source of contamination can be inferred. In some watersheds however, the low frequency of isolation of viruses in drinking water sources imposes a requirement for more frequent, extensive sampling. For *Cryptosporidium*, genotyping may be useful. If Genotype 1 is detected, then human sewage may be the predominant source; however because Genotype 2 can be detected from both humans and animals, discrimination of the source may be possible only by additional sequence analysis. It is worthwhile to develop a database of the national occurrence and distribution of genotypes in water because this will assist in investigations of waterborne outbreaks in the future.

A waterbody that is susceptible to fecal contamination will have potentially multiple human and animal sources of microbial contamination. The goal for source tracking studies should be to also address relative loading. Quantitative information will be an important attribute of a method that could then feed into a total maximum daily load (TMDL; see also Table 1-2) assessment. Although each source of microbial contamination is unique, current source tracking techniques are often isolate-based and do not address the distribution and relative frequency of the various sources. A combination of technologies that would employ (sample) concentration, population assessment, and quantitative measurement of key targets associated with sources should be employed.

Because of its ubiquitous nature as intestinal flora, as discussed in Chapter 4, *Escherichia coli* is the most commonly used organism in surface water source tracking investigations. For example, if *E. coli* is isolated frequently from a surface water source, the approach for assessing relevant health issues will depend on the exact source of the *E. coli*. If the intended use is for drinking water, then the weight or significance of isolating *E. coli* will be lower if the source is birds, gulls, or geese. That is, although bird feces may contain enteric bacteria that are human pathogens, these are easily controlled during conventional treatment. If the source is cattle or sheep, which may also carry *Cryptosporidium* oocysts in their feces, the level of health concern would be higher. The reason for this is that these animals carry *Cryptosporidium* Genotype 2, a zoonotic pathogen known to infect man (see Chapter 3), and because *Cryptosporidium* is more difficult to control with conventional water treatment practices. There would be an even greater health concern if the source is wastewater effluents because these may contain high concentrations of several human pathogens. Enteric viruses and protozoa are of greater concern in non-nitrified, chlorinated, secondary wastewater effluent due to their relatively higher resistance to disinfection compared to enteric bacteria. Thus, source tracking is a valuable method for supplementing sanitary survey inferences and perhaps identifying a new or unknown source of contamination.

While source tracking methods are evolving, at best the current technology can give some indication of the source as being either animal or human. However, more sophisticated molecular analyses will need to be undertaken for more specific distinction of contamination sources (e.g., septic systems versus sewage discharges; pig farms versus chicken farms). Given the amount of money that may have to be spent on infrastructure changes—changes including ongoing agricultural practices and wastewater treatment—more investment should be made to improve the discriminating power of the source tracking methods available. As mentioned previously, source-specific genetic targets hold great promise but must be field tested, and more than one target will be needed to characterize sources in a surface water body. Microarray technology has the potential to screen multiple targets; thus, once these genetic sequences have been identified, it will be very useful in this application. Lastly, it must be kept in mind that all detection methods must be developed in tandem with water concentration and sample preparation methods.

APPLICATION TO GROUNDWATER SOURCES OF DRINKING WATER

When considering the use of indicators and indicator approaches for monitoring the quality of groundwater, important differences between groundwater and surface waters must be understood. The very nature of groundwater makes it a difficult environment to study. Therefore, much less information on the occurrence of indicators and pathogens is available for groundwater than for surface waters. The EPA has recently reviewed the studies that have examined the occurrence of pathogenic microorganisms, generally viruses, and indicators in groundwater systems (EPA, 2000a). It concluded that only one of the studies (Abbaszadegan et al., 1999) was somewhat representative because an attempt was made to collect samples from a variety of hydrogeologic settings. However, even in this study, all of the samples were obtained from community water systems, and water from wells was subject to disinfection prior to distribution for consumption, so it is questionable whether the results of this study are representative of the nation's groundwater. Due to the lack of data, much more effort is needed to screen possible indicators and indicator approaches so that future approaches may suggest which suite of indicators is best for groundwater assessment.

Level A—Screening/Routine Monitoring

In May 2000, the EPA proposed a new regulation, the Ground Water Rule, the purpose of which is to reduce the public health risk associated with the consumption of groundwater that has been contaminated by microorganisms of fecal origin (EPA, 2000a; see also Table 1-1). The approach being taken in the proposed regulation conforms to the Level A—Screening/Routine Monitoring component of the framework shown in Figure 6-1. One of the requirements of the proposed regulation is that systems that draw water from a hydrogeologically sensitive source and do not provide treatment to achieve 99.99 percent virus reduction must monitor their source water for indicator organisms. In the proposal language, utilities may monitor for *Escherichia coli,* enterococci, or coliphages, whichever one is specified by the state.

The decision to require monitoring for indicator organisms rather than directly for enteric pathogens was made on the basis of the impracticality of monitoring for the large number of pathogens that might be present in groundwater that has been contaminated by fecal material (EPA, 2000a). As discussed elsewhere in this report, the analytical methods required for a number of pathogens are expensive, time-consuming, and require a degree of technical expertise beyond that available to many groundwater systems. For some pathogens, analytical methods are not currently available. In addition, many pathogens present a health concern at an extremely low concentration; thus, a large volume of water (e.g., more than 1,000 L) would have to be sampled, which would increase monitoring costs significantly.

The proposed indicator organisms meet many of the desired attributes described in the recommended phased monitoring framework. The indicator organisms can be measured rapidly; they are present in greater numbers than the pathogens; and the analytical methods are amenable to use on large numbers of samples, generally available levels of technical expertise, and can be done at a relatively low cost (i.e., meet the logistical feasibility attribute). In addition, it is not necessary to have quantitative information (i.e., a method that provides presence or absence information would be sufficient for screening purposes). If, however, one of the bacterial indicators is chosen (*E. coli* or enterococci), there are concerns about their similarity to the pathogens, especially viruses, with respect to survival and transport in the subsurface (Gerba, 1984). The difference in size between bacteria (0.2-2 μm) and viruses (0.02-0.1 μm) can result in viruses being transported much greater distances than bacteria, especially in fine-textured soils with small pore sizes (see Chapter 3 for further information). The shorter survival time of bacteria relative to viruses may also be an issue, especially if the time of travel from the surface to the groundwater is long or if the environmental characteristics are harsh. The use of coliphages, or the joint use of coliphages and one of the bacterial indicators, as suggested by the Drinking Water Committee of the EPA's Science Advisory Board[1] (EPA, 2000b), would avoid many of these potential problems. The committee recommends that coliphages be required, in conjunction with bacterial indicators, as indicators of the vulnerability of groundwater to fecal contamination.

Detection of a fecal indicator in groundwater has different implications than detection of indicators in other aqueous environments. Transport of microorganisms through subsurface media is a very complex process; thus, the presence of a

---

[1]See http://www.epa.gov/sab/index.html for further information about the Science Advisory Board.

microorganism at a specific location is subject to a high degree of spatial variability but is often more consistent over time. When detects occur, additional sampling should be conducted to obtain more information about the situation. Likewise, the absence of an indicator in a single subsequent sample does not invalidate the previous positive result. The EPA has determined that between 6 and 18 samples are necessary to determine with 99.9 percent probability that a fecal indicator will be detected in a groundwater that is highly contaminated during at least part of the year (EPA, 2000a).

Depending on the use of such information, it may be sufficient to detect indicators in groundwater. For example, detection of fecal indicators demonstrates that groundwater is vulnerable to contamination by surface sources of fecal material. This may be sufficient to spur action, such as installation of on-site treatment in the case of a drinking water well.

Level B—Investigations to Confirm Health Risk

The detection of indicator organisms in source water during routine monitoring could trigger a second level of study, Level B investigations to confirm health risk. This level of investigation might be desirable if, for example, an assessment of the risk to public health was needed. In the absence of a quantified, documented relationship between the concentrations of indicator organisms and the concentrations of pathogens in water, it would be necessary to collect and analyze samples for specific pathogens. The specific pathogen(s) to study could be determined on the basis of a number of considerations. One such consideration would be the existence and types of possible contaminant sources. If there are no sources of human fecal material in the area, human viruses would likely not be of concern. Another consideration would be the incidence of specific infectious diseases in the source community, because this might indicate specific organisms that should be targeted in the investigation.

The choice of analytical methods in a Level B investigation will be a critical part of the process. For a confirmation of health risk, it is necessary to use a method that provides quantitative information on the pathogens so that an appropriate exposure assessment can be performed. To allow the best input to a risk assessment, the method should also provide information on the viability or infectivity of the pathogens. For viruses, this would require the use of a cell culture system; one limitation of cell culture systems is that information is not obtained on the identity of the virus detected. For bacteria, some cultural technique would be required; depending on the method used, the identity of the microorganisms can be determined. The need to use a culture-based method may limit the number of organisms that can be assessed. For example, no culture-based method has been developed for the detection of noroviruses (formerly called "Norwalk-like viruses"), although they are significant causes of water- and foodborne diseases (CDC, 2003). However, the availability of molecular methods, such as PCR and microarrays, makes it possible to identify specific organisms, including noroviruses, in samples. Microarray technology has already been used to screen a large number of viruses from cell culture samples, and this approach would be quite applicable to the screening of groundwaters (Wang et al., 2002) When molecular methods are combined with cultural methods, information on the presence of specific, infective organisms is obtained.

In the specific case of groundwater, it may be possible to use non-culture-based methods to assess the potential health risk. With respect to groundwater, the concern is whether there is a subsurface pathway through which pathogens in surface sources of contamination can travel and ultimately reach groundwater. Because the microorganisms of concern are not native to groundwater, evidence of their presence, whether in an infective form or not, is proof that the pathway, and thus the potential for groundwater contamination, exists. Therefore, it may be acceptable to use molecular methods such PCR to analyze samples for the presence of pathogens. With the availability of quantitative PCR and information on the ratio between infective and noninfective particles, an estimate of the number of infective microorganisms can be made.

Level C—Source Investigations

In some cases, it may be desirable to determine the source of the contamination so that it can be eliminated. This determination will often require a detailed investigation (Level C). Many enteric pathogens are present in both animal and human wastes. Further complicating matters is that many of the manifestations of infection are relatively nonspecific, such as gastroenteritis, so it is not possible to differentiate sources based on the prevalence of a particular type of illness in the community. In urban areas, there may be multiple point and nonpoint sources of human waste, including septic systems, leaking sewer lines, sites at which reclaimed sewage effluent is being used for irrigation or artificial groundwater recharge, and surface waters receiving treated sewage effluent.

In a situation such as this, it will be necessary to identify the exact source(s) of contamination; thus, traditional methods of microbial identification and enumeration will not be adequate. This can be accomplished by a tracer study, in which some unique substance (such as a colored or fluorescent dye or a microorganism that has been marked with a unique identifying feature) is added to a suspected source and the movement of that substance is followed to determine whether it does, indeed, end up in groundwater. This method can provide useful information; however, it suffers from a number of drawbacks. These include the difficulty of obtaining permission to add tracers to a source, identification of a tracer that has the same properties as the pathogens of interest, and devising an appropriate monitoring scheme. A well-conducted tracer study also requires the involvement of individuals who are highly trained in hydrogeology and/or soil physics; thus, the cost of conducting such a study is very high. In many cases, it may not be practical to conduct a tracer study, especially if extended travel times are involved or if an appropriate network of monitoring wells cannot be installed.

A more common approach for source identification is microbial source tracking (see Box 6-1). Further assessment of coliphage typing as a means to distinguish sewage from other sources should be evaluated since this could be a natural bridge between Level A and Level C investigations. As discussed previously, because of its ubiquitous presence in the intestinal tract of animals, *E. coli* is the most commonly used organism for source tracking. However, detection of *E. coli* in groundwater is much rarer than in surface waters. Thus, all *E. coli* detected in groundwaters should be characterized using a variety of molecular techniques. In groundwater systems the potential for regrowth is high; hence, a marker for biofilm development would be very worthwhile since this may be indicative of other naturally occurring microorganisms (e.g., *Legionella*). Finally, the integration of Level B with Level C studies for direct virus testing would truly indicate a source of human waste as well as human health risk.

IMPEDIMENTS AND DRIVERS TO IMPLEMENTATION

The principal impediments to the development and implementation of a phased monitoring framework for the selection and use of indicators and indicator approaches are the requirements for technical development, the cost of more sophisticated monitoring, and institutional resistance to change. Investment by the government in research designed to develop and standardize new molecular techniques will be an important contributor to resolving the first two impediments. Technical development is important because government investment will be required to make these developments come about in a timely fashion, and cost is critical because investment in methods development and related research will bring down the cost of these methods. Government-funded round-robins and government-sponsored surveys and workshops will also go a long way to overcoming institutional resistance to change. Consequently, as recommended in Chapter 5, EPA should invest in the development of rapid-turnaround biomolecular methods to improve our ability to assess contamination of the nation's water by pathogens.

It is also critical that investments be made in improving or replacing existing methods for collecting and processing samples. Currently, the emphasis is on the development of new detection methodologies that are rapid, sensitive, and specific. However, most of these methods are limited to analyzing very small (microliter-range) sample sizes. Because the presence of a single pathogen in several hundred liters of water may be sufficient to cause a significant public health risk, large samples must be evaluated (see Chapter 5 and Figure 5-5 for further information). The methods that have been developed to concentrate those samples to smaller volumes have generally focused on producing a sample for culture-based analysis. Thus, many substances that are inhibitory to molecular methods are present at highly concentrated levels, which interferes with the analysis. New methods for sampling large volumes must be developed, with the aim that sample analysis be conducted using these new molecular methods. Consequently, the committee also recommends that the EPA invest in research to develop concentration methods designed to vastly increase the size of samples on which biomolecular assay tools can be employed. The effort to develop these concentration and purification methods may cost more and will certainly take longer than development of the assays themselves, but the ultimate payoff in human health protection will be profound.

Evaluation of sources and health risk carries with it issues that may involve legal matters. Communities have the right to know what is in their drinking water supply, even if it is being treated adequately, because some consumers may have cause for concern (see EPA, 1998b). There are issues of who will pay for cleaning up contaminated water systems and how much treatment is needed, given the continued conflict between disinfection and disinfectant by-product formation. As methods are developed, concurrent investment must be made in maintaining health surveillance and addressing risk assessment methods. The level of acceptable risk and the communication of such social values have to be addressed. Better investigation of all waterborne outbreaks with new tools and indicator systems would be very useful for examining relative risks. It may be worthwhile to develop a Risk Advisory committee, perhaps through EPA's Science Advisory Board. The new approaches and new methods will lead to more monitoring data and in some cases, as with microarrays, large databases could be developed very quickly. It is the interpretation of the data and use of the information that will provide the pathway forward.

SUMMARY: CONCLUSIONS AND RECOMMENDATIONS

Microbial indicators are measured to achieve a number of different goals in a variety of water media. No single microbial water quality indicator or even a small set of indicators can meet this diversity of needs. Rather, most appropriate indicators, indicator approaches, and methods depend on the specific application and needs of the situation. Indicator recommendations for selection and use should be developed in the context of a phased monitoring framework. These phases should include screening and routine monitoring (Level A); investigations to confirm health risk (Level B); and detailed investigations for source identification and mitigation (Level C). Under any circumstances, time is of the essence in all three levels of microbial water quality monitoring whenever there exists a possibility that the public may be at risk.

Microbial measurement technology is evolving rapidly and there is an opportunity to leverage these advances toward water quality needs. If sufficient investment is made in the coming decade, indicator systems will undergo a comprehensive evolution, and the correct and rapid identification of waters that are contaminated with pathogenic microorganisms will be substantially enhanced.

Historically, EPA has focused much of its investment on indicators and indicator systems that are used at the screening level (A), but there is an increasing need for national leadership and guidance for the phases of microbial investigation that follow screening. In the limited context of screening, EPA's guidance has been of mixed value, and in this regard the committee concludes that (1) the selection of enterococci for screening at marine recreational beaches is appropriate, because enterococci have been shown to have the best relationship to health risk; (2) existing and proposed monitoring requirements for surface water sources of drinking water are irregular and are not supported by adequate research; and (3) proposed monitoring requirements for groundwater are not adequately protective for viral pathogens.

Based on these conclusions, the committee makes the following recommendations:

• EPA should invest in a long-term research and development program to build a flexible tool box of indicators and methods that will serve as a resource for all three phases of investigation identified in this report.
• That tool box should include the following:

— the development of new indicators, particularly direct measures of pathogens that will enhance health risk confirmation and source identification;
— the use of coliphages, as suggested by EPA's Science Advisory Board, in conjunction with bacterial indicators as indicators of groundwater vulnerability to fecal contamination; and
— the use of routine microbiological monitoring of surface water supplies of drinking water before as well as after treatment.

• A significant portion of that investment should be directed toward concentration methods because existing technology is inadequate to measure pathogens of concern at low concentrations.
• Consistent with previous related recommendations, EPA should invest in comprehensive epidemiologic studies to (1) assess the effectiveness and validity of newly developed indicators or indicator approaches for determining poor microbial water quality and (2) assess the effectiveness of the indicators or indicator approaches at preventing and reducing human disease.
• EPA should develop a more proactive and systematic process for addressing microorganisms on the CCL. The EPA should (1) prepare a review of published methods for each CCL microorganism and groups of related microorganisms; (2) publish those reviews on the Internet so researchers and practitioners can use them and comment on how to improve them; and (3) promote their use in special studies and monitoring efforts.

These conclusions and recommendations should not be taken as an excuse to either cling to or abandon current indicator systems until research develops new approaches. On the contrary, the committee recommends a phased approach to monitoring, as both a means to make existing indicator systems more effective, and to encourage the successive adoption of new, more promising indicator systems as they become available.

REFERENCES

Abbaszadegan, M., P.W. Stewart, M.W. Lechevallier, J.S. Rosen, and C.P. Gerba. 1999. Occurrence of Viruses in Ground Water in the United States. Denver, Colorado: American Water Works Association Research Foundation.

APHA (American Public Health Association). 1998. Standard Methods for the Examination of Water and Wastewater, 20th Edition. Washington, D.C.

Boehm, A.B., S.B. Grant, J.H. Kim, S.L. Mowbray, C.D. McGee, C.D. Clark, D.M. Foley, and D.E. Wellman. 2002. Decadal and shorter period variability and surf zone water quality at Huntington Beach, California. Environmental Science and Technology 36: 3885-3892.

CDC (Centers for Disease Control and Prevention). 2003. Norovirus: Technical Fact Sheet. [On-line]. Available: http://www.cdc.gov/ncidod/dvrd/revb/gastro/norovirus-factsheet.htm.

EPA (U.S. Environmental Protection Agency). 1986. Ambient Water Quality Criteria for Bacteria - 1986. Washington, D.C.: Office of Water. EPA 440-5-84-002.

EPA. 1989a. Drinking Water; National Primary Drinking Water Regulations; Filtration, Disinfection; Turbidity; Giardia lamblia, Viruses, Legionella, and Heterotrophic Bacteria; Final Rule. Federal Register 54: 27486-27541.

EPA. 1989b. Drinking Water; National Primary Drinking Water Regulations; Total Coliforms (Including Fecal Coliforms and E. coli); Final Rule. Federal Register 54: 27544-27568.

EPA. 1998a. Announcement of the Drinking Water Contaminant Candidate List. Notice. Federal Register 64(40): 10274-10287.

EPA. 1998b. Consumer Confidence Reports: Final Rule. Washington, D.C.: Office of Water. EPA 816-F-98-007.

EPA. 1998c. National Primary Drinking Water Regulations: Interim Enhanced Surface Water Treatment; Final Rule. Federal Register 64(241): 69477-69521.

EPA. 1999a. EPA Action Plan for Beaches and Recreational Waters. Washington, D.C.: Office of Research and Development and Office of Water. EPA-600-R-98-079.

EPA. 1999b. Revisions to the Unregulated Contaminant Monitoring Regulation for Public Water Systems: Proposed Rule. Federal Register 64(83): 23397-23458.

EPA. 1999c. Revisions to the Unregulated Contaminant Monitoring Regulation for Public Water Systems: Final Rule. Federal Register 64(180): 50556-50620.

EPA. 2000a. National Primary Drinking Water Regulations: Ground Water Rule; Proposed Rule. Federal Register 65(91): 30193-30274.

EPA. 2000b. Comments on the Environmental Protection Agency's (EPA) Draft Proposal for the Ground Water Rule (GWR). June 30. Washington, D.C.: EPA Science Advisory Board Committee on Drinking Water.

EPA. 2002. Implementation Guidance for Ambient Water Quality Criteria for Bacteria (Draft). Washington, D.C.: Office of Water. EPA-823-B-02-003.

EPA. 2003. National Primary Drinking Water Regulations: Long Term 2 Enhanced Surface Water Treatment Rule; Proposed Rule. Federal Register 68(154): 47640-47795.

Feachem, R.G. 1975. An improved role for fecal coliform to fecal streptococci ratios in the differentiation between human and nonhuman pollution sources. Water Research 9: 689-690.

Gennaccaro, A.L., M.R. McLaughlin, W. Quintero-Betancourt, D.E. Huffman, and J.B. Rose. 2003. Infectious *Cryptosporidium* oocysts in final reclaimed effluents. Applied and Environmental Microbiology 69: 4983-4984.

Gerba, C.P. 1984. Applied and theoretical aspects of virus adsorption to surfaces. Advances in Applied Microbiology 30: 133-168.

Griffith, J.F., S.B. Weisberg, and C.D. McGee. 2003. Evaluation of microbial source tracking methods using mixed fecal sources in aqueous test samples. Journal of Water and Health 1: 141-151.

Haile, R.W., J.S. Witte, M. Gold, R. Cressey, C.D. McGee, R.C. Millikan, A. Glasser, N. Harawa, C. Ervin, P. Harmon, J. Harper, J. Dermand, J. Alamillo, K. Barrett, M. Nides, and G. Wang. 1999. The health effects of swimming in ocean water contaminated by storm drain runoff. Journal of Epidemiology 104: 355-363.

LeChevallier, M.W., W.D. Norton, and R.G. Lee. 1991. Occurrence of *Giardia* and *Cryptosporidium* spp. in surface water supplies. Applied and Environmental Microbiology 57(9): 2610.

Leecaster, M.K., and S.B. Weisberg. 2001. Effect of sampling frequency on shoreline microbiology assessments. Marine Pollution Bulletin 42: 1150-1154.

Malakoff, D. 2002. Microbiologists on the trail of polluting bacteria. Science 295: 2352-2353.

Noble, R.T., D.F. Moore, M. Leecaster, C.D. McGee, and S.B. Weisberg. 2003. Comparison of total coliform, fecal coliform, and enterococcus bacterial indicator response for ocean recreational water quality testing. Water Research 37: 1637-1643.

NRC (National Research Council). 1999. Setting Priorities for Drinking Water Contaminants. Washington, D.C.: National Academy Press.

Prüss, A. 1998. Review of epidemiological studies on health effects from exposure to recreational water. International Journal of Epidemiology 27: 1-9.

Rose, J.B., C.P. Gerba, and W. Jakubowski. 1991. Survey of potable water supplies for *Cryptosporidium* and *Giardia*. Environmental Science and Technology 25: 1393-1400.

Scott, T.M., J.B. Rose, T.M. Jenkins, S.R. Farrah, and J. Lukasik. 2002. Microbial source tracking: Current methodology and future directions. Applied and Environmental Microbiology 68: 5796-5803.

Simpson, J.M., J.W. Santo-Domingo, and D.J. Reasoner. 2002. Microbial source tracking: State of the science. Environmental Science and Technology 36: 5729-5289.

Sorenson, D.L., S.G. Eberl, and R.A. Diksa. 1989. *Clostridium perfringens* as a point source indicator in non-point polluted streams. Water Research 23: 191-197.

Taggart, M. 2002. Factors affecting shoreline fecal bacteria densities around freshwater outlets at two marine beaches. Ph.D. dissertation. University of California, Los Angeles.

Wade, T.J., N. Pai, J.N.S. Eisenberg, and J.M. Colford, Jr. 2003. Do U.S. Environmental Protection Agency water quality guidelines for recreational waters prevent gastrointestinal illness? A systematic review and meta-analysis. Environmental Health Perspectives 111(8): 1102-1109.

Wang, D., L. Coscoy, M. Zylberberg, P.C. Avila, H.A. Boushey, D. Ganem, and J.L. DeRisi. 2002. Microarray-based detection and genotyping of viral pathogens. Proceedings of the National Academy of Sciences 99: 15687-15692.

Appendix A

Emerging and Reemerging Waterborne Pathogens

This appendix addresses a requirement of the committee's statement of task; specifically the requirement to ". . . define currently known waterborne pathogen classes and anticipate those emerging waterborne pathogens that are likely to be of public health concern." For the purposes of this report, emerging and reemerging pathogens can be defined as recently identified waterborne pathogens or those pathogens that were once thought to be under control from a public health perspective but are reappearing and causing increased incidence or geographic range of infections in exposed human populations. In recent years, several such waterborne pathogens have arisen, including recognized pathogens from fecal sources and some "new" pathogens from environmental sources. Several factors contribute to the (re)emergence of waterborne pathogens in the United States (Theron and Cloete, 2002), including the following:

- *Changes in human demographics.* There is an increasing number of "vulnerable subpopulations" in the United States such as infants, children, pregnant women, the elderly, and the immunocompromised (e.g., AIDS patients) who are particularly susceptible to infections resulting from exposure to waterborne pathogens compared to the general populace (see also NRC, 2001).
- *Changes in human behavior.* Urbanization of rural areas allows infections arising in formerly isolated areas, which may once have remained obscure and localized, to reach large and densely populated areas. The use of heated drinking water with warm water reservoirs also promotes the emergence of waterborne pathogens because these systems are ideal habitats for a number of pathogens of public health concern, such as *Legionella* spp. (Lee and West, 1991).

267

• *Breakdown of public health systems.* Although public health measures such as water and wastewater treatment act to minimize human exposure to waterborne pathogens and reduce the incidence of waterborne disease, these systems can and do fail on occasion—often with extensive public health ramifications. Such breakdowns also provide opportunities for pathogens to reemerge.

• *Microbial adaptation.* Microbes are constantly evolving in response to changing environments and environmental conditions. With the increasing use and release of antibiotics and drugs into our waterways, strains of microorganisms that are antibiotic- or drug-resistant have also been increasingly identified (see Chapter 3 for further information).

• *Changes in agricultural practices.* Intensive farming operations (especially concentrated/confined animal feeding operations, or CAFOs) result in high concentrations of animal wastes, which in turn lead to increased pollution of our nation's waters by runoff and intentional (point source) discharges. This is of public health concern because a number of pathogens (e.g., *Cryptosporidium*) routinely contained in such fecal sources can be transmitted to humans through inadequately treated drinking water or through recreational water exposure.

Throughout this report, waterborne pathogens (including those that can be considered emerging or reemerging) can be categorized into four groups: viruses, bacteria, protozoa, and others. ("Others" include cyanobacterial toxins and protists; however, this group is not discussed extensively in this report for reasons outlined in Chapter 1.) Indeed, Chapter 3 on the ecology and evolution of waterborne pathogens and indicator organisms is divided into separate sections for viruses, bacteria, and protozoa. The issue of new and (re)emerging waterborne pathogens has been reviewed in several published reports and articles (EPA, 1998; LeChevallier et al., 1999; Szewzyk et al., 2000; Theron and Cloete, 2002) from a public health and/or water treatment perspective. Therefore, in response to the statement of task, this appendix includes a brief summary of the health effects and mode of transmission of select emerging and reemerging waterborne pathogens in all four groups taken from these and other sources (see Table A-1). Lastly, several of the waterborne pathogens listed in Table A-1 are discussed to some extent (in some cases extensively) elsewhere in this report.

REFERENCES

EPA (U.S. Environmental Protection Agency). 1998. Announcement of the Drinking Water Contaminant Candidate List; Notice. Federal Register 61(94): 24354-24388.

Hilton, C., K. Holmes, K. Spears, L.P. Mansfield, A. Hargreaves, and S.J. Forsythe. 2000. Arcobacter, newly emerging food and waterborne pathogens. Presentation at SGM Warwick, April 12.

LeChevallier, M.W., M. Abbaszdegan, A.K. Camper, G. Izaguirre, M. Stewart, D. Naumovitz, M. Mardhall, C.R. Sterling, P. Payment, E.W. Rice, C.J. Hurst, S. Schaub, T.R. Slifko, J.B. Rose, H.V. Smith, and D.B. Smith. 1999. Emerging pathogens: Names to know and bugs to watch out for. Journal of the American Water Works Association 91(9): 136-172.

Lee, J.V., and A.A. West. 1991. Survival and growth of *Legionella* species in the environment. Journal of Applied Bacteriology Symposium Supplement 70: 121S-129S.

NRC (National Research Council). 2001. Classifying Drinking Water Contaminants for Regulatory Consideration. Washington, D.C.: National Academy Press.

Szewzyk, U., R. Szewzyk, and K.H. Schleifer. 2000. Microbiological safety of drinking water. Annual Review of Microbiology 54: 81-127.

Theron, J., and T.E. Cloete. 2002. Emerging waterborne infections: Contributing factors, agents, and detection tools. Critical Reviews in Microbiology 28: 1-26.

TABLE A-1 Emerging and Reemerging Waterborne Pathogens of Public Health Concern

Waterborne Pathogen	Health Effects	Mode of Transmission	References and URLs
Viruses			
Adenoviruses	Respiratory infections; gastroenteritis; febrile (fever-related) disease with conjunctivitis	Fecal-oral transmission; waterborne transmission	http://www.ce.berkeley.edu/~nelson/ce210a/ Adenovirus/CE210a-adenovirus-k.htm; http://www.cdc.gov/ncidod/dvrd/revb/respiratory/eadfeat.htm; EPA, 1998
Astrovirus	Diarrhea	Fecal-oral transmission	Szewzyk et al., 2000
Coxsackieviruses	Diarrhea and vomiting; skin rashes; myocarditis and pericarditis (inflammation of the heart tissue and the membranous sac enveloping the heart, respectively); aseptic meningitis	Person-to-person transmission; fecal-oral transmission; contaminated water, particularly recreational water	http://www.awwarf.com/newprojects/ pathogens/COXSACKI.html; EPA, 1998
Echoviruses	Subclinical infections; myocarditis; aseptic meningitis	Fecal-oral route	EPA, 1998
Enteroviruses	Gastroenteritis; poliomyelitis (inflammation of the gray matter of the spinal cord); meningitis; respiratory disease; diabetes; encephalitis (inflammation of the brain)	Fecal-oral route or respiratory route	LeChevallier et al., 1999; Theron and Cloete, 2002
Hepatitis viruses	Gastroenteritis	Fecal-oral route; fecally contaminated water and food; person-to-person transmission	LeChevallier et al., 1999
Norwalk/Caliciviruses	Diarrhea and vomiting	Contaminated surface water, groundwater, ice; contaminated shellfish; swimming in water containing sewage	EPA, 1998; LeChevallier et al., 1999
Rotavirus	Diarrhea	Fecally contaminated surface water	Szewzyk et al., 2000; Theron and Cloete, 2002

Bacteria

Arcobacter	Diarrhea	Drinking water reservoirs and treatment plants	Hilton et al., 2000
Aeromonas hydrophila	Gastroenteritis	Drinking water biofilms	EPA, 1998; Theron and Cloete, 2002
Campylobacter (including C. jejuni, C. coli, and related species)	Acute gastroenteritis	Contaminated poultry products; unpasteurized milk; water	LeChevallier, et al., 1999; Szewzyk et al., 2000; Theron and Cloete, 2002
Helicobacter pylori	Gastroduodenal disease including chronic active gastritis; peptic and duodenal ulcer disease; gastric cancer	Unknown; fecal-oral transmission is highly probable	EPA, 1998; LeChevallier et al., 1999; Theron and Cloete, 2002
Legionella spp.	Legionnaires disease (severe lung inflammation) and Pontiac fever (influenza-like form of disease)	Source water to drinking water	Szewzyk et al., 2000
Mycobacterium avium complex	Pulmonary disease	Drinking water treatment and distribution system	EPA, 1998; LeChevallier et al., 1999
Pathogenic Escherichia coli	Hemorrhagic colitis (bleeding from inflamed colon) with bloody diarrhea and hemolytic-uremic syndrome (acute illness following a respiratory infection causing bloody diarrhea)	Contact with contaminated beef or dairy products	LeChevallier et al., 1999; Szewzyk et al., 2000; Theron and Cloete, 2002
Pseudomonas aeruginosa	Nosocomial infections (infections associated with hospital stays) in immunocompromised patients and patients with underlying diseases such as wound and urinary tract infections and pneumonia; tracheobronchitis in cystic fibrosis patients	Water exposed to fecal contamination; surface waters influenced by wastewater discharge; nutrient-rich water	Szewzyk et al., 2000
Yersinia enterocolitica	Gastrointestinal infections	Unknown; contaminated food and water are the most likely sources	Szewzyk et al., 2000; Theron and Cloete, 2002

continued

TABLE A-1 Continued

Waterborne Pathogen	Health Effects	Mode of Transmission	References and URLs
Protozoa			
Acanthamoeba	Granulomatous amoebic encephalitis (swelling of the brain characterized by granulation tissue) in immunosuppressed people; keratitis (severe and potentially blinding infection of the cornea)	Contamination of contact lens or storage case	http://www.awwarf.com/newprojects/pathegeons/ACANTHAM.html; EPA, 1998
Cryptosporidium parvum	Subclinical infections and self-limiting diarrhea in healthy adults; severe diarrhea in infants and immunocompromised persons	Drinking water	Szewzyk et al., 2000; Theron and Cloete, 2002
Cyclospora cayetanensis	Diarrhea, abdominal cramping, decreased appetite, and low-grade fever that can last for several weeks	Unknown; fecal-oral route either directly or via water is thought to be highly probable	LeChevallier et al., 1999
Giardia lamblia	Subclinical infections and self-limiting diarrhea in healthy adults; severe diarrhea in infants and immunocompromised persons	Drinking water	Szewzyk et al., 2000; Theron and Cloete, 2002
Microsporidia	Diarrhea and cholangiopathy (infection of bile ducts) in immunocompromised persons	Surface water; sewage-contaminated waters	EPA, 1998; LeChevallier et al., 1999; Szewzyk et al., 2000
Toxoplasma gondii	Flu-like illness and/or swollen glands in the neck, armpits, or groin	Water contaminated by cats	LeChevallier et al., 1999
Other			
Cyanobacteria toxins	Poisoning; gastrointestinal disease	Nutrient-rich surface water	EPA, 1998; LeChevallier et al., 1999; Szewzyk et al., 2000

Appendix B

Review of Previous Reports

Like Appendix A, this appendix addresses a requirement of the committee's statement of task; specifically the requirement to "review and provide perspective on the importance and public health impacts of waterborne pathogens as discussed in previous National Academies' reports and other seminal reports."

NRC REPORTS

Although this is the first National Research Council (NRC) study to focus specifically on indicators for waterborne pathogens, issues surrounding their use in a variety of applications have been discussed in several recent NRC reports, as summarized in Table B-1 beginning with the most recent. In addition, several of these reports include a discussion of the importance of waterborne pathogens to public health, which is reviewed in Chapters 1 and 2 of this report.

OTHER SEMINAL REPORTS

The U.S. Environmental Protection Agency (EPA) and other groups, including the water industry and academia, have addressed the issue of the microbial quality of drinking water and recreational water and its association with various human health effects such as gastroenteritis, ear and eye infections, dermatitis, and respiratory disease. Thus, Table B-1 also summarizes some of the major reports that have been published addressing these concerns.

Many of the reports included in Table B-1 stress the need for better indicator approaches, including those designed to address the greatest public health threats.

For example, in light of the public health importance of viral pathogens in ground and coastal waters (as described in this report), several of the reports suggest that viral indicators, such as coliphage, be implemented. Other reports recommend development and use of molecular strategies so that new or (re)emerging pathogens can be detected in water. However, all reports included in Table B-1 agree that, given the documented public health impacts of waterborne disease, new and improved indicators of the presence of waterborne pathogens are needed.

REFERENCES

ASM (American Society for Microbiology). 1999. Microbial Pollutants in Our Nation's Water—Environmental and Public Health Issues. Washington, D.C.: ASM Press.

AWWARF (American Water Works Association Research Foundation). 2000a. An Epidemiological Study of Gastrointestinal Health Effects of Drinking Water. Denver, Colorado: American Water Works Association Research Foundation.

AWWARF. 2000b. Development of a Decision Process for Prioritization of Emerging Pathogens Research. Denver, Colorado: American Water Works Association Research Foundation.

AWWARF. 2001. Design of Early Warning and Predictive Source-Water Monitoring Systems. Denver, Colorado: American Water Works Association Research Foundation.

EPA (U.S. Environmental Protection Agency). 1998. Improved Indicator Methods of Pathogen Occurrence in Water. Workshop Summary: August 10-11, 1998. Arlington, Virginia: Office of Water.

EPA. 1999a. EPA Action Plan for Beaches and Recreational Waters. Washington, D.C.: Office of Research and Development and Office of Water. EPA-600-R-98-079.

EPA. 1999b. Review of Potential Modeling Tools and Approaches to Support the BEACH Program. Washington, D.C.: Office of Science and Technology. EPA-823-R-99-002.

EPA. 2001a. Developing Strategy for Waterborne Microbial Disease. Presentation at Waterborne Microbial Disease Stakeholder Meeting. November 6, 2001. Washington, D.C.

EPA. 2001b. Protocol for Developing Pathogen TMDLs: First Edition. Washington, D.C.: Office of Water. EPA-841-R-00-002.

EPA Workshop Group. 2001. Proceedings of Workshop on Development of Microbiological Criteria for Drinking Water Sources, Recreational Waters, and Shellfish Growing Waters, August 27-29. Washington, D.C.: Office of Water.

ILSI (International Life Sciences Institute) Risk Science Institute. 1999. Early Warning Monitoring to Detect Hazardous Events in Water Supplies. T.K. Brosnan, ed. Washington, D.C.: International Life Sciences Institute.

NRC (National Research Council). 1996. Use of Reclaimed Water and Sludge in Food Crop Production. Washington, D.C.: National Academy Press.

NRC. 1998. Issues in Potable Reuse. Washington, D.C.: National Academy Press.

NRC. 1999a. Setting Priorities for Drinking Water Contaminants. Washington, D.C.: National Academy Press.

NRC. 1999b. Identifying Future Drinking Water Contaminants. Washington, D.C.: National Academy Press.

NRC. 2000a. From Monsoons to Microbes. Washington, D.C.: National Academy Press.

NRC. 2000b Watershed Management for Potable Water Supply: Assessing the New York City Strategy. Washington, D.C.: National Academy Press.

NRC. 2001. Classifying Drinking Water Contaminants for Regulatory Consideration. Washington, D.C.: National Academy Press.

NRC. 2002a. Biosolids Applied to Land: Advancing Standards and Practices. Washington, D.C.: National Academy Press.

NRC. 2002b. Opportunities to Improve the U.S. Geological Survey National Water Quality Assessment Program. Washington, D.C.: National Academy Press.

TABLE B-1 Selected Studies and Reports that Address Waterborne Pathogens and Their Indicators

Title	Focus of Study	Points of Emphasis Related to Indicators for Waterborne Pathogens and Committee's Charge
Biosolids Applied to Land (NRC, 2002a)	To evaluate the technical approaches used (by EPA) to establish the chemical and pathogen standards for biosolids	Recommendations: • Further development and standardization of methods for measuring pathogens in biosolids are needed • Research that uses improved pathogen detection technology should be promoted • Research should be conducted to assess whether other indicator organisms, such as *Clostridium perfringens*, could be used in regulation of biosolids
Opportunities to Improve the U.S. Geological Survey National Water Quality Assessment Program (NRC, 2002b)	To provide guidance to the U.S. Geological Survey on opportunities to improve the National Water Quality Assessment (NAWQA) Program as it enters its second decade of nationwide monitoring	Chapter 3 "NAWQA Cycle II Goals—Status": Includes a brief review of the importance of waterborne pathogens in public health risk Conclusions: • Monitoring of all microbes in water, coliforms and/or *Escherichia coli* is neither practical nor optimal, and a different assessment method is required Recommendations: • Methods should be used to distinguish between live and dead viruses • Coliphage assays should be performed at all groundwater wells (in NAWQA Program)
Classifying Drinking Water Contaminants for Regulatory Consideration (NRC, 2001)	To evaluate and revise conceptual approach to generate future Drinking Water Contaminant Candidate Lists (CCLs) first recommended in *Identifying Future Drinking Water Contaminants* (NRC, 1999b)	Chapter 6 "Virulence-Factor Activity Relationships" (VFAR): Describes what VFARs are and concludes that the development and use of VFARs by EPA to help identify future waterborne pathogens appears to be feasible Recommendations: • A scientific working group on bioinformatics, genomics, and proteomics should be established to inform EPA about, and eventually develop and implement, VFARs for drinking water contaminants

Development of Microbiological Criteria for Drinking Water Sources, Recreational Waters, and Shellfish Growing Waters (EPA Workshop, 2001)	To identify approaches for the development of microbiological criteria for drinking water sources and other ambient water uses	Conclusions and recommendations: • Three approaches for the development of microbiological criteria were recommended: (1) selection of one or more pathogens or their surrogates to serve as an indicator of health risk; (2) selection of a single parameter that best correlates with the health risk, indicates the overall fecal or pathogen contamination concentrations in ambient waters, and helps identify the sources of contamination; and (3) development of an index composed of multiple parameters that will indicate waterborne pathogen concentrations and the health risk of gastrointestinal illness. The single indicator approach was considered the weakest • A national research program can be implemented to develop criteria using any of the three approaches • Development of the criteria should be considered an iterative process. A long-term commitment should be made to develop criteria that will identify both sources of microbial contamination and health risks
Developing Strategy for Waterborne Microbial Disease (draft) (EPA, 2001a)	To develop a strategy that unites the influence of the Safe Drinking Water Act and the Clean Water Act	Conclusions: • Research is needed in (1) development of technically sound criteria and risk assessment; (2) development of monitoring tools and diagnostic techniques to rapidly and accurately measure pathogens in different media and determine the potential causes and sources of contamination; and (3) development of modeling tools for forecasting impacts of controlling pathogens through alternative protection and restoration strategies

continued

TABLE B-1 Continued

Title	Focus of Study	Points of Emphasis Related to Indicators for Waterborne Pathogens and Committee's Charge
		Recommendation: • Development of an integrated strategy to reduce the negative impact of microbiological contamination of U.S. waters. Water resource managers must be provided with a sound scientific basis so that they can protect and restore water bodies from microbial contamination via point and nonpoint source discharges by using cost-effective and readily applicable techniques
Design of Early Warning and Predictive Source-Water Monitoring Systems (AWWARF, 2001)	To provide water utilities with information to help them better assess the needs, options, design, and operation of early-warning and source water monitoring programs	Conclusions: • Source water contamination is a significant issue that should be addressed through improved early warning systems; sophisticated systems are the exception rather than the rule • Even with advanced systems, the risk of contaminants in the source water getting into the water supplies exists Recommendations: • A two-step risk-based process for making decisions related to early warning systems. Guidance for the design and operation of early warning systems was provided
Protocol for Developing Pathogen TMDLs (First Edition) (EPA, 2001b)	To assist the development of rational, science-based assessments and decisions that will lead to understandable and justifiable pathogen total maximum daily loads (TMDLs)	Provides a step-by-step description of the TMDL process for pathogens and includes case studies and hypothetical examples to illustrate the major points in the process Recommendations: • States, territories, and tribes should establish TMDLs that will meet water quality standards for each listed water, considering seasonal variations and a margin of safety that accounts for uncertainty

• TMDL submittals should include a monitoring plan to determine whether the TMDL has resulted in attaining water quality standards and to support any revisions to the TMDL that might be required
• Potential indicator organisms for TMDL development include the following:
— Viruses: F1 coliphage; MS2 bacteriophage; poliovirus type 1 strain Lsc2ab; enteroviruses
— Coliform bacteria: total and fecal coliform; *E. coli*; *Klebsiella* spp.
— Enterococcal bacteria: *Streptococcus faecalis*; *S. faecium*
— Protozoa: *Cryptosporidium* spp.; *Giardia* spp.

Findings:

• There were more illnesses among tap water consumers than among subjects in the purified (bottled) water group, suggesting a potential adverse effect originating from the plant or the distribution system. Children were consistently more affected than adults

Conclusion:

• Studies are needed to determine the source of these illnesses (i.e., source water quality, efficacy of treatment, distribution system integrity, and population immunity levels to various pathogens)

Findings:

• A computerized Emerging Pathogens Decision Support System (EPDSS) was developed to evaluate and prioritize waterborne organisms of concern. EPDSS uses quantitative or semiquantitative processes of risk, uncertainty, and decision analysis

An Epidemiological Study of Gastrointestinal Health Effects of Drinking Water (AWWARF, 2000a)

To evaluate drinking water-related gastrointestinal illnesses in a population consuming tap water of a quality that meets current water regulations; determine the source of these illnesses should health effects be observed; provide regulatory agencies and public health authorities with information on health risks associated with drinking water; and find suitable risk indicators

Development of a Decision Process for Prioritization of Emerging Pathogens Research (AWWARF, 2000b)

To develop a decision support tool to assist expert panels in the allocation of limited resources for research related to emerging waterborne pathogens

continued

TABLE B-1 Continued

Title	Focus of Study	Points of Emphasis Related to Indicators for Waterborne Pathogens and Committee's Charge
		• Broad categories of information to be considered in evaluating a pathogen for future research funding should include public health significance; occurrence and ecology; effective or current water treatment; and adequacy of current analytical methods • The structure of EPDSS has two levels of detail, with the second level nested inside the upper level to ensure consistency. Four sets of basic questions are addressed: (1) Are any health effects associated with exposure to this microbe (hazard identification)? (2) To what concentrations are people exposed through all routes considered (exposure assessment)? (3) What is the relationship between exposure to the microbe and probability, severity, or frequency of illness (exposure-response assessment)? (4) What is the overall risk to public health from this microbe? What is the variability of this risk? What is the uncertainty in risk estimates (risk characterization)?
From Monsoons to Microbes (NRC, 2000a)	Overview document based in part on a workshop to elucidate the ocean's role in human health and suggest directions for future efforts to respond to health needs and threats	Chapter 2 "Infectious Diseases" Lists: • Major agents of waterborne disease conveyed by the coastal ocean and their usual routes of transmission to humans • Problems associated with fecal coliforms, as well as *E. coli* and enterococci, as indicators include little association with disease-causing organisms, survival for long times in aquatic habitats, and so on Discussions: • Problems associated with detection and prevention; for example, in seawater, many organisms remain in a viable but non-culturable state

Watershed Management for Potable Water Supply: Assessing the New York City Strategy (NRC, 2000b)

Evaluates the New York City Watershed Memorandum of Agreement (MoA), a comprehensive watershed management plan that allows the City to avoid filtration of its large upstate surface water supply

- Emerging indicator approaches, such as coliphage, gene probes, and immunoassays, were presented

Chapter 5 "Sources of Pollution in the New York City Watershed" emphasizes the public health importance of protozoan parasites *Cryptosporidium* and *Giardia* and their current and future regulatory importance under the federal Safe Drinking Water Act

Chapter 6 "Tools for Monitoring and Evaluation" Recommends:

- Determine the lowest incidence of waterborne disease that can be detected by the City's then current outbreak detection program and increase the sensitivity by studying specific populations
- A *Cryptosporidium* risk assessment be performed on a periodic basis for New York City
- An ongoing program of risk assessment should be used as a complement to active disease surveillance

Appendix C "Microbial Risk Assessment Methods" emphasizes the potential usefulness of linking microbial risk assessment with an epidemiologic surveillance program.

Chapter 3 "Review of Methods for Assessing Microbial Pathogens"

Conclusions:

- There are deficiencies in fecal coliform use as an indicator; for example, it does not account for survival or transport of viral and protozoan pathogens, and pathogens originate from places other than human fecal material

Setting Priorities for Drinking Water Contaminants (NRC, 1999a)

First report to establish a process for setting priorities for an existing CCL; to establish framework for deciding what to regulate, monitor, or study further

continued

TABLE B-1 Continued

Title	Focus of Study	Points of Emphasis Related to Indicators for Waterborne Pathogens and Committee's Charge
		• Limitations in data on health effects of emerging waterborne pathogens make it difficult to establish specific priorities for future regulation, and the expertise is also limited
Identifying Future Drinking Water Contaminants (NRC, 1999b)	Short committee report to identify emerging drinking water contaminants and develop a database to support future decision making on these, as well as help develop an approach to determine future CCLs Report includes three individually authored chapters on (1) historical overview of drinking water contaminants and public water utilities, (2) emerging pathogens, and (3) methods to identify and detect microbial contaminants in drinking water	Emphasizes the inclusion of microbes and all other types of potential contaminants on preliminary and final CCLs Recommendations: • Use common mechanisms of pathogenicity among contaminants in order to include them on future CCLs (analogous to using chemical structure-activity relationships [SARs] for CCLs)
Early Warning Monitoring to Detect Hazardous Events in Water–Supplies (ILSI Workshop, 1999)	To define the state of the science for early-warning monitoring systems for drinking water; to identify strengths and weaknesses of existing technologies and strategies; to raise consciousness regarding the potential for the occurrence of transient hazardous events; and to promote research into prevention, detection, and mitigation or treatment of these events	Conclusions and recommendations: • The detection of hazardous events in source water, especially rivers, is a primary concern • Conventional indicators of water quality are of little value in situations of intentional acts such as the introduction of microbial pathogens or biotoxins into a water system; new and more accurate methods are needed to detect intentional and unintentional microbial hazards introduced into the system • Ideal microbial monitoring methods should be rapid, providing results in 2 hours or less

Microbial Pollutants in our Nation's Water— Environmental and Public Health Issues (ASM, 1999)	To focus attention on the risk of microbial pollutants in water	• Several new approaches were identified, including several optical and assay technologies, but most are still in the research and development phase. Monitoring of health effects in the community using enhanced surveillance strategies to provide early detection of epidemics should be considered because no monitoring system can be constructed to detect all threats Conclusions and recommendations: • Current evidence indicates that microbial pollutants in water, when compared to chemicals, pose far greater risks to communities • EPA needs to focus more on microorganisms in its initiative to study the nation's watersheds. Current indicators of watershed health do not include microbial contaminants of public health concern. The Clean Water Act has to be changed to include more emphasis on microbial threats. Coliform bacteria are not useful for assessing risks due to viruses and protozoa • A task force comprised of EPA, the Centers for Disease Control and Prevention (CDC), the National Institute for Environmental Health Sciences, and other federal agencies, as well as universities and other nongovernmental groups, should be established to outline a national initiative with the goal of protecting the public from waterborne pathogens • An independent scientific assessment should be initiated to address the microbial safety of the nation's water
Action Plan for Beaches and Recreational Waters (BEACH Program) (EPA, 1999a)	To reduce the risk of infection to users of recreational waters through improvements in recreational water programs, communication, and scientific advances; to describe EPA's actions to improve and assist state, tribal, and	Recommendations: • EPA should sponsor conferences and meetings with federal, state, tribal, and local representatives to identify the needs and deficiencies of recreational water quality monitoring programs

continued

TABLE B-1 Continued

Title	Focus of Study	Points of Emphasis Related to Indicators for Waterborne Pathogens and Committee's Charge
	local implementation of recreational water monitoring and public notification programs	• EPA should strengthen water quality standards and implementation programs by developing policies and assisting local managers in their transition to recommend criteria (i.e., *E. coli* and enterococci indicators rather than total and fecal coliforms) • EPA should conduct national Beach Health Surveys annually to collect detailed data on state and local monitoring efforts, applicable standards, water quality communications methods, the nature and extent of contamination problems, and any protection activities • A risk-based evaluation and classification process should be developed, and beaches should be ranked as high, medium, and low • Three broad areas of scientific research should be addressed to improve the science that supports recreational water monitoring programs: (1) water quality indicators research; (2) modeling and monitoring research; and (3) exposure and health effects research
Review of Potential Modeling Tools and Approaches to Support the BEACH Program (EPA, 1999b)	To provide an inventory of predictive models or tools currently in use by agencies responsible for evaluating the need for closing beaches or issuing advisories and warnings	Conclusions: • Waterborne pathogens contaminating recreational areas can originate from various sources located either within the proximity of the beach or at upstream locations within the drainage area or watershed • Beach advisories or closures are issued when pathogen concentrations exceed the water quality standard or local action level; several agencies use mathematical models to predict increased pathogen concentrations

- Selection of the appropriate model for beach advisories will depend on the site conditions of the waterbody of concern. The report reviews the selection of the appropriate model based on a list of screening factors
- Water quality predictive tools that could be applied to beach advisories but are not currently in use by local agencies include pathogen loadings from point- and nonpoint sources and pathogen fate and transport

Conclusions and recommendations:

- *E. coli* should replace total coliforms as an indicator for recreational, raw, surface, and agricultural waters
- *E. coli* should be used not as a single indicator to be applied universally but as an indicator that covers a large part of the field
- A combination of biological and nonbiological measurements (e.g., fecals, turbidity, customer complaints) for intrusion should be used
- EPA should use *E. coli* as the premier indicator for health threats and should reduce the percentages in regulations because regrowth occurs
- Information on the number of fecal coliform violations in the United States that involved *E. coli* should be collected
- *E. coli* data should be added to existing data on coliforms so that historical information and databases do not have to be discarded

Chapter 3 "Microbial Contaminants in Reuse Systems"
Recommendations:

- Facilities should report on effectiveness of treatment processes in removing microbial pathogens

Improved Indicator Methods of Pathogen Occurrence in Water (EPA Workshop, 1998)

To provide scientific and technical guidance on current and alternative indicators and methods; to suggest ways for indicator technology to improve health protection; and to discuss approaches for future dialogues to further the process

Issues in Potable Reuse (NRC, 1998)

To assess public health implications of using reclaimed water as a component of the potable water supply

continued

TABLE B-1 Continued

Title	Focus of Study	Points of Emphasis Related to Indicators for Waterborne Pathogens and Committee's Charge
		• EPA should support research and development on methods for detecting emerging pathogens in environmental samples • Both industry and the research community should develop performance goals appropriate for planned potable reuse Chapter 5 "Public Health Concerns about Infectious Disease Agents" outlines leading pathogens associated with raw sewage and emphasizes the need to be able to effectively monitor for treatment efficacy and, therefore, the need for effective indicator organisms or direct pathogen detection Conclusions: • Coliforms are not reliable for direct pathogen detection; *Clostridium perfringens* should be used instead • There is potential in emerging immunological and molecular biological methods
Use of Reclaimed Water and Sludge in Food Crop Production (NRC, 1996)	To examine the use of treated municipal wastewater and sludge in the production of crops for human consumption	

Appendix C

Detection Technologies

This appendix provides a synopsis of key items related to various widely used methods and technologies for the detection, isolation, enumeration, characterization, and identification of viruses, bacteria, yeasts, molds, and higher organisms in a variety of water samples.[1] For more complete descriptions of detection technologies, readers should consult *Standard Methods for the Examination of Water and Wastewater* (APHA, 1998), *Compendium of Methods for the Microbiological Examination of Foods* (APHA, 2001), *Automated Microbial Identification and Quantitation: Technologies for the 2000's* (Olson, 1996), *Introduction to Bioanalytical Sensors* (Cunningham, 1998), *Biosensors* (Eggins, 1996), *Microarray Biochip Technology* (Schna, 2000), *DNA Microarrays and Gene Expression: From Experiments to Data Analysis and Modeling* (Baldi and Hatfield, 2002), and a review article, "Rapid Methods and Automation in Microbiology" in the inaugural issue of *Comprehensive Reviews in Food Science and Food Safety* published by Institute of Food Technologists (Fung, 2002).

WATER SAMPLING METHODS AND TECHNOLOGIES

The most commonly used method for recovery and concentration of microorganisms in water involves filtration of various volumes through an array of different filter formats. Procedures frequently used for detection of indicator or

---

[1]Mention of commercial systems in this appendix is for illustration purposes and does not imply endorsement by the authors or the National Academies or the exclusion of other commercial systems not specifically mentioned.

enteric bacteria in water involve filtration of 100 mL sample volumes through 0.45-μm-porosity, 47-mm-diameter nitrocellulose membranes and plating on selective media. Membrane filtration is commonly used for indicator microorganisms such as total coliforms, fecal coliforms, enterococci (fecal streptococci), and aerobic spore formers, and for detection of specific organisms or pathogens such as *Escherichia coli*, *E. coli* O157:H7, *Shigella*, *Salmonella*, *Yersinia*, and others. Membrane filtration of turbid surface water can be problematic and a labor-intensive Multiple Tube Fermentation method is often used. Detection of enteric protozoa and viruses requires analysis of much larger volumes of water than the 100 mL typically used for bacterial analysis because of their lower expected concentrations. In contrast to detection of enteric bacteria, conventional filtration approaches for protozoa and viruses require recovery of the organisms from the filtration matrix and further processing for analysis. Elution and dissolution of the filter matrix are commonly used. The selection of elution procedures and reagents is critical if cultural or infectivity assays are the end point analyses because some methods reduce the viability of the recovered organisms. Membrane dissolution methods often result in loss of infectivity, so these procedures are not suitable for assays in which viability or infectivity measurements are crucial.

Concentration methods for microorganisms in water have typically been optimized for a specific pathogen or at best for a limited number of related pathogens (i.e., the enteroviruses). Although there have been attempts to search for a single concentration method for an array of microorganisms in water, to date such efforts have produced mixed results. Development of new technology for concentrating pathogens in water is tedious and requires extensive testing with a variety of organisms and water matrices. Membrane concentration is perhaps the most explored concentration technique, and its advantages and disadvantages have been documented. Membrane concentration procedures are suitable for implementation in the field and for rapid throughput in the laboratory. Hollow-fiber and various ultrafiltration formats have been explored as sampling approaches for multiple bacteria, protozoa, and viruses in water (e.g., Sobsey et al., 1996).

Sampling Methods for Bacteria

Presnell and Andrews (1976) described a combined membrane filter-most probable number (MPN) procedure to increase the amount of water that could be passed (typically 4-5 L of surface water) through the membrane filter. The filters were subsequently washed and MPN analysis was conducted. Van Sluis and Yanko (1977) developed a concentration procedure at the County Sanitation Districts of Los Angeles County (CSDLAC) laboratory to detect the occurrence of *Salmonella* in disinfected effluents and receiving water. The procedure utilized Whatman glass-fiber filters overlaid with filter aid and a pressure filtration apparatus to concentrate 20-liter samples of surface water. The glass-fiber filter and filter aid were then emulsified in diluent. *Salmonella* were detected at concentrations as low as 0.06 MPN per liter. These approaches suggest that reasonably simple procedures can be used to increase the potential detection of pathogenic bacteria in water.

Sampling Methods for Viruses

The most widely used virus sampling and concentration method for large volumes of surface, ground, or finished drinking water is adsorption-elution of virus from microporous filters. Although the pore sizes of the filters (0.2-8.0 μm) are considerably larger than the diameter of the viruses (20-90 nm), viruses are concentrated via adsorption mediated by electrostatic and hydrophobic interactions. However, if the viruses are associated with larger nonviral particles in the water, they will be mechanically strained out in the filter. The accumulated viruses are then desorbed or eluted from the filters with a small volume of eluent fluid. Viruses in this fluid may be assayed directly, but they are usually further concentrated by acid precipitation or by polyethylene glycol precipitation.

Two classes of adsorbent filters have been used to concentrate enteric viruses from freshwater and sewage effluents: negatively charged and positively charged filters. Negatively charged filters are more effective in virus concentration after the water is conditioned by decreasing the pH (3.5) and adding polyvalent cations. Positively charged filters (e.g., Virosorb 1MDS) are commonly used because they have the advantage of adsorbing viruses efficiently over a wide pH range without added polyvalent cations. Nonetheless, negatively charged filters are typically used for virus concentration in marine water. Adsorption-elution methods for virus concentration have limitations such as interference. Dissolved and colloidal substances in water, especially organic matter such as humic and fulvic acids, can interfere with virus adsorption to filters by competing with viruses for adsorption sites. In addition, adsorption-elution efficiencies depend on the enteric virus. For example, Sobsey and Glass (1980) reported overall virus recovery in 1.3 liters of dechlorinated tap water samples containing poliovirus 1, echovirus 1, and reovirus 3 using the electropositive Virosorb 1MDS filter at 57, 53, and 19 percent, respectively. Poliovirus concentration from 1,000 liters of water using Virosorb 1MDS filters is 48 percent with 10^8 plaque-forming units (PFU) of input virus and 24 percent with 200 to 400 PFU of input virus. Virus concentration efficiency will differ with the source of water. A determination of recovery efficiency for seeded virus in the sample water is recommended in the seventeenth edition supplement of *Standard Methods for the Examination of Water and Wastewater* (APHA, 1989b). Despite limitations, virus adsorption-elution from microporous filters is the method of choice for concentrating enteric viruses from large volumes of water.

Viruses concentrated on the adsorbent filters are eluted with beef extract at pH 9.5 followed by pH neutralization prior to assay or storage. Modifications to this procedure have generally involved alternative methods for concentrating viruses recovered in the beef extract eluent. Most modifications have been pursued to adapt this step for use in molecular-based assays. Although beef extract has been the eluent of choice for the past 15 years, it contains reverse transcription polymerase chain reaction (RT-PCR) inhibitors, and the inhibitory effect of beef extract is exacerbated during flocculation procedures for second-step virus concentration (Schwab et al., 1993). Alternative second-step concentration approaches include the use of precipitation agents such as polyethylene glycol and ProCipitate and/or antibody capture (Sobsey et al., 1996). Use of prescreened lots of beef extract and polyethylene glycol (PEG) precipitation for second-step virus concentration results in virus concentrates with lower PCR inhibitors. PEG precipitation has been found to be effective for polioviruses, hepatitis A, and noroviruses (Schwab et al., 1993). Nonetheless, PEG precipitation has to be optimized for the type of water to be assessed.

Various ultrafiltration procedures including tangential flow filtration, vortex flow filtration, and filtration through hollow fibers have also been used to recover and concentrate human viruses and coliphage indicators from up to 100 L of environmental water samples (Jiang et al., 2001). Nonetheless, each one of these approaches has to be optimized for compatibility with cell culture or molecular assays.

Sampling Methods for Protozoa

Early methods for recovering protozoa from water involved passing large volumes of water (up to 1,000 L) through polypropylene yarn-wound filter cartridges with a nominal porosity of 1 μm. More recently developed methods entail passing 10-1,000 L of water through 1-μm absolute porosity pleated membrane capsules, and subsequent concentration and purification of recovered protozoa by immunomagnetic separation (EPA, 1999; McCuin et al., 2001). Other filtration formats used for waterborne protozoa include flat membranes of various diameters (up to 293 mm) and composition (typically polycarbonate or cellulose acetate), compressed foam disks, and hollow-fiber ultrafilters (Clancy et al., 2000; Kuhn and Oshima, 2001).

CONVENTIONAL AGAR PLATE METHOD, MPN METHODS, AND IMPROVED METHODS FOR MONITORING MICROORGANISMS IN WATER

Viable Cell Count Methods and Related Technologies

Viable Cell Count

Use of selective and nonselective agars for growing live bacteria, yeasts, and molds requires water sampling, sample dilution, application of samples into petri dishes, pouring melted agar, incubation of solidified agar samples for a specified time at a specified temperature, and enumeration of colony forming units (CFU) per milliliter depending on the agar used and on the color, shape, size, and fluorescence characteristics of the microorganisms. Further tests of a group or type of microorganism in the water sample (i.e., total viable count, coliform count, fecal coliform count, streptococci count, enterococci count, *Clostridium perfringens* counts) can also be determined. Indeed, myriad combinations of selective and non-selective agars, time and temperature of incubation, aerobic versus anaerobic conditions, volume of sample plated, amount of agar and so forth have been used in performing viable cell counts of water. *Standard Methods for the Examination of Water and Wastewater* (APHA, 1989a) was published to provide consistency of methods between laboratories nationally and, hopefully, internationally (see also Chapter 5 for further discussion).

Improvements of the Conventional Viable Cell Count Method

These methods were developed for efficient operation of the conventional viable cell count method. Many of the methods described here were validated by the Association of Analytical Communities (AOAC International, 2002), a volunteer organization that approves proposed new methods, such as the following:

- Spiral Plating Method (Spiral Biotech, Bethesda, Md.), an automatic plating system on surfaces of agar;
- ISOGRID System (Neogen Incorp., Lansing, Mich.), a membrane filtration system;
- Petrifilm System (3M Co., St. Paul, Minn.), a rehydratable self contained film system;
- Redigel System (3M Co., St. Paul, Minn.), a system that uses pectic gel instead of agar;
- SimPlate System (BioControl, Bellevue, Wash.), a system that uses a round plastic plate with multiple wells as chambers for growth of microbes; and
- Fung's Double Tube, a system for cultivation and enumeration of anaerobic bacteria, especially for *Clostridium perfringens* from food and water (Fung and Lee, 1981).

"Real-Time" Instruments for Viable Cell Count

Viable cell counts can be made with the following real-time instruments:

- direct epifluorescent filter technique involves vital dyes to stain live bacteria for obtaining viable cell counts in about one hour using fluorescent microscopy;

- Chemunex Scan RDI system (Monmouth Junction, N.J.) filters cells on a membrane, stains cells with vital dyes, and then reads the viable cells after 90 minutes in a scanner; and
- MicroStar System (Millipore Corp, Benford, Mass.) utilizes adenosine 5'-triphosphate (ATP) bioluminescence to report microcolonies in a few hours.

Miniaturized Viable Cell Count Methods Using Microtiter Plates and Spot Plating

Miniaturized methods have been developed to reduce the volume of reagents and media used and include the following:

- loop dilution and spot plating of liquid sample on agar surfaces; and
- automated pipette for dilution and spot plating liquid sample on agar surfaces.

Most Probable Number Method

The MPN method is a reliable but laborious method that has been used internationally for about 100 years for coliform enumeration in water. Developments are continuously being made to miniaturize, automate, and computerize the conventional MPN system.

Conventional Five-Tube and Three-Tube MPN Method

Water samples (10 mL, 1 mL, and 0.1 mL) are placed in a series of five tubes or three tubes with nutrients and incubated for a specific time and temperature. Following incubation, the tubes with turbidity and/or gas are recorded and the MPN values are read from established five-tube or three-tube MPN tables.

Miniaturized MPN Modifications in Microtiter Plates

Miniaturized MPN modifications in microliter plates include the following:

- Loop dilution of samples in the wells with nutrients to obtain three-tube MPN (Fung and Kraft, 1969); and
- Automated pipettes to dilute samples in wells and an automated microtiter plate reader and computer to record and interpret MPN of water sample (Irwin et al., 2000; Walser, 2000).

Instrumental Methods for Estimation of Viable Cell Count

Many instruments were developed to measure microbial growth in water samples by monitoring changes such as ATP levels, appearance of specific enzymes, pH, electrical impedance, conductance, capacitance, generation of heat, carbon dioxide, consumption of oxygen, and so on. To obtain useful information, these aforementioned parameters must be related to viable cell count of the same sample series. In general, the larger the number of viable cells in the water sample, the shorter is the detection time of these systems. A scattergram of cells versus detection time is generated that has an inverse slope. By comparing the detection time generated by the microbes of a water sample in these instruments, the initial population can be estimated from the scattergram. The assumption is that as the number of microorganisms in the sample increases, their physical, biophysical, and biochemical activities will also increase. These methods are not suitable for nonviable cells, as injured cells will take much longer to develop a detectable population. The following are selected examples of these instruments.

Impedance and Conductance Methods

Instruments that measure the change of impedance, conductance, or capacitance of the liquid over time as microbes grow include the following:

- Bactometer (bioMerieux, Hazelwood, Mo.), which measures impedance;
- RABIT system (BioScience International, Bethesda, Md.), which measures impedance; and
- Malthus System (Crawley, U.K.), which measures conductance.

ATP Methods

Instruments that measure the increase of microbial ATP over time include the following:

- Lumac (Landgraaf, the Netherlands);
- Biotrace (Neogen, Lansing, Mich.);
- Lightning (BioControl, Bellevue, Wash.);
- Hy-Lite (EM Science, Darmstadt, Germany);
- Charm 4000 (Charm Sciences, Malden, Mass.);
- Celsis System (Cambridge, U.K.);
- Zylux (Maryville, Tenn.); and
- Profile (New Horizon, Columbia, Md.) among others.

A common problem with these systems is the inability to separate background ATP from microbial ATP and to distinguish one type of microbe (e.g., bacteria) from another type (e.g., yeasts) in the same sample.

Turbidity and Color Measurements

Instruments that measure turbidity change or color change include the following:

• Omnispec Bioactivity Monitor System (Wescor, Inc., Logan, Utah), which measures tri-stimulus reflectance colorimetry by monitoring dye pigmentation changes mediated by microbial activities in the water along with appropriate dyes; and
• BioSys (BioSys, Ann Arbor, Mich.), which measures the color change of an agar plug at the bottom of a chamber containing nutrient, dye, and the water sample; the uniqueness of the system is that the color compounds developed during microbial growth are diffused into the agar plug and the automatic measurement of color change by the instrument is done without the interference of particles in the sample.

Measure of Gases

Instruments that measure specific gas development include the following:

• Bactec (Johnston Laboratories, Inc., Cockeysville, Md.), which measures carbon dioxide gas developed by microbial growth in the liquid samples by infrared measurement or radioactivity; and
• BacT/alert Microbial Detection System (Oragnon Teknika/bioMerieux, Hazelwood, Mo.), which measures carbon dioxide development in a liquid sample using a sensor that changes color as carbon dioxide reaches a defined concentration.

IDENTIFICATION AND CHARACTERIZATION OF MICROOORGANISMS FROM WATER

A comprehensive discussion of this subject is beyond the scope of this appendix because there are literally thousands of physical, chemical, biochemical, serological, and immunological tests for the identification and characterization of microorganisms isolated from water. The conventional single-tube test method has largely been replaced by many diagnostics kits developed in the past 30 years. Moreover, many of these tests have been automated, and identification procedures have been computerized to increase the efficiency of operation and eliminate human errors. The following sections briefly identify and describe some detection methodologies for pathogens and indicator organisms that can be considered innovative.

Immunochemical Techniques

A powerful scheme for detecting specific antigens, including both soluble proteins and whole microorganisms, is immunoassays. In this approach, antibodies (either polyclonal or monoclonal) are obtained with specificity toward a particular antigen. The only limitation of this approach is that the antigen must possess a molecular weight greater than 1,000 Daltons. There are a wide variety of different formats for immunoassays. The most useful and sensitive may be the enzyme-linked immunosorbent assay (ELISA; see Figure C-1). In this approach, a first antibody is immobilized on a surface. When a sample containing a particular antigen is exposed to this surface, it binds with a high degree of specificity to the immobilized antibody. After the residual sample is washed to remove any non-specifically bound material from the surface, a second antibody with specificity toward the same antigen, but for a different epitope (binding site), is exposed to the substrate. The second antibody has an enzyme or other group conjugated to it. If the second antibody contains the latter, subsequent exposure to a third solution will direct an enzyme to the complex. The end result is the formation of a "sandwich" in which the initial binding to the first antibody results in the

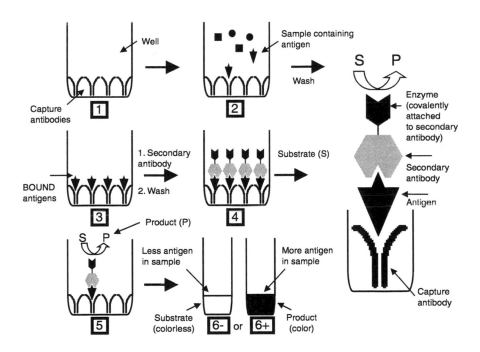

FIGURE C-1 Overview of the ELISA assay.

attachment of an enzyme reporter. A fluorogenic or chromogenic substrate solution is then added to the surface with an appropriate buffer. The intensity of color or fluorescence that forms is proportional to the amount of antigen present in the initial sample.

In addition to immunoassays, immunohistochemical approaches can be used to detect pathogens. In this method, an antibody conjugated to a fluorescent dye is incubated with a sample. The antibody is selected to bind to a particular hapten (site) on the cell, phage, or particle of interest. If binding occurs, the antibody labels the cell so that the cell can be detected more readily. In some cases, multiple labels are employed to label several agents of interest simultaneously. Specific labeling or patterns of labeling can be used to differentiate between cell types.

Nucleic Acid Detection

Genetic methods, based on DNA or RNA analysis, have been rapidly developing in all areas of applied microbiology which include food, clinical, industrial, environmental, and water in the past 20 years.

DNA Sequences

A variety of strategies can be employed for DNA sequence analysis. The simplest approach, analogous to a single sensor, would employ a unique sequence characteristic of the organism of interest. Such an approach would enable the user to identify the pathogen unequivocally because its sequence would not appear in any other organism. The success of this strategy relies on the sequencing of a large number of both pathogenic and nonpathogenic organisms to build up a sufficient database to enable sequence selection. An alternative approach would involve the examination of multiple sequences spread over the genome. In this approach, multiple single nucleotide polymorphisms (SNPs) would be examined. The pattern of expression of the SNPs enables the unique identification of the particular species and strain. This approach can potentially identify patterns of expression that distinguish pathogens from nonpathogens; such distinction would allow development of alarm-type sensors that are based on identifying what the microbe will do rather than what strain or species it is.

Another approach that combines the two methods described involves identifying specific loci with several alleles at each locus. By measuring the expression of the various alleles at a number of loci, it should be possible to generate a unique pattern, a genetic fingerprint or barcode, to minimize the number of regions of the genome that would be examined.

There are various methods for conducting DNA analysis. Sequencing is performed by commercially available instruments. These instruments operate on the basis of a four-color labeling reaction (one color for each base) and read out the sequence from the migration patterns of the DNA sequencing reaction after the reaction products are separated on a polymer gel. They are robust, high-throughput instruments that were used in sequencing the human genome. Although these instruments are fully automated, the time between sample collection and sequence analysis tends to be long. Thus, they are useful primarily for obtaining de novo sequence data rather than for diagnostic or routine analyses.

Once sequence data are available, there are a variety of ways to determine whether a specific sequence is present. DNA arrays operate on the basis of hybridization. These arrays are referred to as microarrays or gene chips (see more below). They are prepared by attaching a single-stranded DNA sequence, the probe, to a surface. Platforms that exist for attaching DNA include photolithographic arrays, spotted arrays, fiber-optic arrays, nanoparticle arrays, and electrochemically addressable arrays. The operating principle for all of these platforms is the same. If the target sequence of interest is present in the sample, it hybridizes via Watson-Crick nucleotide base pairing (i.e., to the complementary probe DNA sequence attached to the surface). The stringency of hybridization conditions can be adjusted so as to enable hybridization only to the perfectly complementary target sequence, and even a single-base mismatch in the DNA target will not result in binding to the array. By attaching many different probe sequences to the array, it is possible to interrogate many (up to hundreds of thousands) sequences simultaneously. In order to detect DNA from a sample, it is first necessary to perform a series of purifications and amplifications. In the purification step, cells must be disrupted and the cell debris, including proteins and membrane fragments, must be removed to isolate the nucleic acid fraction.

DNA is amplified via PCR, which expands the sequences of interest exponentially while all the other sequences grow arithmetically. During the amplification step, specific primers are employed that bracket the DNA sequence to be amplified. These primers often have fluorescent dyes attached to them so that the amplified sequences become labeled for easy detection. Other methods of amplification include rolling circle amplification, ligase chain reaction, and strand displacement assays. The two important aspects of amplification are (1) provision of a detectable amount of DNA sample to employ for subsequent analysis and (2) provision of a specific selection step so that only the specific DNA sequence(s) of interest is amplified.

Basically, a DNA molecule (double helix) of a target pathogen (e.g., *Salmonella*) is first denatured at about 95°C to form two single strands, then the temperature is lowered to about 55°C for two primers (small oligonucleotides specific for *Salmonella*) to anneal to specific regions of the two single stranded DNA molecules. The temperature is then increased to about 70°C for a special heat stable polymerase, the TAQ enzyme from the thermophilic microorganism *Thermus aquaticus*, to add complementary bases (A,T,G,C) to the single-stranded DNA molecule and complete the extension to form a new double strand of DNA. This is called a thermal cycle. After this, the tube is again heated to 95°C for the next cycle. After one thermal cycle, one copy of DNA will become two complete copies. After about 21 cycles and 31 cycles, 1 million and 1 billion copies of the DNA will be formed, respectively (see Figure C-2). The entire process can be accomplished in less than an hour in an automatic thermal cycler. After PCR reactions are complete however, one still needs to detect the presence of the PCR products to indicate the presence of the pathogen.

The first generation of the BAX system for screening family of PCR assays for pathogens (Qualicon, Inc., Wilmington, Del.) used the time consuming and laborious electrophoresis to detect PCR products. Recently, the new BAX system combines DNA amplification and automated homogeneous detection to determine the presence or absence of specific targets like *Salmonella*. All primers, polymerase, and bases necessary for PCR as well as a positive control and an intercalating dye are incorporated into a single tablet. The system works directly from an overnight enrichment of the target organisms and no DNA extraction is necessary. Assays are available for *Salmonella*, *E. coli* O157:H7, *Listeria monocytogenes*, and *Campylobacter jejuni/coli*. The system uses an array of 96

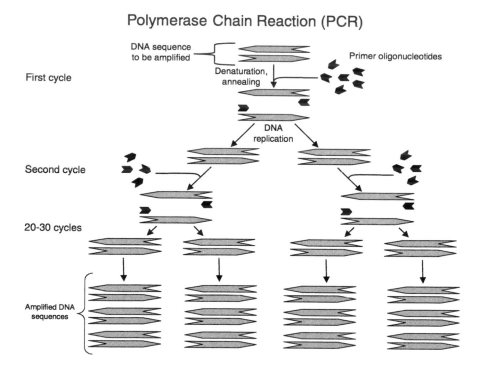

FIGURE C-2 Overview of polymerase chain reaction.

blue light emitting diodes (LEDs) as excitation sources and a photomultiplier tube to detect the emitted fluorescent signal. The integrated system improves the ease of use of the assay and thus, no electrophoresis of PCR products is necessary in this new version of BAX. Notably, a BAX system that detects waterborne *Cryptosporidium parvum* is commercially available.

So called "real time PCR" is a solution method in which a fluorescent signal grows in solution if the target sequence is present. It relies on the use of fluorescent molecules and has the ability to directly measure the amplification products while amplification is taking place. The more target DNA in a broth, the sooner the number of PCR products will reach the detection threshold and be detected since fewer thermal cycles are needed compared with a broth with less target DNA. Thus, the methods are not "real time" per se because it still takes time to reach the threshold for a particular fluorescence level to be detected.

In one manifestation of real time PCR, called TaqMan (Applied Biosystems, Foster City, Calif.), a DNA polymerase enzyme abbreviated Taq obtained from a thermophilic organism is employed. TaqMan possesses exonuclease activity and the TaqMan probe (20-30 base pairs) hybridizes to its complementary target sequence (if present), and consists of a site-specific probe sequence labeled with a fluorescent reporter dye and a fluorescent quencher dye. During the PCR reaction, the hybridized probe is degraded due to the exonuclease activity of the Taq polymerase, thereby separating the quencher from the reporter during extension and resulting in an increase in the fluorescence intensity of the reporter dye. During the PCR amplification, the light emission from the reporter increases exponentially if the target sequence is present—the final level being measured by spectrophotometry after termination of the PCR. Because amplification occurs only in the presence of a particular target sequence, TaqMan assays are of particular value for the detection of specific sequences and are useful for waterborne pathogen detection.

Another real time PCR system called molecular beacon technology (iQ-check system by Bio-Rad, Hercules, Calif.) is now available and all reactions are performed in the same tube. A molecular beacon is a tailor-made hairpin-shaped hybridization probe that is used to specifically bind to target PCR products as it becomes amplified during PCR reactions within the system. On one end of the probe is attached a fluorophore and on the other end a quencher of the fluorophore is attached. In the absence of the target PCR product the beacon forms a hair-pin shape and there is no fluorescence due to quenching of the fluorophore by the quencher. However, during PCR reactions and the generation of target, the beacons will bind to the amplified target and cause the hair-pin molecule to unfold. As the quencher moves away from the fluorophore, fluorescence will occur and can be measured when a threshold of fluorescence is reached. Again, the more target DNA in the broth, the shorter will be the detection time of fluorescence. By using molecule beacons containing different fluorophores, one can detect different PCR products in the same reaction tube, thus performing "multiple" tests of several target pathogens such as *Salmonella*, *E. coli* O157:H7, and *Listeria monocytogenes* in the same broth.

Advantages of real-time PCR in food and water quality monitoring include faster results, no post-PCR analysis, fewer handling steps, no risk of contamination due to a closed tube analysis, and multiplexing capability. Traditional PCR requires thermal cycling to amplify DNA molecules. An isothermal system called nucleic acid-sequence based amplification (NASBA) has recently been developed which has other advantages over the traditional PCR technologies. Since the target is RNA, NASBA can be used to detect RNA viruses and functional mRNA targets.

Conducting multiple PCR reactions enables the simultaneous interrogation of specific nucleic acid sequences, thereby providing excellent specificity for many different microorganisms of interest. Nucleic acid technologies are becoming increasingly sophisticated and powerful tools for applied microbiology and the detection of waterborne pathogens and indicator organisms.

Microarray Technology

Recently, much attention has been directed to the field of "gene chips" and "microchips," in which microarrays of sensors or detection elements are used to detect a great variety of molecules including waterborne and foodborne pathogens and indicator organisms (Ramsay, 1998). Because of ongoing advances in miniaturization technology, as many as 50,000 individual "spots"—with each spot containing millions of copies of a specific DNA probe—can be immobilized on a specialized microscope slide or other substrate. Fluorescently-labeled targets can be hybridized to these spots and be subsequently detected. Microarray technology has been recently discussed as a method for mass testing and diagnostics for viruses and intestinal bacteria in various clinical samples (Petrik, 2001; Sengupta et al., 2003; Wang et al., 2002a,b; Wilson et al., 2002).

Environmental applications of microarray technology have also emerged in recent years. Chips for the detection of a variety of bacteria have been developed which exploit the sequence variability inherent in rDNA, spacer region, and virulence and functional genes, which can help with the identification (taxonomy) of bacteria to the genus and species. The key to these advances has been the validation of these types of microarrays for some 15 groups or genera/species of targeted bacteria (Denef et al., 2003). Addressing functional genes via microarray technology, important specifically for environmental *Pseudomonas*, has also been recently reported (Musarrat and Hashsham, 2003). Recent and forecasted advancements in microarray technology will help make these detection technologies increasingly applicable to the identification and study of waterborne pathogens and indicator organisms.

Ribotyping

Ribotyping involves isolating DNA from a particular microbial strain and fragmenting it with a restriction endonuclease. The resulting DNA is run on a gel, and then probes containing rRNA gene sequences are used to label the gel. The gel labeling pattern can then be used to identify the strain.

Pulsed-Field Gel Electrophoresis (PFGE)

PFGE is similar to ribotyping in that DNA is separated on a gel and labeled. In PFGE, entire chromosomes are separated on the basis of their molecular weights, and the gel pattern distinguishes the different strains from one another.

A newer method for performing DNA fragment analysis involves capillary electrophoresis. In this approach, a glass capillary is employed and the surface charge on the glass serves to separate different DNA lengths from one another. As the different DNA strands elute from the capillary, they are exposed to a laser and the time of elution can be correlated with their position on a hypothetical gel. Capillary electrophoresis systems are much faster than traditional gel-based electrophoresis. In addition, they are amenable to implementation in "lab-on-a-chip" devices.

Mass Spectrometry

An important new area for pathogen detection is mass spectrometry. In this approach, samples, including whole viruses or bacteria, are introduced into the mass spectrometer and ionized and the fragmentation pattern is detected. The ability to detect macromolecules is enabled by developments in sample introduction through ionization including matrix-assisted laser desorption/ionization (MALDI) and electrospray ionization (ESI). The fragmentation pattern, or characteristic peaks in the mass spectrum, provides signatures for the organisms of interest (Demirev, 2001a,b).

Miscellaneous Methods

Aptamers

A recent addition to the armament of specific receptors is the use of aptamers. These materials are isolated from combinatorial oligonucleotide libraries. Antigens are immobilized on a solid support and panned with the oligonucleotide library. Those oligonucleotides that bind with any affinity are retained on the support and those with no binding affinity are washed away. The retained oligonucleotides are displaced from the support, amplified, and panned iteratively, with the higher-affinity binders being preferentially retained and expanded at each step. After several iterations, the oligonucleotides with the highest affinity are isolated, amplified, and sequenced. The resulting aptamers are oligonucleotides (RNA or DNA) that have affinities for the antigen in the micro- to nanomolar range.

Phage Display Libraries

A similar approach employs phage libraries. Phage libraries have combinatorial peptide libraries inserted into a protein that is expressed on the surface of the phage particle. Each phage contains only one peptide insert. Using a similar affinity panning procedure to the one described above for aptamers, successive iterations of binding and expansion of the bound phage lead to high-affinity binders.

Both aptamers and phage display approaches provide new binding receptor entities around which assays can be developed.

Sensors and Sensor Strategies

Sensors collect data and provide information. The information required determines the complexity of the data that must be collected as well as the output of the sensor. If a simple yes or no answer is required, indicating the presence or absence of a particular analyte, the sensor would operate in a fashion akin to a smoke alarm. An example of an alarm sensor would be a trigger sensor employed for detection of various biological agents. Triggers used for bio-detection typically comprise an ultraviolet light source used to excite an air or water sample. Fluorescence emitted by the sample in a particular spectral region signifies the presence of a biological material such as bacteria or spores. Such a sensor is not specific and gives a reading for virtually any biological material. A positive reading causes a sample to be taken and analyzed more thoroughly, hence the word "trigger."

As more information is required, the sophistication of the sensor increases. It may be necessary to determine only that a particular species is there. For example, the presence of a highly infective pathogen would have no threshold concentration; it would be important to know whether or not it was present in the sample. On the other hand, the presence of a particular species may be insufficient information and the abundance of the particular species may be required. Quantitative measurements are critical for most environmental samples in which thresholds are set and regulated. Thus, as one moves from alarm sensors, through sensors that tell one what is there, to sensors that quantify how much of a substance is present, the sophistication of the measurement increases.

Chemical sensors operate on the basis of four generic transduction mechanisms including electrochemical, optical, thermal, and mass measurements. Sensors generally consist of a selective binding layer coupled to a transducer. The binding layer provides selectivity and specificity for the sensor, and the transducer measures a signal resulting from the binding event.

Most sensors are designed using the traditional biological paradigm in which a receptor binds a ligand and generates a signal, referred to as transduction. Many receptor classes exist including ionophores, dyes, and chemically designed binding agents. Developing receptors for the plethora of analytes constitutes a major field of endeavor and is broadly categorized as molecular recognition.

Biosensors employ a biological recognition element such as an enzyme, DNA sequence, receptor, or antibody to perform the recognition and are powerful tools in applied microbiology. Ivnitski et al. (1999) provided a comprehensive overview of different physicochemical instrumental techniques for direct and indirect identification of microorganisms such as infrared and fluorescence spectroscopy, flow cytometry, chromatography, and chemiluminescence techniques as a basis for biosensor construction. The basic concept of a biosensor is simple but their actual construction and operation are quite complex. Basically, a biosensor is a molecule or a group of molecules of biological origin attached to a signal recognition material. When an analyte comes in contact with the biosensor the interaction will initiate a recognition signal that can be transduced and read by an instrument. Many types of biosensors have been developed such as enzymes, antibodies, nucleic acids, cellular materials, and so on. Analytes detected by biosensors include microbial toxins, specific pathogens, carbohydrates, insecticides and herbicides, and antibiotics. The recognition signals used include electrochemical (e.g., potentiometry, voltage changes, conductance, impedance, light addressable, etc.); optical (e.g., UV, bioluminescence, chemiluminescence, fluorescence, laser scattering, reflection and refraction of light, surface plasmon resonance, polarized light); and miscellaneous transducers (e.g.. piezoelectric crystals, thermister, acoustic waves, quartz crystals). Several excellent review articles and books concerning biosensors are available, including Eggins (1997) and Cunningham (1998).

In some cases, biosensors may employ whole cells or tissues for detection. These types of sensing systems receive a significant amount of attention because they enable functional assays to be performed. For example, a heavy metal chemical sensor designed to measure mercury may give a positive readout. On the other hand, a cell-based biosensor that detects mercury by exhibiting toxic effects may not respond in some cases even when mercury is present. This feature underscores an important aspect of cell-based sensors—they have the ability to discriminate between different forms of mercury and can report on their bioavailablity. There is also the possibility to use cell-based biosensors for detecting waterborne pathogens, indicator organisms, and toxins. Although they have not been extensively validated with relevant environmental conditions, this area of emerging technology has significant potential benefit and relevance to waterborne bioagent detection (e.g., see Baeumner, 2003; Belkin, 2003). In all of these approaches, the key is to couple the binding or recognition event to signal generation.

Artificial or electronic "noses" are based on a cross-reactive design model. In this approach, an array of broadly responding sensors is employed. Each sensor responds to a wide number of species rather than to only a single substance. The specificity is encoded in the pattern of response; for example, each substance generates a unique array response pattern that can be distinguished from the pattern for other substances. In this approach, the arrays must be trained on each substance of interest, and a computational pattern recognition program is employed to learn the responses of the array to each substance and then recognize that pattern upon subsequent exposure. Such systems are attracting increasing attention due to their broad specificity and to their ability to respond to substances without the need to design specific receptors.

SUMMARY

Detection technologies concerning fecal indicator microorganisms in water can be addressed in the following three areas:

1. Sampling and Concentration Technologies. The main goal of sampling and concentration technologies is to capture and present target cells, cell components, and metabolites in a large enough number and in a suitable form such that subsequent analysis can be efficiently made. Concentration of target microorganisms can be done by filtration (membrane, ultrafiltration, reverse osmosis, fiber filters with positive or negative charges), centrifugation, flocculation, immunocapture (solid support or immunomagnetic separation), DNA/RNA hybridization, and so on.

2. Enumeration of Target Microbes. This can be done by direct microscopic observation of total cell numbers by differential staining techniques to enumerate live and dead cells. Also, live and injured cells can be grown and enumerated in nonselective agar or selective agar media. MPN procedures can be performed in aliquots of liquid media (non-selective and selective liquid media using the conventional five-tube or three-tube MPN methods and miniaturized MPN procedures using 96-well microtiter plates and computer interphase). Pre-enrichment and enrichment liquid media can also recover and grow target microbes to large numbers for subsequent analysis but will not provide the actual number of cells in the original water sample.

3. Detection and Monitoring of Target Microbes. This can be done by measuring changes in turbidity, color, fluorescence, impedance, conductance, capacitance, heat, pH, specific enzyme activities, gas, ATP levels, and so forth, in the water samples supplemented with appropriate growth-promoting ingredients. More specific information can be obtained by immunochemistry, DNA/RNA hybridization and probes, sequencing of DNA and RNA by sophisticated instruments, PCR and detection of PCR products using electrophoresis or fluorescence technologies. Pulsed-field gel electrophoresis, ribotyping, and related technologies can provide species and subspecies information about target microbes. Target microbes and molecules can be separated and detected by microfluidic, proteomic, genomic, microarray, and microchip technologies. Biosensors and mass spectrometry technologies can also be used to detect many target whole cells and cell components.

To achieve full usefulness of these detection technologies, trained scientists and well designed procedures and instruments must be available and utilized. Final data must be presented logically and scientifically such that informed, appropriate, and defensible decisions can be made by responsible local, regional, national, and indeed, international authorities to promote water safety and protect public health of citizens of the nation and the world.

REFERENCES

AOAC (Association of Analytical Communities) International. 2002. Official Methods of Analysis of AOAC International, 17th Edition, Volumes I and II. Arlington, Virginia.

APHA (American Public Health Association). 1989a. Standard Methods for the Examination of Water and Wastewater. Washington, D.C.

APHA. 1989b. Supplement to Standard Methods for the Examination of Water and Wastewater. Washington, D.C.

APHA. 1998. Standard Methods for the Examination of Water and Wastewater, 20th Edition, Washington, D.C.

APHA. 2001. Compendium of Methods for the Microbiological Examination of Foods, 4th Edition, F.P. Downes and K. Ito, eds. Washington, D.C.

Baeumner, A.J. 2003. Biosensors for environmental pollutants and food contaminants. Analytical Chemistry and Bioanalysis 377(3): 434-445.

Baldi, P., and G.N. Hatfield. 2002. DNA Microarrays and Gene Expression: From Experiments to Data Analysis and Modeling. New York: Cambridge University Press.

Belkin, S. 2003. Microbial whole-cell sensing systems of environmental pollutants. Current Opinion in Microbiology 6(3): 206-212.

Clancy, J.L., Z. Bukhari, T.M. Hargy, J.R. Bolton, B.W. Dussert, and M.M. Marshall. 2000. Using UV to inactivate *Cryptosporidium*. Journal of the American Water Works Association 92(9): 97-104.

Cunningham, A.J. 1998. Introduction to Bioanalytical Sensors. New York: John Wiley & Sons, Inc.

Demirev, P., J.S. Lin, F.J. Pineda, and C. Fenselau. 2001a. Bioinformatics and mass spectrometry for microorganism identification: Proteome-wide post-translational modifications and database search algorithms for characterization of intact *Helicobacter pylori*. Analytical Chemistry 73: 4566-4573.

Demirev, P., J. Ramirez, and C. Fensulau. 2001b. Tandem mass spectrometry of intact proteins for characterization of biomarkers from *Bacillus cereus* T spores. Analytical Chemistry 73: 5725-5731.

Denef, V., J. Park, J.L.M. Rodrigues, T. Tsoi, S. Hashsham, and J. Tiedje. 2003. A more sensitive method enabling the use of spotted oligonucleotide DNA microarrays to study bacterial community functioning. Environmental Microbiology 5(10): 933-943.

Eggins, B. 1997. Biosensors. New York: John Wiley & Sons, Inc.

EPA (U.S. Environmental Protection Agency). 1999. Method 1623. Office of Water. Washington, D.C.: EPA-821-R-99-006.

Fung, D.Y.C., and A.A. Kraft. 1969. Rapid evaluation of viable cell counts using the Microtiter system and MPN techniques. Journal of Milk Food Technology 32: 408-409.

Fung, D.Y.C., and C.M. Lee. 1981. Double-tube anaerobic bacterial cultivation system. Food Science 7: 209-213.

Fung, D.Y.C. 2002. Rapid methods and automation in microbiology. Comprehensive Reviews in Food Science and Food Safety 1: 3-22.

Irwin, P., S.I. Tu, W. Damert, and J. Phillips. 2000. A modified Gauss-Newton algorithm and ninety-six well micro-technique for calculating MON using EXCEL spreadsheet. Journal of Rapid Methods and Automated Microbiology 8: 171-191.

Ivnitski, D., I. Abdel-Hamid, P. Atanasov, and E. Wilkins. 1999. Biosensors for detection of pathogenic bacteria. Biosensors and Bioelectronics 14: 599-624.

Jiang, S.C., R. Nobel, and W. Chu. 2001. Human adenoviruses and coliphage in urban runoff-impacted coastal waters of Southern California. Applied and Environmental Microbiology 67: 179-184.

Kuhn, R.C., and K.H. Oshima. 2001. Evaluation and optimization of a reusable hollow fiber ultrafilter as a first step in concentrating *Cryptosporidium parvum* oocysts from water. Water Research 35(11): 2779-2783.

McCuin, R.M., Z. Bukhari, J. Sobrinho, and J.L. Clancy. 2001. Recovery of *Cryptosporidium* oocysts and *Giardia* cysts from source water concentrates using immunomagnetic separation. Journal of Microbiology Methods 45(2): 69-76.

Musarrat, J., and S.A. Hashsham. 2003. Customized cDNA microarray for expression profiling of environmentally important genes of *Pseudomonas stutzeri* strain KC. Teratogenesis Carcinogenesis and Mutagenesis Supplement 1: 283-294.

Olson, W.P., ed. 1996. Automated Microbial Identification and Quantitation: Technologies for the 2000s. Buffalo Grove, Illinois: Interpharm Press, Inc.

Petrik, J. 2001. Microarray technology: The future of blood testing? Vox Sanguinis 80(1): 1-11.

Presnell, M.W., and W.H. Andrews. 1976. Use of membrane filter and filter aid for concentrating and enumerating indicator bacteria and *Salmonella* from estuarine waters. Water Research 10: 549-554.

Ramsay, G. 1998. DNA chips: State of the art. Nature Biotechnology 16(1): 40-44.

Schna, M., ed. 2000. Microarray Biochip Technology. Sunnyvale, California: TeleChem International, Inc.

Schwab, K.J., R. De Leon, and M.D. Sobsey. 1993. Development of PCR methods for enteric virus detection in water. Water Science and Technology 27: 211-218.

Sengupta, S., K. Onodera, A. Lai, and U. Melcher. 2003. Molecular detection and identification of influenza viruses by oligonucleotide microarray hybridization. Journal of Clinical Microbiology 41(10): 4542-4550.

Sobsey, M.D., and J.S. Glass. 1980. Poliovirus concentration from tap water with electropositive adsorbent filters. Applied and Environmental Microbiology 40(2): 201-210.

Sobsey, M.D., K.J. Schwab, R. De Leon, and Y.-S.C. Shieh. 1996. Enteric Virus Detection in Water by Nucleic Acid Methods. Denver, Colorado: American Water Works Association.

Van Sluis, R.J., and W.A. Yanko. 1977. The Fate of *Salmonella* sp. Following Dechlorination of Tertiary Sewage Effluents. Project report submitted in partial fulfillment of EPA Contract No. 14-12-150. Dechlorination of Wastewater: State-of-the-Art Field Survey and Pilot Plant Studies.

Walser, P.E. 2000. Using conventional microtiter plate technology for the automation of microbiological testing of drinking water. Journal of Rapid Methods and Automated Microbiology 8: 193-207.

Wang, R.F., M.L. Beggs, L.H. Robertson, and C.E. Cerniglia. 2002a. Design and evaluation of oligo-nucleotide-microarray method for the detection of human intestinal bacteria in fecal samples. FEMS Microbiology Letters 213(2): 175-182.

Wang, D., L. Coscoy, M. Zylberberg, P.C. Avila, H.A. Boushey, D. Ganem, and J.L. DeRisi. 2002b. Microarray-based detection and genotyping of viral pathogens. Proceedings of the National Academy of Sciences 99: 15687-15692.

Wilson, W.J., C.L. Strout, T.Z. DeSantis, J.L. Stilwell, A.V. Carrano, and G.L. Andersen. 2002. Sequence-specific identification of 18 pathogenic microorganisms using microarray technology. Molecular and Cellular Probes 16: 119-127.

Appendix D

National Research Council Board Membership and Staff

Staff

FRANCES E. SHARPLES, Director
ROBIN A. SCHOEN, Senior Staff Officer
ROBERT T. YUAN, Senior Staff Officer
KERRY A. BRENNER, Staff Officer
MARILEE K. SHELTON, Staff Officer
EVONNE P.Y. TANG, Staff Officer
ADAM P. FAGEN, Postdoctoral Research Assistant
DENISE GROSSHANS, Financial Associate
SETH H. STRONGIN, Project Assistant
BRENDAN BRADLEY, Intern

WATER SCIENCE AND TECHNOLOGY BOARD

R. RHODES TRUSSELL, *Chair*, Trussell Technologies, Inc., Pasadena,
    California
GREGORY B. BAECHER, University of Maryland, College Park
MARY JO BAEDECKER, U.S. Geological Survey (retired), Reston, Virginia
JOAN G. EHRENFELD, Rutgers University, New Brunswick, New Jersey
DARA ENTEKHABI, Massachusetts Institute of Technology, Cambridge
GERALD GALLOWAY, Titan Corporation, Reston, Virginia
PETER GLEICK, Pacific Institute for Studies in Development, Environment,
    and Security, Oakland, California
CHARLES N. HAAS, Drexel University, Philadelphia, Pennsylvania
KAI N. LEE, Williams College, Williamstown, Massachusetts
CHRISTINE L. MOE, Emory University, Atlanta, Georgia
ROBERT PERCIASEPE, National Audubon Society, Washington, D.C.
JERALD L. SCHNOOR, University of Iowa, Iowa City
LEONARD SHABMAN, Resources for the Future, Washington, D.C.
KARL K. TUREKIAN, Yale University, New Haven, Connecticut
HAME M. WATT, Independent Consultant, Washington, D.C.
CLAIRE WELTY, University of Maryland, Baltimore County, Baltimore
JAMES L. WESCOAT, JR., University of Illinois at Urbana-Champaign

Staff

STEPHEN D. PARKER, Director
LAURA J. EHLERS, Senior Staff Officer
JEFFREY W. JACOBS, Senior Staff Officer
WILLIAM S. LOGAN, Senior Staff Officer
LAUREN E. ALEXANDER, Staff Officer
MARK C. GIBSON, Staff Officer

STEPHANIE E. JOHNSON, Staff Officer
M. JEANNE AQUILINO, Administrative Associate
ELLEN A. DE GUZMAN, Research Associate
PATRICIA JONES KERSHAW, Study/Research Associate
ANITA A. HALL, Administrative Assistant
DOROTHY K. WEIR, Senior Project Assistant

Appendix E

Committee Biographical Information

Mary Jane Osborn, *Chair*, is professor and head of molecular, microbial, and structural biology at the University of Connecticut Health Center. Her fields of specialization are microbial biochemistry, microbiology, and molecular biology. Current research interests include mechanism of cell division in *Escherichia coli*. She was elected to the National Academy of Sciences (NAS) in 1978. Dr. Osborn has served on numerous distinguished committees, including the National Science Board, the President's Committee on the National Medal of Sciences, the Advisory Council of the National Institutes of Health's Division of Research Grants (chair, 1992-1994), the Advisory Council of the Max Planck Institute of Immunobiology, the Board of Scientific Advisors for the Roche Institute for Molecular Biology (chair, 1983-1985), and the NAS Council. Dr. Osborn received her B.A. in physiology from the University of California, Berkeley and her Ph.D. in biochemistry from the University of Washington, Seattle.

R. Rhodes Trussell, *Vice Chair*, is the President of Trussell Technologies, Inc. in Pasadena, California. Dr. Trussell is chair of the Water Science and Technology Board. He previously served on the U.S. Environmental Protection Agency (EPA) Science Advisory Board and chaired that Board's Drinking Water Committee. Dr. Trussell has served on several National Research Council (NRC) committees including, most recently, the Committee on Drinking Water Contaminants, and is a member of the National Academy of Engineering (NAE). He received his B.S. in civil engineering and his M.S. and Ph.D. in sanitary engineering from the University of California, Berkeley.

Ricardo De Leon is the laboratory manager of the microbiology unit in the Water Quality Section of the Metropolitan Water District of Southern California in La Verne, California. His current activities and research include the development and implementation of new technology for the detection of infectious *Cryptosporidium* and other pathogens in water; disinfection of enteric organisms by oxidation and UV light; use of bacterial spores as surrogates for treatment process evaluation; impact of body contact recreation on water quality; and methods development for emerging waterborne pathogens. Dr. De Leon is a member of the Drinking Water Committee of EPA's Science Advisory Board. He received a B.S. and Ph.D. in microbiology from the University of Arizona.

Daniel Y.C. Fung is professor of animal sciences and industry and professor of food science at Kansas State University. His current research focuses on rapid methods and automation in microbiology related to food, environmental, industrial and medical specimens and microbiology of food processing. Dr. Fung's research interests include the rapid detection of harmful and beneficial microorganisms in food and the environment and the control of pathogenic organisms by physical and chemical methods, and fermentation procedures. He is the editor of the *Journal on Rapid Methods*. Dr. Fung has been elected a fellow of the American Academy of Microbiology, Institute for Food Technologists, International Academy of Food Science and Technology, and Institute of Food Science and Technology (U.K.). Dr. Fung received his B.A. in biological sciences from International Christian University in Tokyo, Japan, M.S. in Public Health from the University of North Carolina, and Ph.D. in food technology/food microbiology from Iowa State University.

Charles N. Haas is the Betz Chair Professor of Environmental Engineering at Drexel University. He was formerly a professor and acting chair in the Department of Environmental Engineering at the Illinois Institute of Technology. Dr. Haas' areas of research involve microbial and chemical risk assessment, hazardous waste processing, industrial wastewater treatment, waste recovery, and water and wastewater disinfection processes. He is a member of the Water Science and Technology Board, a fellow of the American Academy of Microbiology, a councilor of the Society for Risk Analysis, and a director of the Association of Environmental Engineering and Science Professors. Dr. Haas has served on several NRC committees, most recently including the Committee on Toxicants and Pathogens in Biosolids Applied to Land, and NRC's Panel on Water System Security Research for its Review of EPA Homeland Security Efforts. He received a B.S. in biology and an M.S. in environmental engineering from the Illinois Institute of Technology and a Ph.D. in environmental engineering from the University of Illinois.

Deborah A. Levy is a commissioned corps officer with the U.S. Public Health Service and a senior epidemiologist with the Division of Healthcare Quality Promotion (DHQP) at the Centers for Disease Control and Prevention (CDC) National Center for Infectious Diseases (NCID). Her current work focuses on emergency response and preparedness for infectious diseases within the healthcare industry. Dr. Levy is a member of CDC's Severe Acute Respiratory Syndrome (SARS) Task Force and is currently coordinating DHQP's SARS preparedness activities. She is also a member of CDC's infection control team that is responding to the influenza outbreak of 2003-2004. Dr. Levy previously served at the CDC in NCID's Division of Parasitic Diseases as the project officer for two interagency agreements with the EPA: to develop a national estimate of waterborne disease, and to conduct research on EPA's Drinking Water Contaminant Candidate List microorganisms. She was also responsible for overseeing CDC's National Waterborne Diseases Outbreak Surveillance System. Dr. Levy joined the CDC in 1996 as an Epidemic Intelligence Service officer. She obtained a B.A. in psychology and an M.P.H. in epidemiology from the University of California, Los Angeles, and a Ph.D. in epidemiology from Johns Hopkins University.

J. Vaun McArthur is senior scientist and professor of research at the University of Georgia Savannah River Ecology Laboratory and an adjunct assistant professor in the Department of Entomology at Clemson University. His current research interests include aquatic microbial ecology with emphasis on factors controlling the distribution and abundance of aquatic bacteria and/or their genes; indirect selection of antibiotic resistance traits by native bacteria; aquatic ecology; stream community metabolism; and macroinvertebrate ecology. He received his B.S. and M.S. in zoology from Brigham Young University and his Ph.D. in biology/ecology from Kansas State University.

Joan B. Rose is an international expert in water pollution microbiology and holds the Homer Nolin Endowed Chair for Water Research at Michigan State University. She recently completed service as the vice chair of the Water Science and Technology Board and also served on the Board on Life Sciences. Previous NRC service includes membership on the Committee on Wastewater Management for Coastal Urban Areas, Committee on Drinking Water Contaminants, and Committee on the Evaluation of the Viability of Augmenting Potable Water Supplies with Reclaimed Water. Dr. Rose was named one of the top 21 most influential people in water in the twenty-first century by Water Technology in 2000 and was awarded the 2001 Clarke Water Prize. She received a B.S. in microbiology from the University of Arizona, an M.S. in microbiology from the University of Wyoming, and a Ph.D. in microbiology form the University of Arizona.

Mark D. Sobsey is a professor of environmental microbiology in the Department of Environmental Sciences and Engineering at the University of North Carolina, Chapel Hill. His research interests lie in the occurrence, distribution, survival, fate, and risk assessment of enteric viruses and other microorganisms in water, soil, shellfish, and other foods. Dr. Sobsey directs the Environmental Virology and Microbiology Laboratory, a coastal North Carolina facility in which students and faculty conduct research into the microbiological quality of drinking water, bathing water, irrigation water, soil, coastal waters and shellfish. He received a B.S. in biology and an M.S. in hygiene from the University of Pittsburgh and a Ph.D. in environmental health sciences from the University of California, Berkeley.

David R. Walt is the Robinson Professor of Chemistry at Tufts University and served as department chairman from 1989 through 1996. His areas of research involve the application of enzymes to organic synthesis, preparation of fiber-optic chemical sensors, immunochemistry, polymeric microstructures, artificial sensing systems, micro- and nanosensors, cell-based biosensors, and combinational chemistry. Some of his work in these areas is focused on the real-time detection of pathogens and toxic chemicals in the environment. Dr. Walt has served on several NRC Committees, including the Committee on Oceanic Carbon and the Panel on Carbon Dioxide. He chaired the NRC's Panel on New Measurement Technologies for the Oceans. Dr. Walt received his B.S. in chemistry from the University of Michigan, and his Ph.D. in chemical biology from the State University of New York at Stony Brook.

Stephen B. Weisberg is executive director of the Southern California Coastal Water Research Project where he specializes in the design and implementation of environmental monitoring programs. Dr. Weisberg serves as chair of the Southern California Bight Regional Monitoring Steering Committee, which is responsible for developing regional integrated coastal monitoring for the Southern California Bight. He also serves on the Steering Committee for the U.S. Global Ocean Observing System, the Alliance for Coastal Technology Stakeholder's Council, and the State of California's Clean Beaches Task Force. Dr. Weisberg's current research focuses on evaluating new technologies for enhancing beach water quality monitoring programs, including use of rapid measurement methods and source tracking technology. He received his B.S. in biology from the University of Michigan and his Ph.D. in biology from the University of Delaware.

Marylynn V. Yates is professor of environmental microbiology in the Department of Environmental Sciences and associate executive vice chancellor at the University of California, Riverside. Dr. Yates conducts research in the area of water and wastewater microbiology. Her current research focuses on contamination of water by human pathogenic microorganisms, especially through use of reclaimed water and biosolids; developing and improving methods to detect microorganisms in environmental samples; persistence of pathogenic microorganisms in the environment, including groundwater; and efficacy of water, wastewater, and biosolids treatment processes to inactivate pathogenic microorganisms. Dr. Yates previously served on the NRC Committee to Improve the U.S. Geological Survey National Water Quality Assessment Program. She received a B.S. in nursing from the University of Wisconsin, Madison; an M.S. in chemistry from the New Mexico Institute of Mining and Technology; and a Ph.D. in microbiology from the University of Arizona.

STAFF

Mark C. Gibson is a program officer at the NRC's Water Science and Technology Board (WSTB) and was responsible for the completion of this report. After joining the NRC in 1998, he directed the Committee on Drinking Water Contaminants, which released three reports, culminating with *Classifying Drinking Water Contaminants for Regulatory Consideration* in 2001. He is also directing the Committee on Water Quality Improvement for the Pittsburgh Region and the Committee on Assessing and Valuing the Services of Aquatic and Related Terrestrial Ecosystems for the WSTB. Mr. Gibson received his B.S. in biology from Virginia Polytechnic Institute and State University and his M.S. in environmental science and policy in biology from George Mason University.

Jennifer Kuzma was a senior program officer for the Board on Life Sciences until January 2003. Dr. Kuzma joined the NRC in January 1999 and served as study director for the NRC report, *Genetically Modified Pest-Protected Plants* (2000). She obtained her Ph.D. in biochemistry from the University of Colorado at Boulder where she worked on the enzymology and molecular biology of isoprene biogenesis from plants and microorganisms.

Seth H. Strongin is a project assistant with the NRC's Board on Life Sciences. He joined the NRC in 2002 and has supported several NRC committees on projects related to bioterrorism, genomics, and agricultural biotechnology, including the 2002 report *Countering Agricultural Bioterrorism*. Mr. Strongin received his B.A. in biology, with a concentration in environmental sciences, from American University.